T0201616

COMPUTATIONAL PHOTONICS

COMPUTATIONAL PHOTONICS

Salah Obayya

University of Glamorgan, UK

A John Wiley and Sons, Ltd., Publication

Registered office
John Wiley & Sons Ltd, The Atrium, Southern Gate, Chichester, West Sussex, PO19 8SQ, United Kingdom

For details of our global editorial offices, for customer services and for information about how to apply for permission to reuse the copyright material in this book please see our website at www.wiley.com.

Library of Congress Cataloging-in-Publication Data

Obayya, Salah.
  Computational photonics / Salah Obayya.
    p. cm.
  Includes bibliographical references and index.
  ISBN 9780470688939 (cloth)
  1. Optoelectronic devices–Mathematical models.   2. Photonics–Mathematics.   I. Title.
  TK8304.O23 2010
  621.36–dc22

                                                                      2010019376

A catalogue record for this book is available from the British Library.

Print ISBN: 978-0-470-68893-9 (H/B)
ePDF ISBN: 9780470667071
oBook ISBN: 9780470667064

Typeset in 11/13pt Times by Aptara Inc., New Delhi, India
Printed and bound in Singapore by Markono Print Media Pte Ltd.

All Praise is due to Allah, and peace and blessings be upon. Prophet Muhammad and upon his family and his Companions.
This work is dedicated to my mother and my late father who raised me up to the good values of hardworking with full devotion.
I would like to dedicate this work also to my wife, Mona, whose love, support, patience, understanding are beyond any scope. In a word, to Mona who means everything to me.

Contents

Preface

Photonics as a field today covers a huge range from communications to science and technology applications, including laser manufacturing, biological and chemical sensing, display technology and optical computing.

For a number of reasons ranging from a higher bandwidth to the inexpensiveness of optical-fibre materials, components based on electric currents are more and more complemented or replaced by technology based on light. Relatively cheap, compact and highly sensitive photonic devices have already been commercialised for a variety of areas, and are expected to be basic blocks for all-optical integrated circuits. As the fabrication of photonic devices is still costly and essentially time consuming, the development of science and technology has grown parallel to the development of accurate numerical techniques able to perform "modelling" of the devices. Besides analysis of the structures, the numerical simulation of the performance of the device makes it possible to perform operations of optimisation and design at an early stage with significant cost savings. The field of numerical modelling has become so important that developing the innovative potential of optics and photonics relies today on sophisticated simulation techniques.

Up to now, there have been many books in the market that deal with numerical modelling techniques. However, the majority only concentrate on one or two numerical methods. This book, on the other hand, covers a comprehensive state-of-the art of many modelling possibilities, ranging from the most innovative techniques in the frequency domain (e.g. the bidirectional beam propagation method), to ones in time domain (e.g. the multiresolution time domain and finite volume time domain methods), for future photonic devices.

Through an extensive number of simulation models and solutions for generic and more specific problems, this book focuses on equipping the reader with the sophisticated concepts of computational modelling in a new easy way to build his/her own codes. Furthermore, it comes with a CD that shows samples of codes, and some examples to enable the reader to understand various computational modelling cases.

This book consists of 11 chapters.

The introduction to the book is given in Chapter 1, where a general overview of modern photonics and recent advances in computational modelling for new photonic devices are given.

From Chapter 2 to Chapter 4, beam propagation method (BPM)-based numerical techniques in the frequency domain are considered. Among them, Chapter 2 presents the governing equations for the full-vectorial BPM, including the finite-element analysis, the perfectly matched layer (PML) scheme for the treatment of boundary conditions and the imaginary-distance BPM that provides the mode solver. Full-vectorial BPM is then assessed in Chapter 3, where modal analysis of rectangular waveguides is given, together with analysis of photonic crystal fibres and liquid-crystal-based photonic crystal fibres.

The bidirectional BPM is presented in Chapter 4. The chapter focuses on the optical waveguide discontinuity problem, giving the formulation of a numerical method able to perform its simulation. After deriving the governing equations, the bidirectional BPM is assessed in several optical examples.

From Chapter 5 onwards, numerical techniques belonging to the area of the time-domain schemes are presented. Chapter 5 starts with a recent innovative modification of the conventional finite-difference time-domain (FDTD) method: the complex-envelope alternating-direction-implicit FDTD. After derivation of the fundamental equations to solve Maxwell's equations, assessment of the technique is given in the specific problem of photonic crystal cavities.

In Chapter 6, the finite-volume time-domain method is proposed as a novel alternative technique to FDTD for the study of electromagnetic problems. The scheme is presented in detail, providing mathematical formulations, analysis of numerical stability and dispersion, and an efficient scheme for the treatment of the boundary conditions. The FVTD is then assessed in Chapter 7 where both linear and nonlinear devices are investigated.

Further modelling possibilities are given in Chapter 8 with the multiresolution time domain (MRTD) method. Basic concepts of the multiresolution analyis are given and its application to the solution of the Maxwell's equations is explained. An accurate and innovative extension of the method to the analysis of second-order nonlinear effects is given. Assessment of the MRTD scheme is presented in Chapter 9, providing code validation in linear photonic devices. Assessment of the technique for the study of second-order nonlinear photonic devices follows in Chapter 10. Chapter 11 shows improvement of the nonlinear-MRTD scheme with the inclusion of auxiliary differential equations that enable the accurate analysis of the linear dispersion of the media. Chapter 11 also presents validation of the code with the generation of second harmonics in a planar waveguide and in a one-dimensional photonic crystal.

Salah Obayya

Acknowledgements

The author would like to acknowledge the people whose help and efforts played an important role in making this book a reality. In particular, I express sincere gratitude to some of my group members; namely Dr. Rosa Letizia, Dr. Domenico Pinto and Dr. Mohamed Farahat.

I wish to mention also the students I taught through the years. Their questions inspired me to undertake this project in the first place.

The author would like to thank the John Wiley team, Sarah Tilley, Jasmine Chang and Jo Tyszka for their patience and support throughout this project.

Last but not least, the author would like to express his gratitude to Prof H. A. Elmikati, Dean of Faculty of Engineering, Mansoura University, whom I owe a lot.

1

Introduction

1.1   Photonics: The Countless Possibilities of Light Propagation

Following the advances in semiconductor physics that have allowed us to fully exploit the conducting properties of certain materials, thereby initiating the transistor revolution in electronics, in the last few decades a new frontier has opened up. The goal in this case is to control the optical properties of materials. An enormous range of technological developments would become possible if we could engineer materials that respond to light waves over a desired range of frequencies by perfectly reflecting them, or allowing them to propagate only in certain directions, or confining them within a specified volume.

Optical solutions to engineering problems are being increasingly found in many fields of application, including medicine, communication, entertainment, sensing and homeland security. Already, fibre optic cables, which simply guide light, have revolutionised the telecommunications industry. Laser engineering, high-speed computing and spectroscopy are just a few of the fields next in line to reap the benefits of the advances in optical materials. Photonics as a field covers today a huge range, from communications to science and technology applications, including laser manufacturing, biological and chemical sensing, display technology and optical computing.

For a number of reasons, ranging from a higher bandwidth to the inexpensiveness of optical fibre materials, components based on electric currents are more and more complemented or replaced by technology based on light. Relatively cheap, compact and highly sensitive photonic devices have already been commercialised for a variety of areas, and are expected to be basic blocks for all-optical integrated circuits, truly revolutionising the way we live.

All-optical systems are believed to be the best solution for the realisation of devices capable of meeting the high-performance characteristics required today and by next-generation telecommunications. In recent decades, with the advent of photonic crystals (PhCs) technology, the goal of a system capable of all-optical signal processing

Computational Photonics   Salah Obayya
© 2011 John Wiley & Sons, Ltd

has made a giant step towards becoming a reality. PhC technology has brought the possibility of manipulating the light in a way that is basically not possible with conventional optical technology. Moreover, this capability of manipulating light is also obtained with an efficiency that is far greater than that obtained with conventional optical devices. However, the complex geometry of devices realised with PhC technology is still a challenge for fabrication technology, although significant progress has been made in this field, making it possible to fabricate new devices exploiting such technology. On the other hand, the mass production of these devices is still far from becoming a reality, and this can pose a serious threat to the development of PhC technology and, most of all, to the development of all-optical systems for ultra-fast telecommunication applications. In this aspect, numerical methods can be of vital help. Generally speaking, the possibility to predict the performance of a device before its practical fabrication can be a key factor towards the development of new and innovative technologies. With the exponential growth of the computational capabilities of modern personal computers (PCs), numerical methods have experienced a sudden expansion in all fields of engineering. This has made possible the development of new applications at a rate that was unthinkable just few decades ago.

Up to now, there have been many books on the market that treat numerical modelling techniques. However, the majority only concentrate on one or two numerical methods. The main goal of this book is to present a comprehensive state-of-the art of the latest modelling possibilities for the analysis and design of innovative devices for next-generation optical communications. Through a thorough analysis of the latest advances in numerical methods, weaknesses and drawbacks will be addressed in order to develop new techniques capable of dealing with simulations of optical devices in an innovative fashion, so as to increase either the accuracy or the efficiency. This aim will be reached through a deep theoretical analysis which will involve the physics behind the functionality of the optical device more directly in the core of the numerical techniques. In this way, the obtained numerical tools will be considered to be more state-of-the-art computational techniques than mere 'number-crunching' algorithms for the solution of Maxwell's equations. The gain in terms of efficiency can be usefully employed for the extension of the methods to dealing with three-dimensional (3D) problems without requiring a prohibitive amount of computational resources, making it possible to obtain simulation results closer to reality.

The information in the book will be essential for the reader to perform and excel in the following:

- Understanding the physics of light propagation in various photonic devices.
- Understanding Maxwell's equations, governing the propagation phenomena.
- Grasping basic concepts of general numerical modelling techniques.
- Deriving the equations underlying the numerical modelling techniques.
- Applying the gained knowledge of computational modelling to modelling photonic devices.

These functions range from introductory (understanding the basic concepts) to advanced levels (application of the concepts to the numerical modelling of photonic devices).

1.2   Modelling Photonics

The last decade has witnessed dramatic progress and interest in micro- and nano-fabrication techniques of complex photonic devices. In almost all cases, an accurate quantitative theoretical modelling of these devices has to be based on advanced computational techniques that solve the corresponding, numerically very large linear, nonlinear or coupled partial differential equations.

Photonics is especially suitable for computation because Maxwell's equations are practically exact, the relevant material properties are well known and the length scales are not too small. Therefore, an exciting aspect of this field is that quantitative theoretical predictions can be made from first principles, without any questionable assumptions or simplifications. The results of such computations have consistently agreed with experiments. This makes it possible and preferable to optimise the design of photonic devices on a computer at an early stage, prior the actual fabrication. The computer becomes the pre-laboratory.

1.2.1   History of Computational Modelling

Many standard numerical techniques for the solution of partial differential equations have been applied to electromagnetics (EM), and each has its own particular strengths and weaknesses. High-quality 'black-box' software is widely available, including free, open-source programs. Indeed, computational photonics has matured so much that many are familiar only with the general principles and capabilities of the different tools.

In the last decades, a wide variety of numerical techniques have been developed and utilised for the design and optimisation of optical devices. Most of them have been successfully employed for the simulation of novel devices, mostly in two-dimensional (2D) space domains. Despite their powerful numerical capabilities, all these techniques possess drawbacks which pose limitations in the range of applicability of the methods to specific classes of problems. These drawbacks become more severe once the extension of these techniques to full-vectorial 3D space domains is considered. These limitations are mainly due to the huge growth in the computational burden required for 3D problems. Furthermore, if nonlinear phenomena are also taken into account, the limitations appear to be even more stringent.

Typically, the existing different numerical schemes can be divided into two main categories: time-domain and frequency-domain methods.

1.2.2   Frequency Domain

Before 1960, the principal approaches in the area of frequency-domain (FD) based numerical techniques involved closed-form and infinite-series analytical solutions, with numerical results from these analyses obtained using mechanical calculators. After 1960, the increasing availability of programmable electronic digital computers permitted such FD approaches to increase significantly in sophistication. Researchers were provided with a whole new range of capabilities offered by powerful high-level programming languages, such as Fortran, rapid random-access storage of large arrays of numbers, and computational speeds orders of magnitude faster than possible with mechanical calculators. In this period, the principal computational techniques for Maxwell's equations included high-frequency asymptotic methods and integral equations. However, these FD methods have some difficulties and drawbacks. For instance, while asymptotic analyses are well suited for modelling the scattering properties of large electrical shapes, such approaches are find it difficult to deal with nonmetallic material compositions and the volumetric complexity of a structure. On the other hand, integral equation methods can deal with material and structural complexity, however their need to construct and solve systems of linear equations limits the electrical size of possible models, especially those requiring detailed treatment of geometric details within a volume.

Although significant progress has been made in solving the ultra-large systems of equations generated by these FD integral equations, the capabilities of even the latest of such technologies cannot keep up with many volumetrically complex structures of recent engineering interest. This also holds for FD finite-element techniques, which generate sparse rather than dense matrices. Moreover, properties of material such as nonlinearities cannot be easily incorporated into the FD solutions of Maxwell's equations, which is a severe constraint, as research today is very active in the fields of active electromagnetic/electronic and electromagnetic/quantum-optical systems, such as high-speed digital circuits, and microwave and millimetre-wave amplifiers and lasers.

1.2.3   Time Domain

Since the arrival of the digital computer, which has profoundly changed the possibilities, time-domain (TD) modelling has offered efficient and flexible techniques to study computational electromagnetic propagation in linear and nonlinear optics.

There are two basic reasons for the success of TD over FD modelling: computational efficiency and problem requirements. Generally, when broadband information is analysed, a TD approach is intrinsically a more immediate choice because it provides a transient response whose bandwidth is limited only by the frequency content of the source, and the time and space sampling adopted in the numerical approach. Moreover, the computational efficiency of TD is also derived from its natural ability to adapt to parallel computer architectures. In addition, the TD approach can usually model problems involving time-varying media and components in a more straightforward way.

A representative example of the rapid growth in TD research is the popularity of the finite-difference time-domain (FDTD) method developed by Taflove.

There are general steps that need to be carried out in order to perform numerical modelling in the time domain:

- Develop time-dependent integral or Maxwell's curl equations.
- Discretise the equations in space and in time by means of an appropriate grid in space, and suitable basis and testing functions.
- Derive a set of equations that relate unknown with known quantities (starting from an initial value that usually is given by the source field).
- Generate a numerical solution of this initial-value problem in space and time.

1.2.4 Chapter Overview

Following this Introduction, from Chapter 2 to Chapter 4, beam propagation method (BPM) based numerical techniques in the frequency domain are considered. Amongst them, Chapter 2 presents the governing equations for the full-vectorial BPM, including the finite-element analysis, the perfectly matched layer (PML) scheme for the treatment of boundary conditions and the imaginary-distance BPM that provides the mode solver. Full-vectorial BPM is then assessed in Chapter 3, where modal analysis of rectangular waveguides is given together with analysis of photonic crystal fibres and liquid-crystal-based photonic crystal fibres.

The bidirectional BPM is presented in Chapter 4. The chapter focuses on the optical waveguide discontinuity problem, giving the formulation of a numerical method able to perform its simulation. After deriving the governing equations, the bidirectional BPM is assessed in several optical examples.

From Chapter 5, numerical techniques belonging to the area of time-domain schemes are presented. Chapter 5 starts with a recent innovative modification of the conventional finite-difference time-domain (FDTD) method: complex-envelope alternating-direction-implicit FDTD. After derivation of the fundamental equations to solve Maxwell's equations, assessment of the technique is driven by the specific problem of photonic crystal cavities.

In Chapter 6, the finite-volume time-domain (FVTD) method is proposed as a novel alternative technique to FDTD for the study of electromagnetic problems. The scheme is presented in detail, providing mathematical formulations, analysis of numerical stability and dispersion, and an efficient scheme for the treatment of the boundary conditions. The FVTD is then assessed in Chapter 7, where both linear and nonlinear devices are investigated.

Further modelling possibilities are given in Chapter 8, with the multiresolution time domain (MRTD) method. Basic concepts of the multiresolution analysis are given and its application to the solution of Maxwell's equations is explained. An accurate and innovative extension of the method to the analysis of second-order nonlinear effects is given. Assessment of the MRTD scheme is presented in Chapter 9, providing code validation in linear photonic devices. Assessment of the technique for the study of second-order nonlinear photonic devices follows in Chapter 10. Chapter 11 shows improvement of the nonlinear-MRTD scheme, with the inclusion of auxiliary differential equations (ADE) that enable the accurate analysis of linear dispersion of the media. Chapter 11 also presents validation of the code with generation of second harmonics in a planar waveguide and in a one-dimensional (1D) photonic crystal.

1.2.5 Overview of Commercial Software for Photonics

The continuous advancement of microwave circuits and electromagnetic devices towards increased functionality and performance requires simultaneous development of modelling tools that are able to keep up with the growing level of sophistication. In order to get a deeper insight into the field of computational electromagnetics (CEM) and grasp what is available in the market, the most widely used commercial packages are listed as follows:

– *COMSOL Multiphysics* (formerly *FEMLAB*) is a finite element analysis, solver and simulation software package for various physics and engineering applications, especially coupled phenomena, or multiphysics. COMSOL Multiphysics also offers an extensive interface to MATLAB and its toolboxes for a large variety of programming, pre-processing and post-processing possibilities.
– *FIMMWAVE* is a generic full-vectorial mode-finder for waveguide structures. FIMMWAVE combines both methods based on semi-analytical techniques with other more numerical methods such as finite difference or finite element.
– *CST MICROWAVE STUDIO®* (CST MWS) offers five solver modules; the Transient, Eigenmode, Frequency Domain, 'Resonant: Fast S-Parameter', 'Resonant: S-Parameter, Fields' (formerly known as Modal Analysis), and the Integral Equation Solver, each offering distinct advantages in their own domains. There are numerical advantages offered by the method used in most of the solvers, the finite integration technique (FIT).
– *CrystalWave* is a design environment for the layout and design of integrated optics components optimised for the design of photonic crystal structures. It is based on both FDTD and finite-element frequency-domain (FEFD) simulators and includes a masque file generator carefully optimised for planar photonic crystal structures.
– *Optiwave* is a suite of engineering design tools. Amongst these, there are *OptiFDTD*, which is based on the FDTD algorithm with second-order numerical accuracy

and the most advanced boundary condition – the uniaxial perfectly matched layer (UPML) boundary condition, and *OptiBPM*, which is based on the BPM.

– *RSoft's Photonic Component Design Suite* allows the design and the simulation of both passive and active photonic devices for optical communications, optoelectronics and semiconductor manufacturing. *FullWAVE* is a simulation tool for studying the propagation of light in a wide variety of photonic structures, including integrated and fibre-optic waveguide devices, as well as circuits and nanophotonic devices, such as photonic crystals. The software employs the FDTD method for the full-vector simulation of photonic structures. *BandSOLVE* is a design tool for the calculation of photonic band structures for all photonic crystal (PC) devices which employs the plane wave expansion (PWE) algorithm.

Indeed, commercial EM simulation programs available today provide powerful design tools, however no single numerical method provides a universal solution, and commercial codes often fail to accurately simulate high-end problems found in cutting-edge research. Therefore, research in computational EM is still essential to keep up with the increasing complexity of devices throughout the EM spectrum and has to develop parallel to fabrication technologies to allow the full establishment of the new photonic solutions in the market.

2

Full-Vectorial Beam Propagation Methods

2.1 Introduction

Numerical simulations play an important role for the design and modelling of guided-wave optoelectronic devices. There are various modelling methods in which not only a full-vector model, but also an approximate scalar model, are used. In this chapter, an overview of beam propagation methods (BPMs) [1–4] is introduced. In addition, the formulation of the wave equations in terms of the electric and magnetic fields is included. Moreover, this chapter includes an introduction to finite-element analysis, followed by the derivation of the finite-element BPM. Also, the formulation of the imaginary-distance full-vectorial finite element BPM scheme proposed in [5] is extended further to fully treat the vectorial complex modes.

2.2 Overview of the Beam Propagation Methods

Analysis and simulation of electromagnetic wave propagation are essential in the modelling and design of optical waveguide devices. The BPM [1–4] has been one of the most popular techniques for modelling and simulation of such optical devices. The major concept of the BPM is the development of a formula that permits the propagation of an initial field distribution along the axial direction by steps of sufficiently small length, as shown in Figure 2.1 [4].

Early publications were focused on the solution of the scalar paraxial wave equation by means of fast Fourier transform (FFT) [6]. However, the formulation of the FFT-BPM is derived under the assumption that the refractive-index difference in the transverse direction is very small, therefore the FFT-BPM cannot be applied to structures with large refractive-index discontinuities. In addition, the FFT-BPM can be

Computational Photonics   Salah Obayya
© 2011 John Wiley & Sons, Ltd

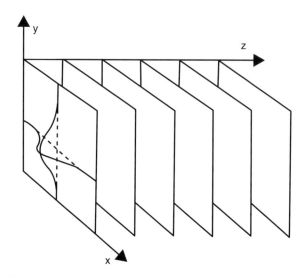

Figure 2.1 Propagation of an initial field distribution along the axial direction.

only used to study the scalar wave propagation, therefore the vectorial properties of
the guided wave cannot be described.

The finite-difference (FD) method was first introduced by Hendow and Shakir [7]
to solve the paraxial scalar wave equation through a cylindrically symmetric structure. Then, Chung and Dagli [8] introduced the FD-BPM to the Cartesian coordinate
system. The scalar FD-BPM has advantages in terms of efficiency and broader applicability [9, 10]. However, the most serious drawback of the scalar FD scheme is the
complete absence of the vector characteristics that are inherent in light propagation
through inhomogeneous and/or anisotropic media. This was partly removed by the
semi-vectorial BPM (SVBPM) that distinguishes two orthogonal, but otherwise completely uncoupled, states of polarisation (TE and TM) [2, 11]. The first approach to
consider the vectorial nature of light was presented in [12] via a FD Crank–Nicolson
scheme. Even though the formulation framework was quite general, the applicability
was restricted only to planar straight waveguides. Since then, many full vectorial BPM
approaches based on the popular finite-difference method (FDM) have been reported
[13–16]. Due to the inefficient discretisation associated with finite differences, the
FDBPM needs large computational resources, especially in simulating nonuniform
optical waveguides.

Due to its numerical efficiency and versatility, some full-vectorial BPM algo-
rithms have been formulated, based on the finite-element method (FEM) [17–21].
Polstyanko *et al.* [17] investigated the full vector finite-element BPM (FE-BPM)
with electric filed formulation. However, Obayya *et al.* [21] formulated the vector
FE-BPM using the transverse magnetic field components. In [1], an efficient vector
FE-BPM for transverse anisotropic material was reported in terms of the transverse

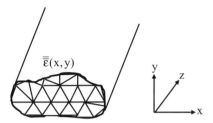

Figure 2.2   Schematic of three dimension optical waveguide.

magnetic-field components, with perfectly matched layer boundary conditions and wide-angle approximation.

2.3   Maxwell's Equations

Maxwell's equations are a set of four partial differential equations which describe the properties of the electric and magnetic fields through the medium. In the frequency domain, the propagation of electromagnetic waves through the waveguide, as shown in Figure 2.2, is governed by Maxwell's equations, which can be written as follows

$$\nabla \times E = -j\omega\mu H \tag{2.1}$$

$$\nabla \times H = j\omega\bar{\bar{\varepsilon}} E \tag{2.2}$$

$$\nabla.(\bar{\bar{\varepsilon}} E) = 0 \tag{2.3}$$

$$\nabla.\mu H = 0 \tag{2.4}$$

where the vector quantities E and H are electric and magnetic field vectors, respectively, $\bar{\bar{\varepsilon}} = \varepsilon_0\bar{\bar{\varepsilon}}_r$ and $\mu = \mu_0\mu_r$. The quantities $\bar{\bar{\varepsilon}}$ and μ define the electromagnetic properties of the medium and are the permittivity tensor and the permeability of the waveguide material, respectively. $\varepsilon_0 = 8.854 \times 10^{-12}$ F/m is the permittivity of free space and $\mu_0 = 4\pi \times 10^{-7}$ H/m is the permeability of free space. $\bar{\bar{\varepsilon}}_r$ and μ_r are the relative permittivity tensor and permeability of the waveguide material. In the absence of the magnetic material, μ_r is set to unity.

2.4   Magnetic-Field Formulation of the Wave Equation

The total electromagnetic field that is supported by a waveguide can be expressed in terms of only the electric or magnetic field components to produce wave equations. In order to express the wave equation in terms of the magnetic field only, the electric field is removed from the derivation by taking the curl of Equation (2.2) and substituting using (2.1). The vector wave equation for the magnetic field vector, H, can be written as

$$\nabla \times \left(\bar{\bar{\varepsilon}}_r^{-1} \nabla \times H \right) - k_0^2 H = 0 \tag{2.5}$$

where k_0 is the free space wave number $k_0^2 = \omega^2 \mu_0 \varepsilon_0$. The anisotropic material is assumed to have one of its principal axes points in the direction of the waveguide. Under this assumption [22], the permittivity tensor takes the form

$$\bar{\bar{\varepsilon}} = \varepsilon_0 \bar{\bar{\varepsilon}}_r = \varepsilon_0 \begin{pmatrix} \varepsilon_{xx} & \varepsilon_{xy} & 0 \\ \varepsilon_{yx} & \varepsilon_{yy} & 0 \\ 0 & 0 & \varepsilon_{zz} \end{pmatrix} \tag{2.6}$$

For isotropic waveguides, $\varepsilon_{xx} = \varepsilon_{yy} = \varepsilon_{zz}$ and $\varepsilon_{xy} = \varepsilon_{yx} = 0$. Using the vector notation $(\nabla \times (\varphi A) = \nabla \varphi \times A + \varphi \nabla \times A)$, the vector wave Equation (2.5) can be rewritten as follows

$$\nabla^2 H + k_0^2 \bar{\bar{\varepsilon}}_r H = -\bar{\bar{\varepsilon}}_r^{-1} \nabla \bar{\bar{\varepsilon}}_r \times (\nabla \times H) \tag{2.7}$$

The transverse component of the vector wave Equation (2.7) can be obtained such that

$$\nabla^2 H_t + k_0^2 \bar{\bar{\varepsilon}}_{rt} H_t = -\bar{\bar{\varepsilon}}_{rt}^{-1} \nabla_t \bar{\bar{\varepsilon}}_{rt} \times (\nabla_t \times H_t) \tag{2.8}$$

where the subscript 't' stands for the transverse components and $\bar{\bar{\varepsilon}}_{rt}$ is the transverse component of the dielectric tensor and can be defined such that

$$\bar{\bar{\varepsilon}}_{rt} = \begin{pmatrix} \varepsilon_{xx} & \varepsilon_{xy} \\ \varepsilon_{yx} & \varepsilon_{yy} \end{pmatrix} \tag{2.9}$$

2.5 Electric-Field Formulation of the Wave Equation

The wave equation can also be expressed only in terms of the electric field. In this case, the magnetic field is removed from the derivation by taking the curl of Equation (2.1) and substituting using (2.2). The vector wave equation for the electric field vector, E, can be written as [3]

$$\nabla^2 E + \bar{\bar{\varepsilon}}_r k_0^2 E = \nabla(\nabla . E) \tag{2.10}$$

The transverse component of (2.10) is given by

$$\nabla^2 E_t + \bar{\bar{\varepsilon}}_{rt} k_o^2 E_t = \nabla_t \left(\nabla_t . E_t + \frac{\partial E_z}{\partial z} \right) \tag{2.11}$$

Using Gauss' law $\nabla.(\bar{\bar{\varepsilon}}_r E) = 0$, $\partial E_z / \partial z$ can be calculated as follows

$$\nabla_t.(\bar{\bar{\varepsilon}}_{rt} E_t) + \frac{\partial \varepsilon_{zz}}{\partial z} E_z + \varepsilon_{zz} \frac{\partial E_z}{\partial z} = 0 \tag{2.12}$$

If ε_{zz} is slowly variant in the z-direction, then $\frac{\partial \varepsilon_{zz}}{\partial z} E_z$ is much smaller than the other two terms in (2.12) and therefore one can obtain

$$\frac{\partial E_z}{\partial z} \approx \frac{-1}{\varepsilon_{zz}} \nabla_t.(\bar{\bar{\varepsilon}}_{rt} E_t) \tag{2.13}$$

Using Equations (2.11) and (2.13), the electric-field-dependent wave equation can be derived as

$$\nabla^2 E_t + \bar{\bar{\varepsilon}}_{rt} k_0^2 E_t = \nabla_t \left(\nabla_t . E_t + \frac{-1}{\varepsilon_{zz}} \nabla_t.(\bar{\bar{\varepsilon}}_{rt} E_t) \right) \tag{2.14}$$

2.6   Perfectly Matched Layer

To create a practical solver, the effects of the simulation boundaries should be considered. Basic BPM and FEM boundary conditions set the field just outside the simulation area to zero, simulating a perfectly conducting metal box. The PML [23] is an artificial absorbing layer for wave equations, commonly used to truncate computational regions in numerical methods to simulate problems with open boundaries, especially in the FDM and FEM methods. The key property of a PML that distinguishes it from an ordinary absorbing material is that it is designed so that the waves incident upon the PML from a non-PML medium do not reflect at the interface. This property allows the PML to strongly absorb outgoing waves from the interior of a computational region without reflecting them back into the interior.

The PML was originally formulated by Berenger in 1994 [23] for use with Maxwell's equations, and since that time there have been several related reformulations of PML for both Maxwell's equations and for other wave equations. Berenger's original formulation is called a split-field PML, because it splits the electromagnetic fields into two unphysical fields in the PML region. A later formulation that has become more popular because of its simplicity and efficiency is called uniaxial PML or UPML [24], in which the PML is described as an artificial anisotropic absorbing material. Although both Berenger's formulation and UPML were initially derived by manually constructing the conditions under which incident plane waves do not reflect from the PML interface from a homogeneous medium, both formulations were later shown to be equivalent to a much more elegant and general approach: stretched-coordinate PML [25, 26]. This approach uses a coordinate transformation in which one (or more) coordinate is mapped to a complex number which is actually an analytic continuation of the wave equation into complex coordinates, replacing propagating (oscillating) waves by exponentially decaying waves. This viewpoint allows PMLs to be derived for inhomogeneous media such as waveguides, as well as for other coordinate systems and wave equations.

The wave equation for the magnetic field can be modified to include the parameters for the PML layers as follows

$$\nabla \times \left(\bar{\bar{k}} \nabla \times H\right) - k_0^2 H = 0 \tag{2.15}$$

where $\bar{\bar{k}} = 1/\bar{\bar{\varepsilon}}_r$ and the del operator ∇ in this case is defined as

$$\nabla = \hat{u}_x \alpha_x \frac{\partial}{\partial x} + \hat{u}_y \alpha_y \frac{\partial}{\partial y} + \hat{u}_z \alpha_z \frac{\partial}{\partial z} = \nabla_T + \hat{u}_z \alpha_z \frac{\partial}{\partial z} \tag{2.16}$$

with $\hat{u}_x$, $\hat{u}_y$ and $\hat{u}_z$ are the unit vectors associated with the x-, y- and z-directions, respectively, and α_x, α_y and α_z are parameters associated with the PML boundary conditions. Since the waves are assumed to propagate along the z-direction, the parameter α_z is set to unity, while the other PML parameters have to be determined such that the wave impedance of the PML layer placed around the computational domain is exactly the same as that of the adjacent medium inside the computational domain. Hence, the PML medium perfectly matches the computational domain medium, which will allow the unwanted radiation to leave the computational domain freely without any reflection. The necessary condition can be derived as [19, 23]

$$\alpha_{x,y} = 1 - j \frac{3\lambda \rho^2}{4\pi n d^3} \ln\left(\frac{1}{R}\right) \tag{2.17}$$

where λ is the wavelength, d is the thickness of the PML (kept constant in all directions), n is the refractive index of the adjacent medium, ρ is the distance from the inner PML interface, R is the reflection coefficient and c is the free-space speed of light. The parameters α_x and α_y are set in different regions as follows. Inside the orthodox computational domain, both α_x and α_y are set to unity, while for PML regions normal to the x-direction, α_x is set as indicated in (2.17), while α_y is set to unity, and the situation is reversed for PML regions normal to the y-direction. For corners, both α_x and α_y are set as indicated in (2.17). With these PML arrangements in different regions, any radiation wave will freely leave the computational domain whatever the angle it hits the PML computational domain boundaries.

2.7   Finite-Element Analysis

The finite-element method [27, 28] is a numerical technique for finding approximate solution of partial differential equations and integral equations. It can be used for modelling a wide class of problems by breaking up the computational domain into elements of simple shapes. For this purpose, suitable shape functions are used to approximate the unknown function within each element. The approximating functions are defined in terms of field variables of specified points called nodes. Therefore, in the finite-element analysis the unknowns are the field variables of the nodes. Once the field variables of the nodes are found, the field variables at any point can be obtained by using the shape functions.

2.7.1   Types of Elements

The elements [27, 28] can be classified as 1D elements, 2D elements and 3D elements, as shown in Figure 2.3. The 1D and 2D elements are suitable for the analysis of 1D and 2D problems, respectively. The three-noded triangular element is an important example of a 2D element, while the tetrahedron is the basic element for 3D problems. The tetrahedron has four nodes, one at each corner, as shown in Figure 2.3(c) [28].

In the finite-element analysis, the selected finite points at which basic unknowns are to be determined are called nodes. The basic unknowns at any point inside the element are determined by using approximating functions in terms of the nodal values of the element. The nodes are either external or internal nodes. The nodes which occur on the edge surfaces of an element are called external nodes, while internal nodes occur inside an element. The nodes which occur at the ends of 1D elements or at the corners in the 2D or 3D elements are also called primary elements. However, the nodes which occur along the sides of an element, but not at the corners, are called secondary nodes.

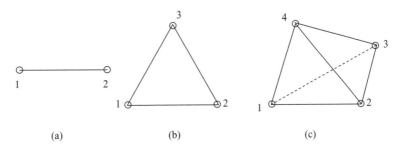

Figure 2.3   (a) 1D (b) 2D and (c) 3D elements.

2.7.2   Shape Functions

The finite-element analysis is used to find the field variables at the nodes. This can be done by assuming that any point inside the element basic variable is a function of values at nodal points of the element. The interpolation or approximating function [28] is the one which relates the field variable at any point within the element to the field variables of nodal points. This is also called shape function. Due to its simple mathematical implementation, polynomial shape functions are commonly used. In addition, any function can be well approximated by using polynomial shape functions. The 1D polynomial shape function of nth order can be given by [28]

$$u(x) = \alpha_1 + \alpha_2 x + \alpha_3 x^2 + \cdots + \alpha_{n+1} x^n \tag{2.18}$$

However, a general form of the 2D polynomial shape function is given by [28]

$$u(x, y) = \alpha_1 + \alpha_2 x + \alpha_3 y + \alpha_4 x^2 + \alpha_5 xy + \alpha_6 y^2 + \alpha_7 x^3 + \cdots + \alpha_m y^n \tag{2.19}$$
$$v(x, y) = \alpha_{m+1} + \alpha_{m+2} x + \alpha_{m+3} y + \cdots + \alpha_{2m} y^n \tag{2.20}$$

It is worth noting that the higher the order of the approximating polynomial, the lower the error in the final solution. In addition, it is required to get expansion polynomials that yield the highest order of approximation with a minimum number of unknowns associated with the element shape.

2.7.3   Galerkin's Procedure

Galerkin's method [28] is used for solving a set of differential equations specified over a region with specific boundary conditions and is briefly explained in this section. Assuming that the governing differential equation of a specified region V takes the following form

$$L(u) = P \tag{2.21}$$

where L is a differential operator on a basic unknown u, the required value of u should satisfy specified values on the boundary of the region. If $\bar{u}$ is taken as approximate solution, then the error $E(x)$ at a point x is given by

$$E(x) = L(\bar{u}) - P \tag{2.22}$$

where $E(x)$ is the residual at point x. In order to get the required solution, the residual relative to weighting function w_i is set to zero, that is

$$\oint_v w_i(L\bar{u} - P)dV = 0 \quad \text{for } i = 1 \text{ to } n \tag{2.23}$$

There are different approaches, depending upon the selection of the weighting function. In Galerkin's method [28], Equation (2.23) is taken as

$$\oint_v \psi(L\bar{u} - P) = 0 \tag{2.24}$$

where ψ is also chosen from the basis function used for constructing the approximate solution function $\bar{u}$. Let

$$\bar{u} = \sum_{i=1}^{n} Q_i G_i \tag{2.25}$$

where Q_i is the basic unknown vector and G_i are basis functions. G_i are usually polynomial in the space coordinates x, y and z. Then, in Galerkin's method, the weighting function ϕ is taken as

$$\phi = \sum_{i=1}^{n} \psi_i G_i \tag{2.26}$$

In the above equation ϕ_i are arbitrary, except at the points where boundary conditions are satisfied. Since ψ is constructed similar to $\bar{u}$, Galerkin's method leads to a simplified method. Thus, in Galerkin's method, the basis function G_i is chosen and ϕ_i is determined in $\bar{u} = \sum_{i=1}^{n} Q_i G_i$ to satisfy $\oint_v (L\bar{u}) - PdV = 0$, where coefficient ϕ_i are arbitrary except at specified boundaries.

2.8 Derivation of BPM Equations

2.8.1 Slowly Varying Envelope Approximation

The slowly varying approximation assumes [15] that, since the simulation follows the propagation of light in the structure, the optical field can be defined in terms of its envelope and rapid phase components, as shown in Figure 2.4 [4], that is

$$H(x, y, z) = h(x, y, z)e^{-jk_o n_o z} \tag{2.27}$$

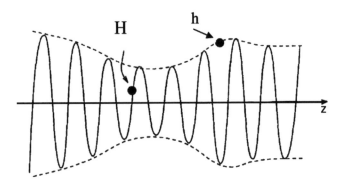

Figure 2.4 Envelope and rapid phase of optical field.

where $h(x, y, z) = h_T(x, y, z) + h_z(x, y, z)$ is the magnetic field's envelope, h_T and h_z are the magnetic field's transverse and axial components, which are given by $h_T = h_x \hat{u}_x + h_y \hat{u}_y$ and $h_z = h_z \hat{u}_z$, respectively. In addition, n_0 is the reference index which is used to satisfy the slowly varying envelope approximation.

Using the divergence relation $\nabla.H = 0$ and the wave equation (2.15), one can obtain, after some mathematical treatments, the following vectorial wave equation in terms of the transverse component [1]

$$\bar{\bar{k}}_a \frac{\partial^2 h_T}{\partial z^2} - 2\gamma \bar{\bar{k}}_a \frac{\partial h_T}{\partial z} - \bar{\bar{k}}_b \nabla_T (\nabla_T.h_T) - \nabla_T \times k_{zz} \nabla_T \times h_T + (\bar{\bar{k}}_c + \gamma^2 \bar{\bar{k}}_a) h_T$$
$$+ \frac{\partial \bar{\bar{k}}_a}{\partial z} \frac{\partial h_T}{\partial z} + \gamma^{-1} \frac{\partial \bar{\bar{k}}_a}{\partial z} \nabla_T \frac{\partial h_z}{\partial z} = 0 \tag{2.28}$$

where $\gamma = jk_o n_o$, $\bar{\bar{k}}_a$, $\bar{\bar{k}}_b$ and $\bar{\bar{k}}_c$ are the transverse tensors given by

$$\bar{\bar{k}}_a = \begin{bmatrix} k_{yy} & -k_{yx} \\ -k_{xy} & k_{xx} \end{bmatrix} \tag{2.29}$$

$$\bar{\bar{k}}_b = \gamma^{-1} \frac{\partial \bar{\bar{k}}_a}{\partial z} - \bar{\bar{k}}_a \tag{2.30}$$

$$\bar{\bar{k}}_c = k_o^2 - \gamma^{-1} \frac{\partial \bar{\bar{k}}_a}{\partial z} \tag{2.31}$$

Using the slowly varying approximation $\left\| \frac{\partial \bar{\bar{k}}_a}{\partial z} \right\| \ll \left\| \frac{\partial h_T}{\partial z} \right\| \ll \left\| \frac{\partial h_z}{\partial z} \right\|$, (2.28) can be simplified to

$$\bar{\bar{k}}_a \frac{\partial^2 h_T}{\partial z^2} - 2\gamma \bar{\bar{k}}_a \frac{\partial h_T}{\partial z} - \bar{\bar{k}}_b \nabla_T (\nabla_T.h_T) - \nabla_T \times k_{zz} \nabla_T \times h_T + (\bar{\bar{k}}_c + \gamma^2 \bar{\bar{k}}_a) h_T = 0$$
$$\tag{2.32}$$

2.8.2 Paraxial and Wide-Angle Approximations

As shown in Equation (2.32), there is a second-order derivative to be solved in the x-, y- and z-directions. The calculation of the second derivative in the z-direction is difficult, therefore a Padé approximation is used. The simplest method is to assume that $\frac{\partial^2}{\partial z} \approx 0$, which is called the Padé approximation of order zero. This approximation is accurate when the envelope of the electric field changes slowly in the z-direction. In addition, it can be used for waveguides which are weakly guiding. However, this approximation is not accurate in the case of structures which guide the light at a large angle to the assumed propagation direction. Therefore, a higher order of $\frac{\partial^2}{\partial z}$ is required, which is called the higher-order Padé approximation (wide-angle situations).

In order to improve the accuracy of the calculation of the propagation of light at large angles to the assumed propagation direction, an approximation of $\frac{\partial^2}{\partial z}$ should be taken into account. However, the memory requirement and computational time for the simulation are increased.

2.8.3 Discretisation by the Finite-Element Method

The FEM is applied to the transverse variation of Equation (2.32) and the cross section of the computational domain Ω is divided into N el triangles with unknowns N_p over the corresponding nodes. In addition, a set of Lagrangian polynomial basis functions of first or second order $\{\psi_j\}$, $j = 1, \ldots, N_p$ are introduced. Moreover, $h_T(x, y, z)$ is defined, such that [1]

$$h_T(x, y, z) = \sum_{j=1}^{N_{px}} h_{xj}(z)\psi_j(x, y)\hat{u}_x + \sum_{j=N_{px}+1}^{N_p} h_{yj}(z)\psi_j(x, y)\hat{u}_y \qquad (2.33)$$

where the coefficients h_{xj} and h_{yj} are the unknown field values on the partition nodes. Then Equation (2.32) can be rewritten in a matrix form, as follows

$$[M]\frac{\partial^2\{h_T\}}{\partial z^2} - 2\gamma[M]\frac{\partial\{h_T\}}{\partial z} + ([K] + \gamma^2[M])\{h_T\} = \{0\} \qquad (2.34)$$

Where $\{h_T\}$ represents a column vector containing the unknowns h_{xj} and h_{yj}, $\{0\}$ is the null column vector and $[M]$ and $[K]$ are the global matrices, given by [1]

$$[M]_{ij} = \int_\Omega \bar{\bar{k}}_a \vec{\psi}_j . \vec{\psi}_i d\Omega \qquad (2.35)$$

$$[K]_{ij} = -\int_\Omega (\bar{\bar{k}}_{zz}\nabla_T \times \vec{\psi}_j).(\nabla_T \times \vec{\psi}_i)d\Omega + \int_\Omega (\nabla_T \times \vec{\psi}_j)\nabla_T.(\bar{\bar{k}}_b^T \vec{\psi}_i)d\Omega$$

$$- \oint_{\partial\Omega} (\nabla_T.\vec{\psi}_j)(\bar{\bar{k}}_b^T \vec{\psi}_i).\hat{n}d\ell + \int_\Omega \bar{\bar{k}}_c \vec{\psi}_j . \vec{\psi}_i d\Omega \qquad (2.36)$$

where $(\ )^{\mathrm{T}}$ denotes the transpose operation, $d\Omega$ introduces all boundaries over the cross-sectional domain Ω, $\hat{n}$ is the outward normal unit vector linked to those boundaries and

$$\vec{\psi}_j = \vec{\psi}_j \hat{u} \tag{2.37}$$

where $\hat{u} = \hat{u}_x$, for $j = 1, \ldots, N_{px}$, and $\hat{u} = \hat{u}_y$, for $j = N_{px} + 1, \ldots, N_p$. Matrices $[M]$ and $[K]$ can also be expressed as a summation of element matrices linked to the x and y coordinates, over all elements e as follows [1]

$$[M] = \sum^e \begin{bmatrix} [M^e_{xx}] & [M^e_{xy}] \\ [M^e_{yx}] & [M^e_{yy}] \end{bmatrix} \tag{2.38}$$

$$[K] = \sum^e \begin{bmatrix} [K^e_{xx}] & [K^e_{xy}] \\ [K^e_{yx}] & [K^e_{yy}] \end{bmatrix} \tag{2.39}$$

where

$$\left[M^e_{xx}\right] = k^e_{yy}\left[S^e_1\right] \tag{2.40}$$

$$\left[M^e_{xy}\right] = -k^e_{yx}\left[S^e_1\right] \tag{2.41}$$

$$\left[M^e_{yx}\right] = -k^e_{xy}\left[S^e_1\right] \tag{2.42}$$

$$\left[M^e_{yy}\right] = k^e_{xx}\left[S^e_1\right] \tag{2.43}$$

$$\begin{aligned}\left[K^e_{xx}\right] &= -\alpha^2_y k^e_{zz}\left[S^e_3\right] - \alpha^2_x k^e_{bxx}\left[S^e_2\right] - \alpha_x\alpha_y k^e_{bxy}\left[S^e_4\right] \\ &\quad - \alpha_x(k^e_{bxx}n^e_x + k^e_{bxy}n^e_y)\left[L^e_1\right] + k^e_{cxx}\left[S^e_1\right]\end{aligned} \tag{2.44}$$

$$\begin{aligned}\left[K^e_{xy}\right] &= -\alpha_x\alpha_y k^e_{zz}\left[S^e_4\right] - \alpha_x\alpha_y k^e_{bxx}\left[S^e_4\right]^T - \alpha^2_y k^e_{bxy}\left[S^e_3\right] \\ &\quad - \alpha_y(k^e_{bxx}n^e_x + k^e_{bxy}n^e_y)\left[L^e_2\right] + k^e_{cxx}\left[S^e_1\right]\end{aligned} \tag{2.45}$$

$$\begin{aligned}\left[K^e_{yx}\right] &= -\alpha_x\alpha_y k^e_{zz}\left[S^e_4\right]^T - \alpha^2_x k^e_{byx}\left[S^e_2\right] - \alpha_x\alpha_y k^e_{byy}\left[S^e_4\right] \\ &\quad - \alpha_x(k^e_{byx}n^e_x + k^e_{byy}n^e_y)\left[L^e_1\right] + k^e_{cyx}\left[S^e_1\right]\end{aligned} \tag{2.46}$$

$$\begin{aligned}\left[K^e_{yy}\right] &= -\alpha^2_x k^e_{zz}\left[S^e_2\right] - \alpha_x\alpha_y k^e_{byx}\left[S^e_4\right]^T - \alpha^2_y k^e_{byy}\left[S^e_3\right] \\ &\quad - \alpha_y(k^e_{byx}n^e_x + k^e_{byy}n^e_y)\left[L^e_2\right] + k^e_{cyy}\left[S^e_1\right]\end{aligned} \tag{2.47}$$

where k^e_{zz}, k^e_{rs}, and k^e_{lrs} are the average values of the components k_{zz}, k_{rs} and k_{lrs}, respectively over the element e. The sub-indexes (r, s) are the coordinate pair (x, y) and sub-index l represents b or c. In addition, the auxiliary element matrices

$[S^e_{1,2,3,4}]$ and $[L^e_{1,2}]$ are taken as

$$[S^e_1] = \int_{\Omega^e} \{\psi^e\}\{\psi^e\}^T d\Omega \qquad (2.48)$$

$$[S^e_2] = \int_{\Omega^e} \frac{\partial\{\psi^e\}}{\partial x}\frac{\partial\{\psi^e\}^T}{\partial x} d\Omega \qquad (2.49)$$

$$[S^e_3] = \int_{\Omega^e} \frac{\partial\{\psi^e\}}{\partial y}\frac{\partial\{\psi^e\}^T}{\partial y} d\Omega \qquad (2.50)$$

$$[S^e_4] = \int_{\Omega^e} \frac{\partial\{\psi^e\}}{\partial y}\frac{\partial\{\psi^e\}^T}{\partial x} d\Omega \qquad (2.51)$$

$$[L^e_1] = \int_{\partial\Omega^e} \{\psi^e\}\frac{\partial\{\psi^e\}^T}{\partial x} d\ell \qquad (2.52)$$

$$[L^e_2] = \int_{\partial\Omega^e} \{\psi^e\}\frac{\partial\{\psi^e\}^T}{\partial y} d\ell \qquad (2.53)$$

where $\{\psi^e\}$ represents a column vector containing the corresponding shape functions. In addition, Ω^e and $\partial\Omega^e$ are the element's area and boundary respectively and n^e_x and n^e_y are the x and y components of the outward normal unit vector, respectively, linked to $\partial\Omega^e$.

Assuming the Padé approximation, Equation (2.34) can be transformed into the following matrix equation

$$[\tilde{M}]\frac{d\{h_T\}}{dz} + [K]\{h_T\} = \{0\} \qquad (2.54)$$

where

$$[\tilde{M}] = [M] - \frac{1}{4\gamma^2}([K] + \gamma^2[M]) \qquad (2.55)$$

The θ-finite-difference equation can be obtained as

$$([\tilde{M}(z)] + \theta\Delta z[K(z)])\{h_T(z+\Delta z)\} = ([\tilde{M}(z)] - (1-\theta)\Delta z[K(z)])\{h_T(z)\} \qquad (2.56)$$

where Δz is the step size along the propagation direction. Equation (2.56), for transverse magnetic fields, can be solved by an iterative procedure to get the required magnetic fields. At each iteration step, a matrix equation can be solved by employing the ORTHOMIN algorithm [29]. The parameter θ is introduced to control the scheme used to solve the finite difference equations. This scheme is unconditionally stable if $\theta \geq 0.5$.

2.9   Imaginary-Distance BPM: Mode Solver

For modal solution purposes, the imaginary-distance full-vectorial FE-BPM scheme proposed in [5] is extended further to fully treat the vectorial complex modes, that is, for the ℓth mode whose effective index and field distribution are assumed to be $n_{eff,\ell}$ and $\{h_{t,l}\}$, respectively. By relaxing the derivative terms in Equation (2.34) to zero, the following modal analysis matrix equation for the ℓth mode can be written as

$$[K]\{h_{t,l}\} = -\gamma^2[M]\{h_{t,l}\} \tag{2.57}$$

Following the kth propagation step, and using Equations (2.56) and (2.57), the field distribution of the ℓth mode yields

$$\{h_{t,l}\}_{k+1} = \frac{-2\gamma - 0.5\Delta z k_o^2 \left(n_{eff,\ell}^2 - n_o^2\right)}{-2\gamma + 0.5\Delta z k_o^2 \left(n_{eff,\ell}^2 - n_o^2\right)}\{h_{t,l}\}_k \tag{2.58}$$

Assuming that there exist m modes (including radiation modes) in the waveguide, the propagating field, $\{h_t\}_k$, at the kth propagation step can be written as

$$\{h_t\}_k = \sum_{\ell=1}^{m} C_{\ell,k}\{h_{t,l}\} \tag{2.59}$$

where $C_{\ell,k}$ is the complex amplitude of the ℓth mode at the kth propagation step. If the propagation step size, Δz, is selected as

$$\Delta z \approx j\frac{4n_o}{\left(n_{eff,\ell}^2 - n_o^2\right)k_o} \tag{2.60}$$

then, for a sufficiently large number of propagation steps, $\{h_t\}_k$ will converge to the ℓth mode eigenvector $\{h_{t,l}\}_k$ and its effective index, $n_{eff,\ell}$, can be obtained using

$$n_{eff,\ell,k}^2 = \frac{\{h_t\}_k^*[K]_k\{h_t\}_k}{k_o^2\{h_t\}_k^*[M]_k\{h_t\}_k} \tag{2.61}$$

at the kth propagation step, where the symbol $(\ )^*$ denotes complex conjugate and transpose. However, at the start of propagation the value of the effective index of the desired mode is not known, this being required to determine the step size Δz using

Equation (2.60). So, at the beginning, Δz is calculated with the effective index taken as the largest index of refraction in the structure and, with the iterative adaptation at each propagation step, the propagating field and the calculated effective index are eventually seen to be converging to the desired mode. The value of the reference index of refraction, n_0, may be arbitrarily chosen. However, n_0 is chosen as the smallest index of refraction in the structure so as to make the imaginary part of the complex step size always positive, and as a result, convergent propagation toward the desired mode is achieved. Using the above procedure, an arbitrary starting field, $\{h_t\}_k$, will converge to the fundamental TE or TM mode depending on whether it is polarised in the y- or x-directions, respectively. However, to calculate one of the higher-order modes, such as the ith mode, all the lower-order modes $(i - 1)$ should be filtered out from the spectrum of the starting field, to yield

$$\{h_t\}_{1,new} = \{h_t\}_1 - \sum_{\ell=1}^{i-1} \frac{\{h_{t,l}\}^*[M]\{h_t\}_1}{\{h_{t,l}\}^*[M]\{h_{t,l}\}}\{h_{t,l}\} \tag{2.62}$$

Thus, using Equation (2.62) as a new starting field, the propagating field will converge to the desired ith-order mode without converging to any of the lower-order modes.

References

[1] Patrocinio da Silva, J., Hugo, E., Figueroa, H. and Frasson, A.M.F. (2003) Improved vectorial finite-element BPM analysis for transverse anisotropic media. *J. Lightwave Technol.*, **21** (2), 567–576.
[2] Huang, W.P., Xu, C., Chu, S.T. and Chaudhuri, S.K. (1992) The finite difference vector beam propagation method: Analysis and assessment. *J. Lightwave Technol.*, **10**, 295–305.
[3] Xu, C.L., Huang, W.P., Chrostowski, J. and Chaudhuri, S.K. (1994) A full-vectoriai beam propagation method for anisotropic waveguides. *J. Lightwave Technol*, **12** (11) 1926–1931.
[4] Ma, F., Xu, C. L. and Huang, W. P. (1996) Wide-angle full vectorial beam propagation method. *IEE Proc. Optoelectron.*, **143** (2) 139–143.
[5] Obayya, S.S.A., Azizur Rahman, B.M., Grattan, Kenneth T.V. and El-Mikati, H.A. (2002) Full vectorial finite-element-based imaginary distance beam propagation solution of complex modes in optical waveguides. *J. Lightwave Technol.*, **20**, 1054.
[6] Feit, M.D. and Fleck, J.J.A. (1978) Light propagation in graded-index optical fibers. *Appl. Opt.*, **17** (24), 3990–3998.
[7] Hendow, S.T. and Shakir, S.A. (1986) Recursive numerical solution for nonlinear wave propagation in fibers and cylindrically symmetric systems. *Appl. Opt.*, **25** (11), 1759–1764.
[8] Chung, Y. and Dagli, N. (1990) An assessment of finite difference beam propagation method. *IEEE J. Quantum Electron.*, **26** (8), 1335–1339.
[9] Yevick, D. and Hermansson, B. (1990) Efficient beam propagation techniques. *IEEE J. Quantum Electron.*, **26**, 109–112.
[10] Lee, P.C., Schulz, D. and Voges, E. (1992) Three-dimensional finite difference beam propagation algorithm for photonic devices. *J. Lightwave Technol.*, **10**, 1832–1838.

[11] Liu, P.L., Yang, S.L. and Yuan, D.M. (1993) The semivectorial beam propagation method. *IEEE J. Quantum Electron.*, **29**, 1205–1211.

[12] Clauberg, R. and von Allmen, P. (1991) Vectorial beam-propagation method for integrated optics. *Electron. Lett.*, **27** (8), 654–655.

[13] Chund, Y., Dagli, N. and Thylen, L. (1991) Explicit finite difference vectorial beam propagation method. *Electron. Lett.*, **27** (23), 2119–2121.

[14] Huang, W.P., Xu, C.L. and Chaudhuri, S.K. (1992) A finite-difference vector beam propagation method for three-dimensional waveguide structures. *IEEE Photon. Technol. Lett.*, **4**, 148–151.

[15] Huang, W.P. and Xu, C.L. (1993) Simulation of three-dimensional optical waveguides by a full-vector beam propagation method. *IEEE J. Quantum Electron.*, **29** (10), 2639–2649.

[16] Xu, C. and Huang, W.P. (1995) Finite-difference beam propagation method for guided-wave optics. *Prog. Electromag. Res., PIER*, **11**, 1–49.

[17] Polstyanko, S.V., Dyczij-Edlinger, R. and Lee, J.F. (1996) Full vectorial analysis of a nonlinear slab waveguide based on the nonlinear hybrid vector finite-element method. *Opt. Lett.*, **21**, 98–100.

[18] Montanari, E., Selleri, S., Vincetti, L. and Zoboli, M. (1998) Finite element full vectorial propagation analysis for three dimensional z-varying optical waveguides. *J. Lightwave Technol.*, **16**, 703–714.

[19] Koshiba, M., Tsuji, Y. and Hikari, M. (1999) Finite-element beam-propagation method with perfectly mathed layers boundary conditions. *IEEE Trans. Magn.*, **35**, 1482–1485.

[20] Obayya, S.S.A., Rahman, B.M.A. and El-Mikati, H.A. (2000) New Full Vectorial Numerically Efficient Propagation Algorithm Based on the Finite Element Method. *J. Lightwave Technol.*, **18** (3), 409–415.

[21] Obayya, S.S.A., Rahman, B.M.A. and El-Mikati, H.A. (2000) Full vectorial finite element beam propagation method for nonlinear directional coupler devices. *IEEE J. Quantum Electron.*, **36**, 556–562.

[22] Fallahkhair, A.B., Li, K.S. and Murphy, T.E. (2008) Vector finite difference modesolver for anisotropic dielectric waveguides. *J. Lightwave Technol.*, **26** (11), 1423–1431.

[23] Berenger, J. (1994) A perfectly matched layer for the absorption of electromagnetic waves. *J. Comput. Phys.*, **114**, 185–200.

[24] Gedney, S.D. (1996) An anisotropic perfectly matched layer absorbing media for the truncation of FDTD latices. *Antennas Prop., IEEE Trans.*, **44**, 1630–1639.

[25] Chew, W.C. and Weedon, W.H. (1994) A 3d perfectly matched medium from modified Maxwell's equations with stretched coordinates. *Microwave Opt. Technol. Lett.*, **7**, 590–604.

[26] Chew, W.C., Jin, J.M. and Michielssen, E. (1997) Complex Coordinate Stretching as a Generalized Absorbing Boundary Condition. *Microwave and Opt. Technol. Lett.*, **15** (6), 363–369.

[27] Volakis, J.L., Chatterjee, A. and Kempel, L.C. (1998) *Finite Element Method for Electromagnetics: Antennas, Microwave Circuits, and Scattering Applications*, John Wiley & Sons, Inc., Hoboken.

[28] Bhavikatti, S.S. (2005) *Finite Element Analysis*, New Age International (P) Ltd., Publishers, New Delhi.

[29] Behie, A. and Vinsome, P.K.W. (1982) Block iterative methods for fully implicit reservoir simulation. *Soc. Pet. Eng. J.*, **22** (5), 658–668.

3

Assessment of the Full-Vectorial Beam Propagation Method

3.1 Introduction

In this chapter, the numerical precision of the vectorial finite-element beam propagation method (VFEBPM) is demonstrated through the analysis of a single optical waveguide [1], a rectangular directional coupler, an electro-optic modulator and switches. In addition, a comparison between the VFEBPM and the Marcatili analytical approach [2] is investigated thoroughly. The different modal properties of the photonic crystal fibre (PCF) [3,4] is also examined, using the imaginary-distance full-vectorial finite-element beam propagation method (IDVFEBM) [4]. Moreover, the liquid-crystal PCF [5] and its application to polarisation rotator is introduced.

3.2 Analysis of Rectangular Waveguide

3.2.1 Effect of Longitudinal Step Size

To assess the numerical precision of the proposed VFEBPM, it is first applied to a single rectangular guide [1] shown in the inset of Figure 3.1. The refractive index of the core, n_g, is taken as 3.26, while that of the substrate, n_s, is fixed to 3.20 at the operating wavelength of 1.3 μm. In all simulations, the width of the perfectly matched layer (PML) is taken as 1.0 μm, the theoretical reflection coefficient, R, as 10^{-100}, and the reference refractive index, n_0 as the mean value of the core and substrate indexes. The rectangular waveguide cross section is represented by 7200 first-order triangular elements. The waveguide is launched with the fundamental transverse electric (TE) or transverse magnetic (TM) modal field profiles obtained from the vector finite-element

Computational Photonics   Salah Obayya
© 2011 John Wiley & Sons, Ltd

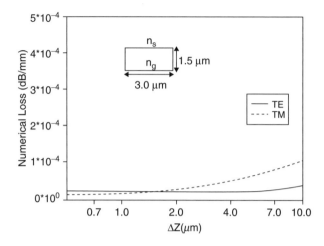

Figure 3.1 Effect of the propagation step size on nonphysical power loss for both TE and TM waves. The inset is a schematic diagram of the rectangular guide under consideration. (Reproduced with permission from Obayya, S.S.A., Rahman, B.M.A. and El-Mikati, H.A. (2000) New full vectorial numerically efficient propagation algorithm based on the finite element method. *IEEE J. Lightwave Technol.*, **18** (3), 409–415. © 2000 IEEE.)

method (VFEM) [6]. Figure 3.1 shows the effect of the propagation step size, Δz, on the level of numerical dissipation for the two polarised modes, TE and TM. The time taken for each simulation point in Figure 3.1 is around 10 min using a SUN 4/85 Workstation. It is suspected that some stable numerical algorithms may not conserve the propagating beam power. However, it is revealed from Figure 3.1 that for the range $\Delta z < 2.0$ µm, the nonphysical power loss is less than 0.000 02 dB/mm for both polarised modes. Therefore, the proposed VFEBPM can be regarded as a stable and power-conserving technique as well.

Next, the capability of the proposed VFEBPM as a 'mode solver' is investigated. Initially, an arbitrary initial field is launched into the considered rectangular waveguide, and allowed to propagate along the imaginary axis. The fundamental TE or TM mode is then evolved [7].

To test the accuracy of the solution, a simple rectangular spatial pulse with a sharp step rise is launched into the waveguide. For TE excitation, the field profile of its dominant H_y component at $z = 50$ µm is shown in Figure 3.2. This field profile resembles the fundamental TE mode after propagating a relatively short distance of 50 µm. The variation of the effective indices, $n_{\text{eff}} = \beta/k$, where β is the mode propagation constant, for both TE and TM modes with the imaginary propagation distance are also calculated. The effective indices for the TE and TM modes are found to be 3.241 25 and 3.241 21, respectively, after propagating a distance of around 40 µm along the imaginary axis.

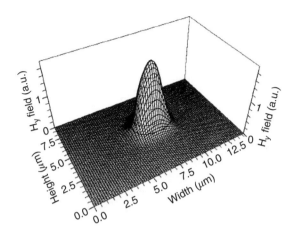

Figure 3.2 The fundamental component H_y field profile for the quasi-TE mode at $z = 50\ \mu\text{m}$. (Reproduced with permission from Obayya, S.S.A., Rahman, B.M.A. and El-Mikati, H.A. (2000) New full vectorial numerically efficient propagation algorithm based on the finite element method. *IEEE J. Lightwave Technol.*, **18** (3), 409–415. © 2000 IEEE.)

The same fundamental TE and TM modes have also been rigorously solved by using the VFEM [6] with 12 800 first-order triangular elements to discretise the waveguide cross section. The variation of the percentage errors in calculating the TE and TM effective indices using the VFEBPM are shown in Figure 3.3. It is revealed from this figure that the percentage errors were only 0.0012 and 0.0018% for the TE and TM modes, respectively. This proves the accuracy of the proposed VFEBPM approach.

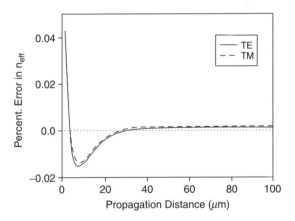

Figure 3.3 Variation of the errors in the n_{eff} calculation of the fundamental TE and TM modes with the longitudinal imaginary distance. (Reproduced with permission from Obayya, S.S.A., Rahman, B.M.A. and El-Mikati, H.A. (2000) New full vectorial numerically efficient propagation algorithm based on the finite element method. *IEEE J. Lightwave Technol.*, **18** (3), 409–415. © 2000 IEEE.)

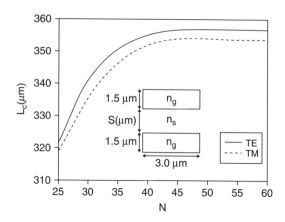

Figure 3.4   Effect of the transverse mesh divisions on the coupling lengths of TE and TM waves. The inset is a schematic diagram of the directional coupler under consideration. (Reproduced with permission from Obayya, S.S.A., Rahman, B.M.A. and El-Mikati, H.A. (2000) New full vectorial numerically efficient propagation algorithm based on the finite element method. *IEEE J. Lightwave Technol.*, **18** (3), 409–415. © 2000 IEEE.)

3.2.2   Analysis of Rectangular Directional Coupler

The inset of Figure 3.4 shows a simple rectangular directional coupler consisting of two vertically coupled rectangular guides. The core index, n_g, and the substrate index, n_s, of the directional coupler are taken as 3.26 and 3.20, respectively, and the waveguide separation, S, is fixed at 0.8 μm. In addition, the operating wavelength is equal to 1.3 μm. The effect of the transverse mesh divisions on the two polarised modes, TE and TM, polarised waves is first studied. Initially, one of the guides with its isolated TE or TM fundamental modal field profile is launched. The coupling length can be defined as the minimum distance at which a maximum power transfer between the guides occurs. The effect of transverse mesh divisions on the coupling length for both TE and TM polarised waves is shown in Figure 3.4, where N is the number of divisions in either the x- or y-directions. In this case, Δz is taken as 0.5 μm. It is evident from Figure 3.4 that for $N > 45$, the coupling lengths settle to 357 μm and 353 μm for TE and TM polarised waves, respectively. For $N = 60$, in both transverse directions, the effect of Δz on the coupling length for both TE and TM polarised waves is shown in Figure 3.5. As shown from this figure, the coupling lengths for the two polarised waves are nearly constant for $\Delta z < 1.0$ μm.

In order to test the accuracy of the VFEBPM in calculating the coupling length, the same directional coupler structure has also been analysed using the VFEM [6] with 12 800 first-order triangular elements. The coupling length L_c can be calculated using the propagation constants of the even and odd supermodes obtained by

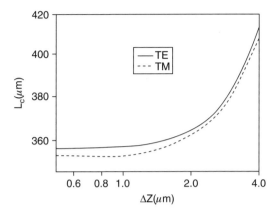

Figure 3.5 Effect of the propagation step size (Δz) on the coupling lengths of the TE and TM waves. (Reproduced with permission from Obayya, S.S.A., Rahman, B.M.A. and El-Mikati, H.A. (2000) New full vectorial numerically efficient propagation algorithm based on the finite element method. *IEEE J. Lightwave Technol.*, **18** (3), 409–415. © 2000 IEEE.)

the VFEM, as follows

$$L_C = \frac{\pi}{\beta^e - \beta^o} \tag{3.1}$$

where β^e and β^o are the propagation constants of the even and odd supermodes, respectively. For this coupler, the coupling lengths obtained by using the VFEM are 356.2 and 354.2 µm for TE and TM polarisations, respectively. The percentage difference between the coupling length results obtained using the VFEBPM and the VFEM are 0.22 and 0.34% for TE and TM polarisations, respectively, which shows the high numerical accuracy of the newly developed VFEBPM in calculating the coupling length for both TE and TM polarisations.

3.2.3 Effect of Structure Geometrical Parameters

The effect of varying the waveguide separation, S, and the core index, n_g, of the rectangular directional coupler, shown in the inset of Figure 3.4, is also considered. Figure 3.6 shows the variation of the coupling length of the two polarised modes with the separation, S, at different core index values. In all cases, it may be noted that the coupling lengths for TE and TM increase exponentially (linear for a semi-log scale) with increasing S. For lower core indices, cases 'a', $n_g = 3.26$ and 'b', $n_g = 3.3$, the polarisation effect on the coupling length is negligible. However, for case 'c', $n_g = 3.4$, the coupling length for TM is relatively higher than its TE counterpart. These features, related to the polarisation effect on the coupling length, cannot be

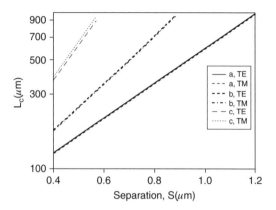

Figure 3.6 Variation of TE and TM coupling lengths with the waveguide separation (*S*) and the index difference (*Δn*) as a parameter. (Reproduced with permission from Obayya, S.S.A., Rahman, B.M.A. and El-Mikati, H.A. (2000) New full vectorial numerically efficient propagation algorithm based on the finite element method. *IEEE J. Lightwave Technol.*, **18** (3), 409–415. © 2000 IEEE.)

accurately predicted using less accurate scalar BPM algorithms. In all cases, the TE and TM coupling lengths have been recalculated using the VFEM with 12 800 first-order triangular elements and the differences between these results and those obtained using the VFEBPM are always less than 0.8%. It is worth noting that the accuracy of the solutions may be improved by using a finer mesh discretisation.

3.2.4   FV-BPM Versus Analytical Marcatili's Approach

The Marcatili [2] analytical approach is a simple method to calculate the effective refractive index of a channel waveguide. In addition, it can be used for studying the coupling coefficients of a simple channel directional coupler. Figure 3.7 shows a cross section of the rectangular directional coupler studied by Marcatili [2]. The coupler cross section is sub-divided into areas of refractive indices from n_1 to n_5. In the vth

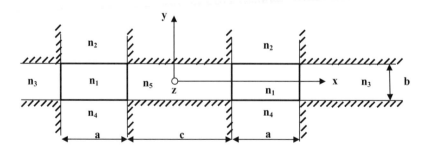

Figure 3.7   Coupler cross section sub-divided for analysis.

medium ($v = 1, 2, 3, 4, 5$), the refractive index is n_v and the propagation constants k_{xv}, k_{yv}, k_z are related by

$$k_{xv}^2 + k_{yv}^2 + k_z^2 = \omega^2 \varepsilon \mu n_v^2 = k_v^2 \tag{3.2}$$

where k_z is the axial propagation constant, while k_{xv} and k_{yv} are the transverse propagation constants in medium v along x and y directions, respectively. The propagation constant k_v of a plane wave in a medium of refractive index n_v can be computed by $k_v = (2\pi/\lambda)n_v$. Marcatili [2] observed that most of the power travels in region 1, a small part travels in regions 2, 3, 4 and 5, and even less travels in the six shaded areas. Therefore, the refractive indices in the six shaded areas are not specified. The coupling length L_c can also be calculated from [2]

$$\frac{\pi}{2L_C} = 2\frac{k_x^2}{k_z}\frac{\xi_5}{a}\frac{\exp(-c/\xi_5)}{1 + k_x^2\xi_3^2} \tag{3.3}$$

where a is the width of the two cores of refractive index, n_1, c is the distance between the two cores and ξ_3 and ξ_5 measure the penetration depths of the field components in media 3 and 5.

For the TM modes of the fundamental component, E_{pq}^y, where the sub-indexes p and q indicate the number of extrema each component has within the guide, k_x, k_y, k_z can be given by [2]

$$k_x = \frac{p\pi}{a}\left(1 + \frac{A_3 + A_5}{\pi a}\right)^{-1} \tag{3.4}$$

$$k_y = \frac{q\pi}{b}\left(1 + \frac{n_2^2 A_2 + n_4^2 A_4}{\pi n_1^2 b}\right)^{-1} \tag{3.5}$$

$$k_z = \left(k_1^2 - \left(\frac{p\pi}{a}\right)^2\left(1 + \frac{A_3 + A_5}{\pi a}\right)^{-2} - \left(\frac{q\pi}{b}\right)^2\left(1 + \frac{n_2^2 A_2 + n_4^2 A_4}{\pi n_1^2 b}\right)^{-2}\right)^{1/2} \tag{3.6}$$

where $A_{2,3,4,5}$ are defined such that

$$A_{2,3,4,5} = \frac{\lambda}{2\left(n_1^2 - n_{2,3,4,5}^2\right)^{1/2}} \tag{3.7}$$

In addition, $\xi_{3,5}$ can be calculated by

$$\xi_{3,5} = \frac{A_{3,5}}{\pi}\left[1 - \left[\frac{pA_{3,5}}{a}\frac{1}{1 + \frac{A_3 + A_5}{\pi a}}\right]^2\right]^{-1/2} \tag{3.8}$$

For the TE modes, E^x_{pq}, k_x, k_y, k_z can be given by [2]

$$k_x = \frac{p\pi}{a}\left(1 + \frac{n_3^2 A_3 + n_5^2 A_5}{\pi n_1^2 a}\right)^{-1} \tag{3.9}$$

$$k_y = \frac{q\pi}{b}\left(1 + \frac{A_2 + A_4}{\pi b}\right)^{-1} \tag{3.10}$$

$$k_z = \left(k_1^2 - \left(\frac{p\pi}{a}\right)^2\left(1 + \frac{n_3^2 A_3 + n_5^2 A_5}{\pi n_1^2 a}\right)^{-2} - \left(\frac{q\pi}{b}\right)^2\left(1 + \frac{A_2 + A_4}{\pi b}\right)^{-2}\right)^{1/2} \tag{3.11}$$

In addition, $\xi_{3,5}$ are defined such that

$$\xi_{3,5} = \frac{A_{3,5}}{\pi}\left[1 - \left[\frac{pA_{3,5}}{a}\frac{1}{1 + \frac{n_3^2 A_3 + n_5^2 A_5}{\pi n_1^2 a}}\right]^2\right]^{-1/2} \tag{3.12}$$

We will now compare the results of the VFEBPM and the Marcatili analytical approach [2] and those reported by Kumar *et al.* [8]. A more accurate analysis than the Marcatili approach [2] has been made in [8], taking the effect of the corner regions into account through first-order perturbation theory. In the comparison, a simple rectangular directional coupler, as shown in the inset of Figure 3.8, is used. The coupling

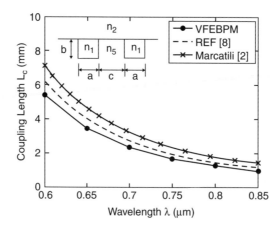

Figure 3.8 Variation of coupling length L_c as a function of wavelength for a rectangular core directional coupler with a = b = c = 2.0 μm.

length L_c for E_{11}^y is calculated with the parameters, $n_1 = 2.211$, $n_2 = 1.0$, $n_3 = n_4 = n_5 = 2.2$ and $a = b = c = 2.0$ μm. Figure 3.8 shows the variation of the coupling length as a function of the wavelength. It is evident from this figure that the results of Marcatili's approach are larger than those reported by [8] and the VFEBPM. In Marcatili's approach [2], the powers in the shaded areas are neglected. In addition, Kumar *et al.* [9] showed that Marcatili's results for a rectangular directional coupler can be improved by taking the effect of corner regions into account. Moreover, they demonstrated that Marcatili's approach is not accurate for small values of depth of channel waveguides [8]. Also, Kuznetsov [10] reported that the expressions of Marcatili overestimate the coupling coefficient, especially in the weaker guiding cases, by as much as a factor of 2.

3.3 Photonic Crystal Fibre

Figure 3.9 shows a schematic diagram of a photonic crystal fibre (PCF), consisting of two rings of arrays of air holes arranged in a silica substrate whose refractive index is taken as 1.45 at a wavelength of 1.55 μm, and where d is the hole diameter and Λ is the hole pitch. Owing to the symmetry of the structure along both the x- and y-directions, only one quarter of the structure has been represented using 16 200

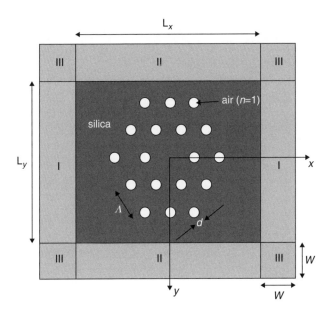

Figure 3.9 Schematic diagram of a two-ring PCF of 18 air holes. (Reproduced with permission from Obayya, S.S.A., Rahman, B.M.A. and Grattan, K.T.V. (2005) Accurate finite element modal solution of photonic crystal fibres. *Optoelectron. IET Proc.*, **152** (5), 241–246. © 2005 IET.)

first-order triangular elements. In all the simulations, the computational window area $L_x/2 \times L_y/2$ has been taken to be 10 μm × 10 μm, and this has been terminated by a PML whose width, W, is 1.0 μm and split into five divisions, where the theoretical reflection coefficient, R, is set to 10^{-8}. The parameters α_x and α_y are set in different regions as follows. Inside the orthodox computational domain, both α_x and α_y are set to unity, while for PML regions normal to the x-direction, α_x is set, as indicated in Equation (2.17), while α_y is set to unity, and the situation is reversed for PML regions normal to the y-direction. For corners, both α_x and α_y are set as indicated in Equation (2.17). Only the fundamental H_{11}^y mode has been considered, as it is degenerate with the H_{11}^x mode, owing to the rotational symmetry of the structure.

3.3.1 Effective Index of Modes

The variation of the real part of the complex effective index of a two-ring PCF with the hole pitch, Λ, at different d/Λ ratios is shown in Figure 3.10 [4]. It is evident from this figure that the effective index of the fundamental mode increases monotonically with increasing hole pitch, Λ, or decreasing d/Λ. Figure 3.11 shows the field profiles of the dominant H_y component of the fundamental H_{11}^y mode for a hole pitch value of

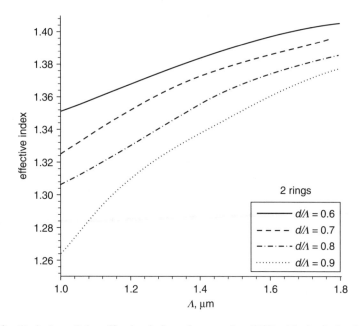

Figure 3.10 Variation of the effective index of a two-ring PCF with the hole pitch, Λ, with d/Λ as a parameter. (Reproduced with permission from Obayya, S.S.A., Rahman, B.M.A. and Grattan, K.T.V. (2005) Accurate finite element modal solution of photonic crystal fibres. *Optoelectron. IET Proc.*, **152** (5), 241–246. © 2005 IET.)

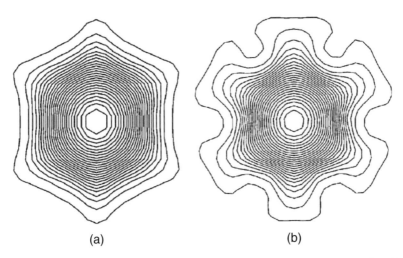

(a)   (b)

Figure 3.11   The dominant H_y field distributions of the fundamental TE mode for a hole pitch value of 1.8 μm and two different values of d/Λ: (a) $d/\Lambda = 0.8$ (b) $d/\Lambda = 0.6$. (Reproduced with permission from Obayya, S.S.A., Rahman, B.M.A. and Grattan, K.T.V. (2005) Accurate finite element modal solution of photonic crystal fibres. *Optoelectron. IET Proc.*, **152** (5), 241–246. © 2005 IET.)

1.8 μm and two different values of the d/Λ ratio, 0.6 and 0.8. It is revealed from this figure that the dominant H_y field component is more confined in the core region at $d/\Lambda = 0.8$ than its counterpart at $d/\Lambda = 0.6$. Such a confinement feature of the mode to the core region is directly linked to how much the mode is 'leaking' into the outer air-hole region. This effect can be shown more clearly by inspecting quantitatively the variation of the confinement loss with the hole pitch for different values of the d/Λ ratio.

3.3.2   Losses

The confinement loss of the fundamental mode can be computed from the imaginary part of the complex effective index n_{eff}

$$\text{Confinement Loss (dB/m)} = 8.686 \times 10^6 \times k_o \times \text{Im}(n_{\text{eff}}) \qquad (3.13)$$

where $\text{Im}(n_{\text{eff}})$ stands for the imaginary part. Figure 3.12 shows the variation of the confinement losses of a two-ring PCF with the hole pitch, Λ, at different d/Λ ratios. As may be noted from Figure 3.12, the confinement loss decreases rapidly with increasing hole pitch Λ or the ratio d/Λ. At $d/\Lambda = 0.9$, as the hole pitch Λ increases from 1.0 to 1.8 μm, the confinement loss drops from a value of nearly 3100 to 0.023 dB/m,

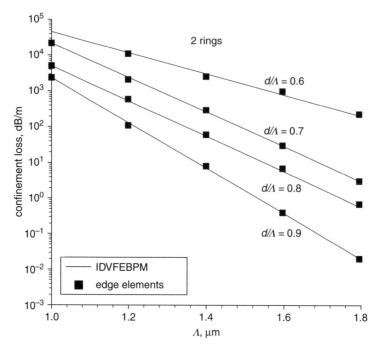

Figure 3.12 Variation of the confinement losses of a two-ring PCF with the hole pitch, Λ, with d/Λ as parameter. (Reproduced with permission from Obayya, S.S.A., Rahman, B.M.A. and Grattan, K.T.V. (2005) Accurate finite element modal solution of photonic crystal fibres. *Optoelectron. IET Proc.*, **152** (5), 241–246. © 2005 IET.)

respectively. Similarly when $\Lambda = 1.8$ μm, the confinement loss reaches a peak of nearly 221 dB/m as the ratio d/Λ is decreased from 0.9 to 0.6.

Next, for three rings of 36 air holes and four rings of 60 air holes, the variations of the confinement loss with the hole pitch, Λ, with the ratio d/Λ as a parameter are shown in Figures 3.13 and 3.14, respectively. It can be observed from these figures that the confinement loss can be improved by using a bigger hole pitch, Λ, and/or a bigger value of d/Λ (i.e. wider air holes). In addition, the confinement loss can be decreased significantly by increasing the number of rings of air holes. In this case, by increasing the number of air holes, the mode tends to become more confined to the core region. It is also worth noting that the variations of the effective index with the hole pitch, Λ, with the ratio d/Λ taken as a parameter are almost independent of the number of rings. The confinement loss results for two and three rings of air holes obtained here using IDVFEBPM are in excellent agreement with their counterparts obtained in [11], using the vector edge elements, and in [12], using the multipole method. The difference between the confinement loss results obtained using the IDVFEBPM and both the multipole and edge-element methods is less than 0.1%. However, the present IDVFEBPM is formulated in terms of only two transverse magnetic-field components.

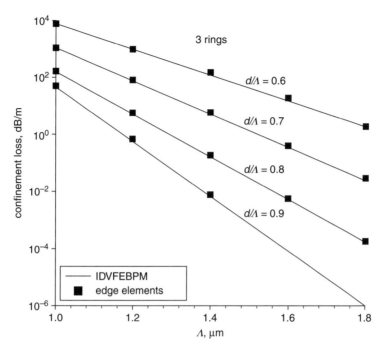

Figure 3.13 Variation of the confinement losses of a three-ring PCF with the hole pitch, Λ, with d/Λ as parameter. (Reproduced with permission from Obayya, S.S.A., Rahman, B.M.A. and Grattan, K.T.V. (2005) Accurate finite element modal solution of photonic crystal fibres. *Optoelectron. IET Proc.*, **152** (5), 241–246. © 2005 IET.)

Therefore, it should be more efficient numerically than the edge-element method that employs three electric-field components [11].

3.3.3 Effective Mode Area

The effects of the hole pitch, Λ, and the ratio d/Λ on the effective mode area are now studied. The effective mode area, A_{eff}, is related to the effective area of the fibre core area, which is computed using [13]

$$A_{\text{eff}} = \frac{\left(\iint\limits_{\Omega} |H_t|^2 \, dx \, dy \right)^2}{\iint\limits_{\Omega} |H_t|^4 \, dx \, dy} \tag{3.14}$$

where H_t is the transverse magnetic field vector and Ω is the area enclosed within the computational window. Figure 3.15 shows the variation of the effective mode area, A_{eff}, of a four-ring PCF with hole pitch, Λ, at different d/Λ ratios. It is quite

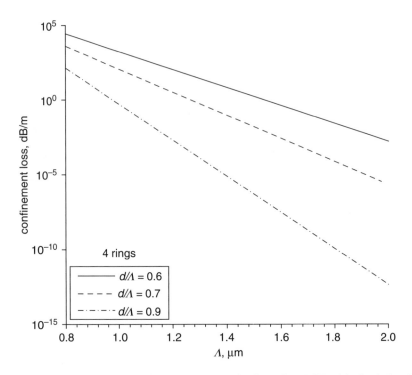

Figure 3.14 Variation of the confinement losses of a four-ring PCF with the hole pitch, Λ, with d/Λ as parameter. (Reproduced with permission from Obayya, S.S.A., Rahman, B.M.A. and Grattan, K.T.V. (2005) Accurate finite element modal solution of photonic crystal fibres. *Optoelectron. IET Proc.*, **152** (5), 241–246. © 2005 IET.)

interesting to note that the effective mode area for a PCF can be tailored easily to suit different applications. For example, a large mode area can be useful in high power transmission with less fibre nonlinearity effects [14]. Alternatively, small mode areas can suit applications where exploitation of enhanced fibre nonlinearity is needed. As may be seen from Figure 3.15, there is a certain range of hole pitch, Λ, for which the effective mode area, A_{eff}, steadily increases with Λ; for example, when the ratio $d/\Lambda = 0.6$, the effective mode area steadily increases with tincreasing Λ, when it is larger than 1.0 μm. However, the effective mode area, A_{eff}, rapidly becomes large as the hole pitch becomes slightly less than 1.0 μm. The same effect can also be observed with $d/\Lambda = 0.7$.

In order to explain this sudden expansion in A_{eff}, the 'cutoff' conditions of the fundamental mode require analysis. The fundamental mode reaches the cutoff line when its effective index becomes equal to the cladding effective index of the fundamental space-filling mode (FSM) [15].

The FSM is defined as the fundamental mode propagating in an infinitely periodic array of air holes with no central defect. In this case, only one periodic cell with

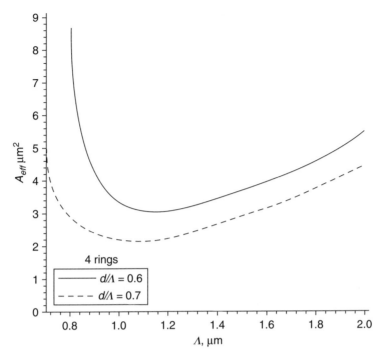

Figure 3.15 Variation of the mode effective area, A_{eff}, of a four-ring PCF with the hole pitch, Λ, at two different values of d/Λ ratio, 0.6 and 0.7. (Reproduced with permission from Obayya, S.S.A., Rahman, B.M.A. and Grattan, K.T.V. (2005) Accurate finite element modal solution of photonic crystal fibres. *Optoelectron. IET Proc.*, **152** (5), 241–246. © 2005 IET.)

the appropriate electric and magnetic wall boundary conditions applied, as shown in Figure 3.16, has been considered using the IDVFEBPM. Figure 3.17 shows the variation of the effective indices of the fundamental mode propagating in our PCF, n_{eff}, and the FSM (without central defect), n_{FSM}, with the hole pitch, Λ, for two values of the ratio d/Λ. As may be noted from this figure, the fundamental mode reaches the cutoff when the hole pitch is below 1.0 and 0.8 μm for $d/\Lambda = 0.6$ and 0.7, respectively. Therefore, as the fundamental mode approaches the cutoff, its effective mode area, A_{eff}, tends to extend to the outer regions of the air holes, which explains the significant increase in A_{eff} in these ranges of the hole pitch, Λ. For a curved PCF, it is not recommended to a use large hole pitch, Λ, to achieve a large mode effective area, as the radiation losses increase in proportion to Λ^3 [15]. Therefore, it is particularly interesting to exploit the behaviour of the effective mode area near the cutoff hole pitch in the design of a curved PCF with a large A_{eff} and keep the radiation losses at moderate levels.

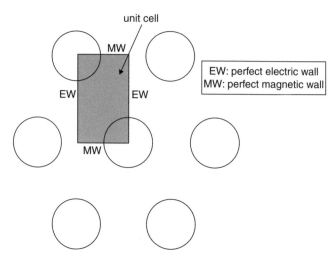

Figure 3.16 Schematic diagram of a unit cell of the infinitely periodic PCF considered in the FSM mode calculation. (Reproduced with permission from Obayya, S.S.A., Rahman, B.M.A. and Grattan, K.T.V. (2005) Accurate finite element modal solution of photonic crystal fibres. *Optoelectron. IET Proc.*, **152** (5), 241–246. © 2005 IET.)

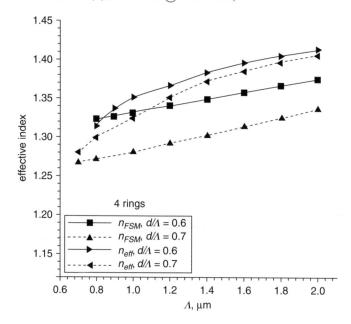

Figure 3.17 Variation of the effective index of the fundamental mode, n_{eff}, and the FSM mode, n_{FSM}, of a four-ring PCF with the hole pitch, Λ, and two values of d/Λ. (Reproduced with permission from Obayya, S.S.A., Rahman, B.M.A. and Grattan, K.T.V. (2005) Accurate finite element modal solution of photonic crystal fibres. *Optoelectron. IET Proc.*, **152** (5), 241–246. © 2005 IET.)

3.3.4 Dispersion

Finally, the chromatic dispersion of the PCF is considered. The chromatic dispersion D of a PCF may be calculated using

$$D = -\frac{\lambda}{c}\frac{\partial^2[\mathrm{Re}(n_{\mathrm{eff}})]}{\partial\lambda^2} \tag{3.15}$$

where Re stands for the real part and c is the speed of light in free space. The dispersion is determined easily using the formula given in (3.15) once the dependence of the effective index on the wavelength λ is determined. The second-order derivative in (3.15) has been approximated using a central finite difference formula. Also, the silica material dispersion, using the formulae given by Sellmeier in [16], has been included in the computations. For four rings of 60 air holes, Figure 3.18 shows the wavelength dependence of the refractive index of the silica material and also the effective index of the fundamental mode for different values of d/Λ. It may be noted from this figure that the effective index steadily decreases as the wavelength is increased. Therefore, it is evident that at shorter wavelengths the mode tends to be more confined, while at longer wavelengths the mode becomes less confined to the core region.

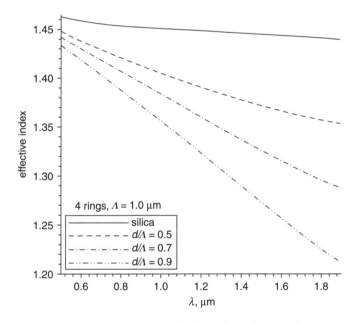

Figure 3.18 Variation of the effective index of a four-ring PCF with the wavelength when the hole pitch, Λ, is 1.0 μm, with d/Λ as parameter. (Reproduced with permission from Obayya, S.S.A., Rahman, B.M.A. and Grattan, K.T.V. (2005) Accurate finite element modal solution of photonic crystal fibres. *Optoelectron. IET Proc.*, **152** (5), 241–246. © 2005 IET.)

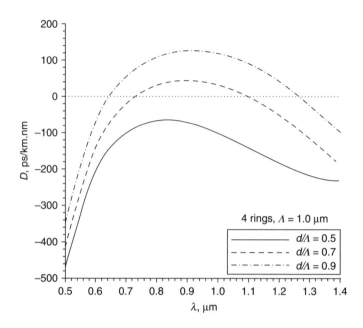

Figure 3.19   Variation of the chromatic dispersion of a four-ring PCF with the wavelength when the hole pitch, Λ, is 1.0 μm with d/Λ as parameter. (Reproduced with permission from Obayya, S.S.A., Rahman, B.M.A. and Grattan, K.T.V. (2005) Accurate finite element modal solution of photonic crystal fibres. *Optoelectron. IET Proc.*, **152** (5), 241–246. © 2005 IET.)

Figure 3.19 shows the dispersion, D, for a four-ring PCF with $\Lambda = 1.0$ μm at three different values of d/Λ. In addition, the dispersion, for a four-ring PCF with $d/\Lambda = 0.9$ at different values of the hole pitch, Λ, is shown in Figure 3.20. As may be observed from these figures, the peak value of the dispersion strongly depends on d/Λ rather than the hole pitch, Λ, itself. It can be also noted from these figures that the zero dispersion point can be obtained at the desired wavelength by changing the geometrical parameters of the PCF.

3.4   Liquid-Crystal-Based Photonic Crystal Fibre

3.4.1   Design

Figure 3.21 shows a cross section of a triangular lattice soft glass nematic liquid-crystal-based PCF [5] (NLC-PCF) whose cladding holes have been infiltrated with a nematic liquid crystal (NLC) of type E7. All the holes have the same diameter d and are arranged with a hole pitch, Λ. The NLCs used are anisotropic materials consisting of rod-like molecules which are characterised by ordinary index n_o and extraordinary index n_e. Moreover, the local orientation of the NLCs, as shown in Figure 3.21, is

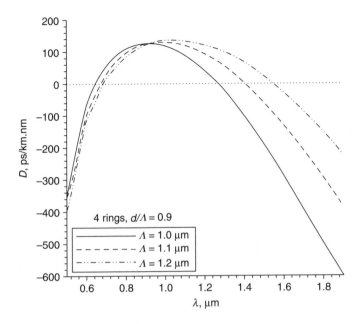

Figure 3.20 Variation of the chromatic dispersion of a four-ring PCF with the wavelength when the hole pitch, d/Λ, is 0.9 µm, with the hole pitch, Λ, as parameter. (Reproduced with permission from Obayya, S.S.A., Rahman, B.M.A. and Grattan, K.T.V. (2005) Accurate finite element modal solution of photonic crystal fibres. *Optoelectron. IET Proc.*, **152** (5), 241–246. © 2005 IET.)

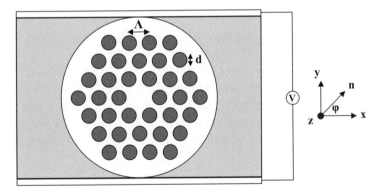

Figure 3.21 Cross section of an NLC-PCF sandwiched between two electrodes and surrounded by silicone oil. The director of the NLC is shown at the right. (Reproduced with permission from Hameed, M.F.O., Obayya, S.S.A. and Wiltshire, R.J. (2010) Beam propagation analysis of polarization rotation in soft glass nematic liquid crystal photonic crystal fibers. *IEEE Photon. Technol. Lett.*, **22** (3), 188–190. © 2010 IEEE.)

described by the director, which is a unit vector n along the direction of the average orientation of the molecules.

Under the application of a static electric field, the director's orientation can be controlled, since the liquid crystal molecules tend to align their axes according to the applied field. This can be achieved successfully with better field uniformity over the fibre's cross section, as described by Haakestad *et al.* [17]. In [17], the fibre is placed between two pairs of electrodes, allowing for the arbitrary control of the alignment of the NLC director via an external voltage, as shown schematically in Figure 3.21. In addition, two silica rods of appropriate diameter are used to control the spacing between the electrodes, and the fibre is surrounded by silicone oil, which has higher dielectric constant than air. Therefore, the external electric field will be uniform across the fibre cross section, which results in good alignment of the director of the NLC. In addition, the nonuniform electric-field region will be at the edges, away from the core regions, thus having little effect on the performance of the proposed fibre. Other layouts, such as those described in [18, 19] can also be used to ensure better field uniformity over the fibre's cross section.

The ordinary n_o and extraordinary n_e indexes of the E7 material were measured by Li *et al.* [20] at different visible wavelengths over a temperature range of 15 to 50 °C with steps of 5 °C. Then the Cauchy model was used to fit the measured n_o and n_e, which can be described as follows [20]

$$n_e = A_e + \frac{B_e}{\lambda^2} + \frac{C_e}{\lambda^4} \tag{3.16}$$

$$n_o = A_o + \frac{B_o}{\lambda^2} + \frac{C_o}{\lambda^4} \tag{3.17}$$

where A_e, B_e, C_e, A_o, B_o and C_o are the coefficients of the Cauchy model. The Cauchy coefficients at $T = 25$ °C are $A_e = 1.6933$, $B_e = 0.0078$ μm^2, $C_e = 0.0028$ μm^4, $A_o = 1.4994$, $B_o = 0.0070$ μm^2 and $C_o = 0.0004$ μm^4. The variation of n_o and n_e of the E7 material with wavelength at different temperatures T from 15 to 50 °C with steps of 5 °C is shown in Figure 3.22. It is evident from this figure that n_e is higher than n_o at the measured temperature values within the reported wavelength range. In the proposed design, the relative permittivity tensor ε_r of the E7 material is taken as [21]

$$\varepsilon_r = \begin{pmatrix} n_o^2 \sin^2 \varphi + n_e^2 \cos^2 \varphi & (n_e^2 - n_o^2) \cos \varphi \sin \varphi & 0 \\ (n_e^2 - n_o^2) \cos \varphi \sin \varphi & n_o^2 \cos^2 \varphi + n_e^2 \sin^2 \varphi & 0 \\ 0 & 0 & n_o^2 \end{pmatrix} \tag{3.18}$$

where φ is the rotation angle of the director of the NLC, as shown in Figure 3.20.

The in-plane alignment of the NLC can be exhibited under the influence of appropriate homeotropic anchoring conditions [21, 22]. Haakestad *et al.* [17]

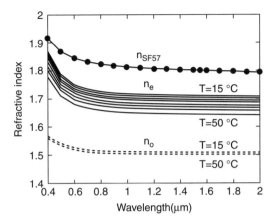

Figure 3.22   Variation of n_o and n_e of E7 material with wavelength at different temperatures, T, from 15 to 50 °C with steps of 5 °C. The solid line with closed circles represents the variation of the refractive index of SF57 material, n_{SF57}, with the wavelength. (Reproduced with permission from Hameed, M.F.O., Obayya, S.S.A. and Wiltshire, R.J. (2010) Beam propagation analysis of polarization rotation in soft glass nematic liquid crystal photonic crystal fibers. *IEEE Photon. Technol. Lett.*, **22** (3), 188–190. © 2010 IEEE.)

demonstrated experimentally that in a strong field limit, the NLC of type E7 is aligned in plane in capillaries of diameter 5 μm. In addition, Alkeskjold and Bjarklev [23] presented experimentally in-plane alignment of the E7 material in PCF capillaries of diameter 3 μm with three different rotation angles, 0°, 45° and 90° using two sets of electrodes.

The substrate of the nematic liquid-crystal PCF (NLC-PCF) is a soft glass of type SF57 (lead silica). The wavelength-dependent refractive index of the SF57 material is also shown in Figure 3.22. It can be seen from this figure that the refractive index of the SF57 material is higher than n_o and n_e of the E7 material which guarantees the index guiding of the light through the high-index core NLC-PCF.

The Sellmeier equation for SF57-type soft glass [24] is given by

$$n_{SF57}^2 = A_o + A_1\lambda^2 + \frac{A_2}{\lambda^2} + \frac{A_3}{\lambda^4} + \frac{A_4}{\lambda^6} + \frac{A_5}{\lambda^8} \qquad (3.19)$$

where n_{SF57} is the refractive index of the SF57 material, $A_o = 3.247\,48$, $A_1 = -0.0096$ μm^{-2}, $A_2 = 0.0494$ μm^2, $A_3 = 0.00294$ μm^4, $A_4 = -1.4814 \times 10^{-4}$ μm^6, and $A_5 = 2.7843 \times 10^{-5}$ μm^8 [24].

The practical techniques that have been used in manufacturing nonsilica PCFs are capillary stacking [25], drilling [26], built-in casting [26] and extrusion [24, 28, 29]. Of these, extrusion offers a controlled and reproducible approach for fabricating complex-structured PCFs with a good surface quality. In addition, extrusion can be used to produce structures that could not be created with capillary stacking approaches. Therefore, most of the nonsilica PCFs in the literature are fabricated by extrusion. The extrusion approach has been recently extended to soft glasses such as lead silicate (SF57 glass) [24, 28, 29] and tellurite [30]. SF57-type soft glass has a low processing temperature of $\sim$520 °C [31], while the softening temperature for silica glass is 1500–1600 °C. Therefore, it is possible to extrude the PCF directly from the bulk glass. In addition, lead silicate glasses offer the highest thermal and crystallisation stability, making them particularly attractive for PCF fabrication.

The filling of PCFs with liquid or liquid-crystalline materials has already been demonstrated in the literature [17, 19, 32–34]. Arc-fusion techniques have been successfully implemented for the infiltration of central defect cores [33], while extensive control of the infiltration process of either core or cladding capillaries can be achieved by using UV-curable polymers [34]. In [17], all the cladding holes of a silica PCF were filled with NLC by capillary forces and electrically tunable photonic bandgap guidance was reported. In addition, a tunable light switch using silica PCF, whose central defect and cladding holes were filled with NLC was studied by Fang *et al.* [19].

3.4.2   Modal Hybridness

All the holes of the NLC-PCF have the same diameter d and are arranged with a hole pitch $\Lambda = 5$ μm and a d/Λ ratio of 0.7. In addition, n_o and n_e of the E7 material are fixed at 1.5024 and 1.6970, respectively, at an operating wavelength $\lambda = 1.55$ μm and a temperature of 25 °C. The rotational angle of the director of the NLC is taken as 45° and n_{SF57} is fixed at 1.802 at $\lambda = 1.55$ μm. The dominant H_y and nondominant H_x field profiles of the quasi-TE mode are shown in Figure 3.23(a) and (b), respectively. It can be observed that the field profiles of the dominant and nondominant components of the quasi-TE mode are very similar. The maximum value of H_x is 0.998, normalised to the maximum value of the dominant H_y component. This means that the proposed NLC-PCF supports highly hybrid modes, which is very useful in designing polarisation conversion devices. To understand the effect of the infiltration of the NLC, a soft glass PCF with air holes will be considered. Figure 3.23(c) shows the dominant H_y field components of the quasi-TE mode for a soft glass PCF with a hole pitch of 5 μm and a d/Λ ratio equal to 0.7. It can be observed from this figure that this field profile is symmetric in nature, as the PCF structure itself is symmetric. The nondominant H_x field components of the quasi-TE mode are shown in Figure 3.23(d), which is clearly antisymmetric in nature and has a maximum magnitude of only 0.019, normalised to the maximum value of H_y. The nondominant H_y field profile of the quasi-TM mode is not shown here, but this profile is similar to the nondominant H_x field profile of the quasi-TE mode. Therefore, it should be noted that the infiltration of the NLC has a great impact on the hybridness of the modes in the suggested NLC-PCF.

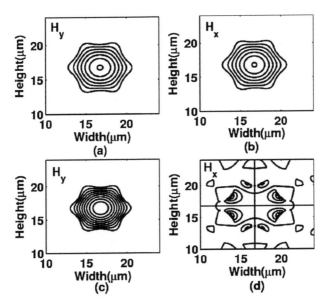

Figure 3.23 Contour plot of the dominant H_y and nondominant H_x field profiles of the fundamental quasi-TE mode for (a and b) NLC-PCF and (c and d) soft glass PCF with air holes. (Reproduced with permission from Hameed, M.F.O., Obayya, S.S.A. and Wiltshire, R.J. (2010) Beam propagation analysis of polarization rotation in soft glass nematic liquid crystal photonic crystal fibers. *IEEE Photon. Technol. Lett.*, **22** (3), 188–190. © 2010 IEEE.)

3.4.3 Effective Index

The soft glass NLC-PCF shown in Figure 3.21 has been analysed and studied thoroughly. All the holes of the NLC-PCF have the same radius r and are arranged with a hole pitch, Λ, of 2.3 µm. In addition, n_o, n_e and n_{SF57} are taken as 1.5024, 1.6970 and 1.802 respectively at $\lambda = 1.55$ µm. Moreover, the rotation angle and the temperature are fixed at 90° and 25 °C respectively. The variation of the real part of the complex effective index of the quasi-TM mode with wavelength at different r, 0.6 µm, 0.7 µm, 0.8 µm and 0.9 µm, is shown in Figure 3.24. It is evident from this figure that the effective index of the quasi-TM mode decreases with increasing wavelength due to less confinement of the mode through the core region at long wavelengths. Moreover, the effective refractive index of the cladding region and hence the effective index of the quasi-TM mode decreases with increasing radius of the infiltrated NLC hole, as revealed in Figure 3.24.

3.4.4 Dispersion

Figure 3.25 shows the variation of the chromatic dispersion of the quasi-TE and TM modes with wavelength at different r values. It is evident from this figure that the dispersion of the quasi-TM modes is negative over all studied wavelength ranges,

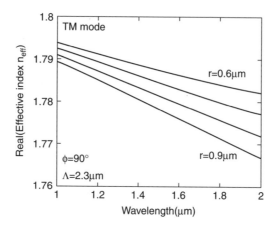

Figure 3.24 Variation of the real part of the complex effective index of the quasi-TM modes of NLC-PCF with the wavelength at different values of hole radius, r.

which can be used for dispersion compensation. It should be noted that the negative dispersion can be obtained using the conventional PCF [35]. However, the negative dispersion of the proposed design can be tuned using the temperature or an external electric field. Figure 3.25 also reveals that the dispersion of the quasi-TM mode is less than that of the quasi-TE mode. This can be explained by analysing the dominant field components of the quasi-TE and TM modes, and the direction of the director of the NLC. The dominant electric-field components of the quasi-TE and TM modes are E_x and E_y, respectively. At $\varphi = 90°$, ε_r of the E7 material the diagonal of $[\varepsilon_{xx}, \varepsilon_{yy}, \varepsilon_{zz}]$, where $\varepsilon_{xx} = n_o^2$, $\varepsilon_{yy} = n_e^2$ and $\varepsilon_{zz} = n_o^2$. It may be observed that ε_{yy} is greater than ε_{xx},

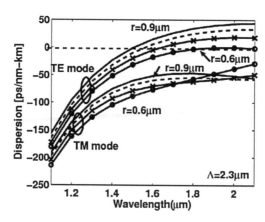

Figure 3.25 Variation of the dispersion of the quasi-TE and TM modes with the wavelength at different values of hole radius, r: 0.6 μm, 0.7 μm, 0.8 μm and 0.9 μm. (Reproduced with permission from Hameed, M.F.O., Obayya, S.S.A., Al-Begain, K., et al. (2009) Modal properties of an index guiding nematic liquid crystal based photonic crystal fiber. *IEEE J. Lightwave Technol.*, **27** (21), 4754–4762. © 2009 IEEE.

therefore the effective refractive index of the cladding region of the quasi-TM mode is greater than that of the quasi-TE mode. As a result, the effective index n_{eff} of the quasi-TM mode is greater than that of the quasi-TE mode. In addition, the change of the effective index n_{eff} of the quasi-TM mode over the reported wavelength range is less than that of the quasi-TE mode. Consequently, the dispersion of the quasi-TM mode is less than that of the quasi-TE mode. It can also be observed from Figure 3.25 that the slope of the quasi-TE dispersion can be adjusted until a flat dispersion is obtained, by changing the radius of the infiltrated NLC cladding holes. It is found that a flat dispersion of ± 3 ps/nm.km is obtained over a wide wavelength range from 1.8 to 2.1 μm for the quasi-TE mode at $r = 0.6$ μm. However, there is a cross over in the dispersion curve of the quasi-TM mode at $r = 0.6$ μm at long wavelengths; the mode starts spreading into the cladding region, which increases the effective mode area. At $r = 0.6$ μm, the quasi-TM mode spreads dramatically in the cladding region, which might lead to an increase in the dispersion, as shown in Figure 3.25.

3.4.5 Tunable Liquid-Crystal-Based Photonic Crystal Fibre Polarisation Converter

As shown in the inset of Figure 3.26, when a TE polarised mode obtained from a soft glass PCF with air holes is launched directly into the NLC-PCF, the input power excites two hybrid modes along the polarisation rotator (PR) waveguide. These two

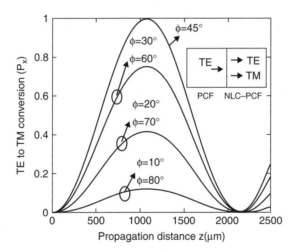

Figure 3.26 Evolution of the TM powers along the propagation direction at different rotation angles of the director of the NLC. The solid black lines represent P_x at $\varphi = 10°$, $20°$, $30°$ and $45°$, while the values of P_x at $\varphi = 60°$, $70°$ and $80°$ are represented by dotted grey lines. However, the hole pitch, Λ, and d/Λ ratio are taken as 5.0 μm and 0.7, respectively. (Reproduced with permission from Hameed, M.F.O., Obayya, S.S.A. and Wiltshire, R.J. (2010) Beam propagation analysis of polarization rotation in soft glass nematic liquid crystal photonic crystal fibers. *IEEE Photon. Technol. Lett.*, **22** (3), 188–190. © 2010 IEEE.)

modes become out of phase at a distance equal to L_π from the beginning of the PR section. Therefore, the H_y component will be cancelled, while the H_x component will be added which produces a nearly pure TM mode. The calculated value of L_π using the FVFEBPM is 1072 µm, which is in excellent agreement with the 1071 µm calculated by the VFEM [6]. The polarisation power factors, P_y and P_x, are defined as the power carried by the H_y and H_x field components, respectively, over the PR waveguide cross section, normalised to the total power. Figure 3.26 shows the variation of the P_x power for the TE input along the axial direction, at different rotation angles of the director of the NLC. It is evident from this figure that for TE excitation, initially P_x is zero and it slowly increases to a maximum value at $z = L_\pi$; if the PR section is not terminated at this position the optical power P_x starts decreasing. It should be noted that nearly 99.813% polarisation conversion can be obtained at $z = 1072$ µm when the rotational angle is equal to 45°. It is also evident from Figure 3.26 that the conversion ratio increases with increasing rotational angle from 0 to 45° until complete conversion occurs at $\varphi = 45°$. The conversion ratio then decreases with increasing rotational angle. In addition, the hybridness and hence the conversion ratios at $\varphi = 10°$, 20° and 30° are approximately equal to the hybridness and hence the conversion ratios at $\varphi = 80°$, 70° and 60°, respectively, as shown in Figure 3.26.

3.5   Electro-Optical Modulators

High-speed integrated electro-optic modulators (EOMs) and switches are the basic building blocks of modern wideband optical communication systems and the future trend of ultrafast signal processing technology. Therefore, a great deal of research effort has been devoted to developing low-loss, efficient broadband modulators in which the RF signal is used to modulate the optical carrier frequency [36]. The design of EOMs usually relies on the use of either directional couplers (DCs) [37] or Mach–Zehnder (MZ) [38] arrangements. In DC-based EOMs, the externally applied electric field affects the refractive index distribution in two coupled waveguides that are used in such a way that this change is asymmetric, and this also affects the light propagation in the two guides, the coupling length, the phase matching and hence the power coupling transfer between them. However, in contrast, in MZ-based EOMs, the two guides are relatively far apart from each other, with either one or both of the two guides being affected by the applied field. So, at the end of the device, the two waves emerging from the two guides are either in phase or in anti-phase, and this gives rise to output switching properties related to the applied electric field.

3.5.1   Design of the Electro-Optical Modulator

For electro-optic semiconductor waveguides [39], the refractive-index distribution becomes a 3 × 3 tensor and depends on the potential distribution as follows

$$\bar{n}(x, y) = \begin{bmatrix} n(x, y) + \Delta n_{xx}(x, y) & \Delta n_{xy}(x, y) & 0 \\ \Delta n_{yx}(x, y) & n(x, y) + \Delta n_{yy}(x, y) & 0 \\ 0 & 0 & n(x, y) + \Delta n_{zz}(x, y) \end{bmatrix}$$

$$(3.20)$$

where $n(x, y)$ is the refractive-index distribution of the structure in the case of no applied modulating field and Δn_{xx}, Δn_{xy}, Δn_{yx} and Δn_{zz} are the changes in the refractive index occurring due to the electro-optic effect, and these are related to the applied modulating electric fields via

$$\Delta n_{xx}(x, y) = -\Delta n_{zz} = \frac{n^3(x, y)}{2} r_{41} E_y(x, y), \Delta n_{yy}(x, y) = 0 \quad (3.21)$$

$$\Delta n_{xy}(x, y) = \Delta n_{yx}(x, y) = \frac{n^3(x, y)}{2} r_{41} E_x(x, y) \quad (3.22)$$

where r_{41} is the electro-optic coefficient for selected semiconductor materials, such as InP and GaAs, and this value has been taken as equal to 1.4×10^{-6} μm/V for GaAs [40].

A schematic diagram of a deeply etched GaAs [41] EOM is shown in Figure 3.27. As shown in this figure, a 0.2 μm 10% AlGaAs layer, a thick GaAs core with height

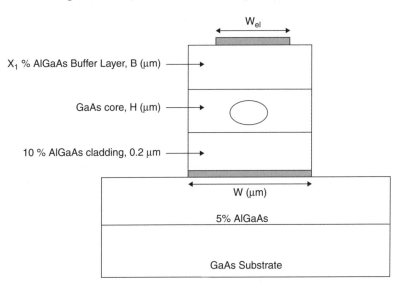

Figure 3.27 Schematic diagram of a deeply etched AlGaAs/GaAs semiconductor EOM. (Reproduced with permission from Obayya, S.S.A., Haxha, S., Rahman, B.M.A. and Grattan, K.T.V. (2003) Optimization of optical properties of a deeply-etched semiconductor optical modulator. *IEEE J. Lightw. Technol.*, **21** (8), 1813–1819. © 2003 IEEE.)

H (μm) and a buffer AlGaAs layer with an aluminium concentration of $x_1\%$ and height B (μm) are all deposited on a 2 μm thick 5% AlGaAs spacer layer. The whole structure is deposited on a very thick insulating GaAs substrate, as shown in Figure 3.27.

The ground electrode, with $V = 0$, is placed between the 10% AlGaAs layer and the substrate, while the hot electrode $V \neq 0$ is deposited on top of the buffer AlGaAs layer. Several major manufacturers are developing high-speed GaAs EOMs using a highly doped layer beneath the lower cladding layer as the lower electrode, which may be connected to a ground metal electrode on the side. In this case a doping density of $10^{18}/\text{cm}^3$ has been considered for the 0.1 μm thick GaAs lower electrode and the 2 μm thick lower spacer layer. The width of the waveguide is W (μm), while the electrode width is W_{el} (μm) and the operating wavelength is 1.55 μm. As a deeply etched waveguide structure suffers from lower bending loss, a more compact system design is possible than for the shallow-etched counterparts. This waveguide structure is theoretically multimoded when the waveguide width is greater than 2.5 μm. However, from a practical point of view, the use of a high-index GaAs substrate leads to a situation where these higher-order modes suffer very high leakage radiation losses into the substrate, while the fundamental mode shows virtually no leakage loss [41,42], and so effectively behaves like a single-moded guide.

Figure 3.28(a) and (b) show the contour plots of the horizontal E_x and vertical E_y modulating electric-field components when the waveguide width is 5 μm, the electrode width is 4.9 μm, the core height is 1.5 μm, the buffer thickness is 1.1 μm and the buffer Al concentration is 30%. In this case, the applied voltage is 5 V. It

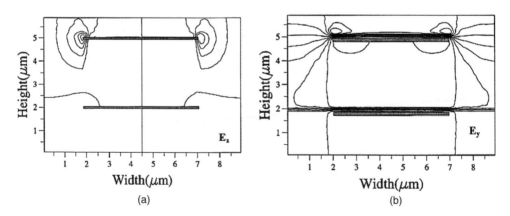

Figure 3.28 Contour plot of the (a) horizontal modulating field component E_x, and (b) the vertical modulating field E_y when $W = 5$ μm, $W_{el} = 4.9$ μm, $B = 1.1$ μm, $H = 1.5$ μm and $x_1 = 30\%$. (Reproduced with permission from Obayya, S.S.A., Haxha, S., Rahman, B.M.A. and Grattan, K.T.V. (2003) Optimization of optical properties of a deeply-etched semiconductor optical modulator. *IEEE J. Lightw. Technol.*, **21** (8), 1813–1819. © 2003 IEEE.)

should be noted that the E_y field is symmetric and nearly uniform in the waveguide core region, whereas the E_x profile is asymmetric and nearly zero in this region, with its two maxima around the left and right corners of the upper electrode. The vector-field profile would be nearly vertical, except near the left and right edges of the upper electrode.

As can be deduced from Equations (3.20) to (3.22), the H_y (or E_x) modes are mainly affected by refractive-index changes in the x direction, namely Δn_{xx}, which are directly proportional to E_y, while the H_x (or E_y) modes are affected by refractive-index changes in the y direction Δn_{yy}, which are zero. Therefore, the H_y modes (quasi-TE) will be considered throughout the rest of the book. However, if an asymmetry has been brought to the structure, the horizontal field component, E_x, will not be symmetric, giving rise to a nonzero Δn_{xy}, an off-diagonal refractive-index component which can cause a coupling between the two orthogonal TE and TM modal states.

3.5.2   The Effect of the Core Height

Figure 3.29 shows the variation of the half-wave voltage length product $V_\pi L$ with the core height H when the buffer thickness, B, is 1.1 μm and the buffer layer Al

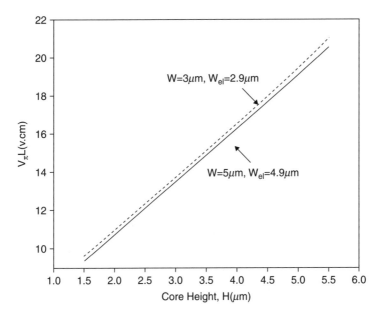

Figure 3.29   Variation of $V_\pi L$ with the core height, H, for two different values of waveguide width, W, and electrode width, W_{el}. (Reproduced with permission from Obayya, S.S.A., Haxha, S., Rahman, B.M.A. and Grattan, K.T.V. (2003) Optimization of optical properties of a deeply-etched semiconductor optical modulator. *IEEE J. Lightw. Technol.*, **21** (8), 1813–1819. © 2003 IEEE.)

concentration, x_1, is 30%, for two cases of the waveguide width W and the electrode width W_{el}. The refractive indexes considered here are 3.232, 3.377, 3.329 and 3.353 for the upper 30% AlGaAs cladding, the GaAs core, the 10% lower AlGaAs cladding and the 5% AlGaAs substrate, respectively. The $V_\pi L$ quantity is a very important parameter in designing EOMs and switches and for MZ [38] structures. It is defined as the product of the voltage at which the phase difference between the two branches of the MZ structure is 180°, and the length of the electrode. Assuming that the two waveguide branches of the MZ arrangement are sufficiently separated not to have coupling between their isolated guided modes, $V_\pi L$ can be calculated using

$$V_\pi L = \frac{\pi V_0}{\Delta \beta}, \Delta \beta = \beta_1 - \beta_0 \qquad (3.23)$$

where V_0 is the applied voltage, and β_1 and β_0 are the propagation constants of the fundamental H_{11}^y modes of the MZ arms with and without the applied voltage, respectively. It may be observed from Figure 3.29, that as the core height H increases, $V_\pi L$ also increases. In particular, as H increases from 1.5 to 5.5 μm, the corresponding value of $V_\pi L$ also increases from nearly 9 to 20.5 V.cm. These results are intuitive and can be justified simply, as follows. For a particular core height, it may be presumed that a specific modulating electric-field magnitude (or the applied voltage) is needed to obtain a phase difference of 180° at the MZ output. However, if the core height is increased, then it may be anticipated from a simple parallel-plate approximation that, in order to maintain the same modulating electric-field strength in the core area, the applied modulating voltage has to be increased. In addition, it can be observed from Figure 3.29 that the variation of $V_\pi L$ with the core height is almost independent of the value of the waveguide width W, as long as the electrode width W_{el} is nearly equal to the corresponding waveguide width, W. In addition, for a core height, H, of 1.5 μm, the total etch depth is 3.0 μm, and the corresponding $V_\pi L$ is 9.6 V.cm. However, if the etching depth is reduced to only 2.6 μm, the value of $V_\pi L$ will increase by 3.6%. This shows that deeply etched waveguides are better than their shallow-etched counterparts, not only due to better light confinement, but also because of the reduced $V_\pi L$ needed to work as modulators.

The variation of $V_\pi L$ with the core height H for three different values of Al concentration x_1 of the buffer AlGaAs layer are shown in Figure 3.30. In this case, the waveguide width, W, is 5 μm, the electrode width, W_{el}, is 4.9 μm and the buffer thickness B is 1.1 μm. As can be seen from this figure, the value of $V_\pi L$ increases as the core height, H, increases for the three different values of x_1. On the other hand, for a particular core height, H, $V_\pi L$ is found to increase slightly as the value of x_1 increases from 10 to 30% and ultimately to 100%. These results can be explained physically as the increase of x_1 leads to an increase in the refractive-index difference between the core and buffer layers in such a way that the optical field of the mode H_{11}^y will be pushed slightly downward and away from the hot electrode. This would

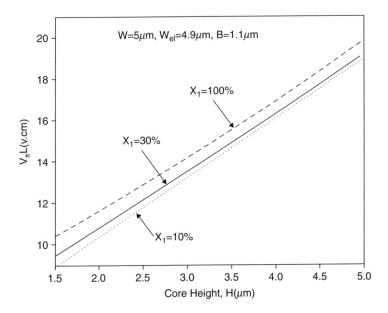

Figure 3.30 Variation of $V_\pi L$ with the core height, H, for three different values of Al concentration of the buffer layer, x_1. (Reproduced with permission from Obayya, S.S.A., Haxha, S., Rahman, B.M.A. and Grattan, K.T.V. (2003) Optimization of optical properties of a deeply-etched semiconductor optical modulator. *IEEE J. Lightw. Technol.*, **21** (8), 1813–1819. © 2003 IEEE.)

lead to slightly less interaction between the modulating electric field and the optical field, which would give rise to a slight increase in $V_\pi L$ as x_1 increases.

3.5.3   The Effect of the Electrode Width

The variation of $V_\pi L$ with the electrode width, W_{el}, for different values of the waveguide width, W, and the core height, H, is shown in Figure 3.31. In these cases, the buffer layer thickness, B, is 1.1 μm, and Al concentration, x_1, is 30%. It can be noted from this figure that for all the three combinations of W and H, as the electrode width is increased, the corresponding value is found to be monotonically reducing, and until W_{el} approaches the value of the full waveguide width, W, the corresponding values of $V_\pi L$ tend to converge to certain values. As a result, the electrode width should not be narrower than the waveguide width, but rather it could be equal to the waveguide width to improve the performance and also to facilitate its fabrication. When W_{el} is small compared to W, the modulating electric-field profile spreads less in the horizontal direction, compared with a greater value of W_{el}, giving a smaller overlap with the optical field profile. This is why the value of the $V_\pi L$ product will be higher in the former case than in the latter. On the other hand, for the same waveguide and electrode

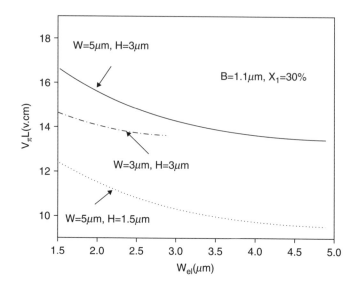

Figure 3.31 Variation of $V_\pi L$ with the electrode width, W_{el}, for three different combinations of waveguide width, W, and core height, H. (Reproduced with permission from Obayya, S.S.A., Haxha, S., Rahman, B.M.A. and Grattan, K.T.V. (2003) Optimization of optical properties of a deeply-etched semiconductor optical modulator. *IEEE J. Lightw. Technol.*, **21** (8), 1813–1819. © 2003 IEEE.)

width values, Figure 3.31 illustrates the difference in the values of $V_\pi L$ when the core height H is 3 µm compared with the case when H is only 1.5 µm. In particular, when $W = 5$ µm and $W_{el} = 4.9$µm, the value of $V_\pi L$ is around 14 V.cm when H is 3 µm, and this value reduces to only 9.5 V.cm when H is 1.5 µm. This reduction in $V_\pi L$ as H is reduced is a direct consequence of the increased electric-field intensity, and hence lower modulating voltages are needed to keep the 180° phase difference condition at the output ports of the MZ structure. It might be envisaged that $V_\pi L$ reduction can be achieved by designing a waveguide with lower values of the core height H; however, care should be taken in doing so, as the mode might reach cut off, or the scattering loss due to roughness or fabrication imperfections around the sidewalls might increase significantly if extremely low values of H are used. In addition, it can be seen from this figure that for the same core height, H, the value of $V_\pi L$ tends to the same value as long as the electrode width, W_{el}, and the waveguide width, W, are nearly similar, irrespective of the value of the waveguide width itself.

3.5.4 The Effect of the Buffer Thickness

Figure 3.32 shows the variations of the $V_\pi L$ product with the buffer thickness, B, for two different values of the core height, H. In this case, the width of the waveguide,

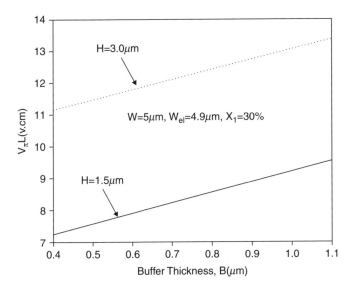

Figure 3.32 Variation of $V_\pi L$ with the buffer thickness, B, for two different values of core height, H. (Reproduced with permission from Obayya, S.S.A., Haxha, S., Rahman, B.M.A. and Grattan, K.T.V. (2003) Optimization of optical properties of a deeply-etched semiconductor optical modulator. *IEEE J. Lightw. Technol.*, **21** (8), 1813–1819. © 2003 IEEE.)

W, the electrode width, W_{el}, and the Al concentration of the buffer layer x_1 are 5 μm, 4.9 μm, and 30%, respectively. As can be seen from this figure, the value of $V_\pi L$ reduces as the buffer thickness, B, decreases. In this case, when the buffer thickness is increased the interaction between the applied modulating field and the optical field will be less, as the optical field is shifted downward and away from the peak modulating field intensity. Hence, as the buffer thickness, B, is increased in order to maintain the 180° phase difference between the two waveguide branches of the MZ, the value of the applied voltage should be increased, leading to a linear increase of the $V_\pi L$, as shown in Figure 3.32. Also, when the core height is 1.5 μm and the buffer thickness, B, lies in the region 0.5 to 0.4 μm, the value of $V_\pi L$ drops dramatically to a level near 7.5 V.cm. Practically speaking, this is a good value for $V_\pi L$; however, the optical losses, as will be seen subsequently, may be large within this range of buffer thickness.

On the other hand, the variation of $V_\pi L$ with the buffer thickness, B, for different values of the Al concentration of the buffer layer x_1 is shown in Figure 3.33, where the value of the waveguide width, W, is 5 μm, the electrode width, W_{el}, is 4.9 μm and the core height, H, is 1.5 μm. For a given buffer thickness, B, the value of $V_\pi L$ is slightly increased as x_1 increases from 30 to 50% and then to 100%. In particular, when the buffer thickness, B, is 0.4 μm, $V_\pi L$ increases from 7.3 to 7.4 V.cm and then to 7.7 V.cm, as x_1 increases from 30 to 50% and then to 100%, respectively, which

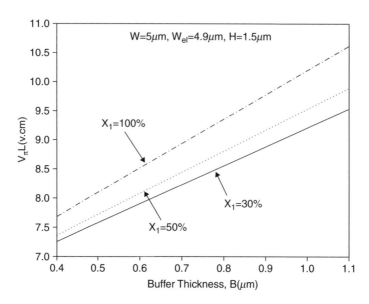

Figure 3.33 Variation of $V_\pi L$ with the buffer thickness, B, for three different values of Al concentration of the buffer layer, x_1. (Reproduced with permission from Obayya, S.S.A., Haxha, S., Rahman, B.M.A. and Grattan, K.T.V. (2003) Optimization of optical properties of a deeply-etched semiconductor optical modulator. *IEEE J. Lightw. Technol.*, **21** (8), 1813–1819. © 2003 IEEE.)

indicates, at this stage, that the aluminium concentration of the buffer layer has only a small influence on the design of optical modulators.

Next, the effects of the buffer layer thickness, B, and the Al concentration x_1 on the optical losses will be thoroughly investigated. For this purpose, the top electrode has been assumed to consist of two metal layers, gold (Au) and Titanium (Ti), each with a thickness t of 0.1 μm. The perturbation technique combined with the vector H-field finite-element modal solution [43] has been utilised to estimate the optical losses due to this imperfectly conducting electrode. The variations of the optical losses with the buffer thickness, B, for different values of the Al concentration x_1 of the buffer layer are shown in Figure 3.34. As may be noted from this figure, the optical losses are drastically reduced as either the buffer thickness or the Al concentration of the buffer is increased, where in either case, the optical field will be more confined to the core and a smaller portion of this field will be concentrated near the electrode region with losses. In particular, for a buffer thickness of 0.5 μm and an Al concentration of 30%, the optical loss has been estimated to be 12 dB/cm, while if the buffer thickness is increased to 1.0 μm, the optical loss is greatly reduced to nearly 1 dB/cm. On the other hand, instead of increasing the buffer thickness (from $B = 0.5$ μm), if the aluminium concentration is increased from 30 to 100%, the optical loss decreases significantly

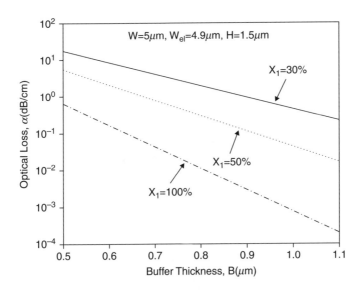

Figure 3.34 Variation of the optical loss, α, in decibels per centimetre with the buffer thickness for three different values of Al concentration of the buffer layer, x_1. (Reproduced with permission from Obayya, S.S.A., Haxha, S., Rahman, B.M.A. and Grattan, K.T.V. (2003) Optimization of optical properties of a deeply-etched semiconductor optical modulator. *IEEE J. Lightw. Technol.*, **21** (8), 1813–1819. © 2003 IEEE.)

from 12 to 0.8 dB/cm. It should be noted that the overall optical loss plays a dominant role in determining the bandwidth of the modulator.

The variation of the microwave index, n_m, and of the microwave characteristic impedance, Z_c, with the buffer thickness, B, for two different values of the core height, H, is given in Figure 3.35. In this case, the waveguide width, W, the electrode width, W_{el}, the Al concentration of the buffer layer x_1 and the electrode thickness t are 5 µm, 4.9 µm, 30% and 0.1 µm, respectively. Although the present model can deal with electrodes of finite conductivities, as a reasonable approximation, however, the doped semiconductor ground lower electrode has been considered as a perfectly conducting metal electrode in order to simplify the calculations. As shown in this figure, the microwave index, n_m, reduces linearly, while the microwave characteristic impedance, Z_c, increases linearly with increasing buffer thickness, for both values of the core height, H. For velocity matching between the microwave and optical signals, the value of n_m should be equal to the value of the effective index of the optical fundamental mode H_{11}^y, (n_{eff}), while for impedance matching purposes, the microwave characteristic impedance should equal to 50 Ω. If the velocity matching is examined, it can be deduced easily from Figure 3.35 that it is not possible to find a range of values of the buffer thickness, B, for which n_m is within the range of n_{eff}, for both values of the core height, H. It should be noted that if the electrode thickness

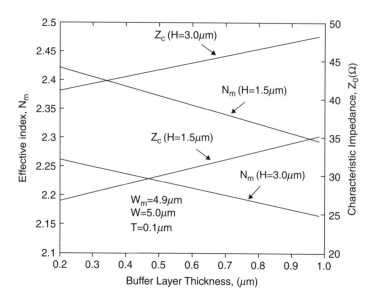

Figure 3.35 Variation of the microwave index, n_m, and the microwave characteristic impedance, Z_c, with buffer thickness, B, for two different values of core height, H. (Reproduced with permission from Obayya, S.S.A., Haxha, S., Rahman, B.M.A. and Grattan, K.T.V. (2003) Optimization of optical properties of a deeply-etched semiconductor optical modulator. *IEEE J. Lightw. Technol.*, **21** (8), 1813–1819. © 2003 IEEE.)

is increased, then the conductor loss reduces; however, the phase matching between the microwave and optical waves would deteriorate. To investigate this further, an additional study of segmented electrodes and/or the use of material such as Ta$_2$O$_5$, which has a high dielectric constant ($\varepsilon_r = 27$) at microwave frequencies and a low refractive index ($n = 2.03$) at optical frequencies, may be considered [44]. However, it is only for $H = 3.0$ µm that values of B higher than 0.9 µm can give reasonably good matching values for Z_c.

3.5.5 The Effect of the Al Concentration of the Buffer Layer

Next, for a waveguide width, W, of 5 µm, an electrode width, W_{el}, of 4.9 µm and a core height, H, of 1.5 µm, Figure 3.36 shows the variation of $V_\pi L$ and the buffer thickness, B, with the Al concentration of the buffer layer, x_1, when the electrode optical losses are fixed at either 0.2 or 0.5 dB/cm. For each value of x_1, an iteration loop has been implemented into the modal solution program in order to find the buffer thickness needed to get the desired level of optical losses. Once the required buffer thickness is obtained, the corresponding value of $V_\pi L$ is then calculated. As may be observed from this figure, for $x_1 = 40\%$ and for a fixed level of optical loss of 0.5 dB/cm, the required buffer thickness, B, is equal to 0.82 µm and the corresponding

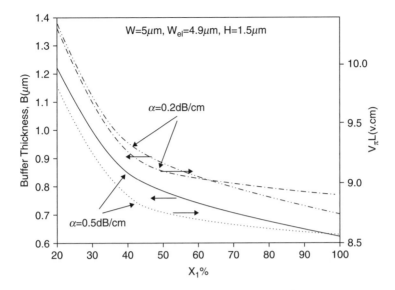

Figure 3.36 Variation of the required buffer thickness, B, and the corresponding $V_\pi L$ with Al concentration of the buffer layer, x_1, when the optical loss, α, is fixed to 0.2 dB/cm or 0.5 dB/cm. (Reproduced with permission from Obayya, S.S.A., Haxha, S., Rahman, B.M.A. and Grattan, K.T.V. (2003) Optimization of optical properties of a deeply-etched semiconductor optical modulator. *IEEE J. Lightw. Technol.*, **21** (8), 1813–1819. © 2003 IEEE.)

$V_\pi L$ product is 8.7 V.cm. However, for the same value of x_1, if the accepted level of losses is reduced to 0.2 dB/cm, the design value of the buffer thickness required would be 0.94 μm and the corresponding value of $V_\pi L$ would increase slightly to 9.0 V.cm. It can be observed that by increasing x_1, the overall $V_\pi L$ can be reduced for a given total optical loss.

3.6 Switches

Finally, as an example of a nonidentical directional coupler, an electro-optic LiNbO$_3$ channel coupler modulator [1], shown in the inset of Figure 3.37, is simulated at a wavelength of 0.85 μm. Without any applied modulating signal, the refractive index of both LiNbO$_3$ guides, n_g, is taken as 2.3, while that of the cladding, n_s, is 2.29. With appropriate electrode design, for nonzero modulating potential, it is assumed that the index in one guide increases, while the other guide decreases by the same amount, $\Delta n/2$, and the cladding index remains constant. The effect of the electro-optically induced index difference, Δn, on the TE coupling length is shown in Figure 3.37, where the results obtained using the VFEBPM and VFEM [45] are in close agreement.

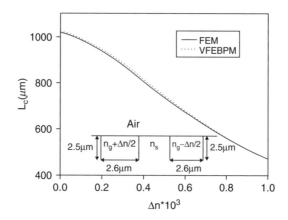

Figure 3.37 Effect of electro-optically induced index difference on the coupling length. The inset is a schematic diagram of the electro-optic LiNbO channel coupler modulator under consideration. (Reproduced with permission from Obayya, S.S.A., Rahman, B.M.A. and El-Mikati, H.A. (2000) New full vectorial numerically efficient propagation algorithm based on the finite element method. *IEEE J. Lightwave Technol.*, **18** (3), 409–415. © 2000 IEEE.)

Without applied modulating potential, $\Delta n = 0$, the two guides are phase-matched, giving rise to a peak coupling length, 1010 μm, which monotonically reduces with increasing Δn as the two guides continue to lose their synchronicity. When the input power is launched into guide 'b', the power evolution in the two guides in the axial direction, z, is shown in Figure 3.38 for two values of Δn. For, $\Delta n = 0$, full power transfer from guide 'b' to guide 'a' can be observed at around 1010 μm, while for

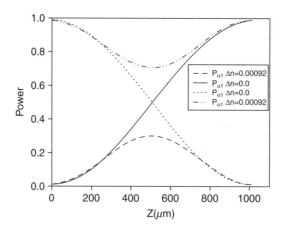

Figure 3.38 The evolution of optical power in the two guides along the axial direction. (Reproduced with permission from Obayya, S.S.A., Rahman, B.M.A. and El-Mikati, H.A. (2000) New full vectorial numerically efficient propagation algorithm based on the finite element method. *IEEE J. Lightwave Technol.*, **18** (3), 409–415. © 2000 IEEE.)

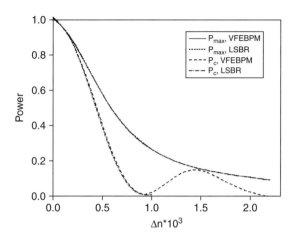

Figure 3.39 Variation of the output (P_o) and maximum powers (P_{max}), with the electro-optically induced index difference. (Reproduced with permission from Obayya, S.S.A., Rahman, B.M.A. and El-Mikati, H.A. (2000) New full vectorial numerically efficient propagation algorithm based on the finite element method. *IEEE J. Lightwave Technol.*, **18** (3), 409–415. © 2000 IEEE.)

$\Delta n = 0.000\ 92$, a fraction of the guide 'b' power is initially transferred to guide 'a' due to their phase mismatch; this power is then returned to guide 'b' with virtually no overall power is transferred to guide 'a'. It can be noted that, the coupling length, L_c, is equal to 505.0 μm, half of the value for $\Delta n = 0$. In this case, at $z = L_c$, the optical power couples back to guide 'b', as shown in Figure 3.38. With guide 'a' as an output port, the variation of the maximum and output powers, P_{max} and P_o, with Δn are shown in Figure 3.39. The maximum power, P_{max}, is the power transferred to guide 'a' at a device length equal to the coupling length at concerned particular value of Δn, while the output power, P_o, denotes the power transferred to guide 'a' at a fixed device length equal to the peak coupling length, $L_{co} = 1010$ μm. The power transfer results obtained using the VFEBPM are in excellent agreement with those obtained using the least-squares boundary residual (LSBR) method [45]. As shown in Figure 3.39, the output power is minimum for Δn values at which L_{co} is an even multiple of the coupling length, while for Δn values at which L_{co} is an odd multiple of the coupling length, the output power reaches a maximum value, but not the unity peak, as the two guides are strongly phase-mismatched and hence, the full power transfer is inhibited.

References

[1] Obayya, S.S.A., Rahman, B.M.A. and El-Mikati, H.A. (2000) New full vectorial numerically efficient propagation algorithm based on the finite element method. *J. Lightwave Technol.*, **18** (3), 409–415.

[2] Marcatili, E.A.J. (1969) Dielectric rectangular waveguide and directional coupler for integrated optics. *Bell Syst. Tech. J.*, **48**, 2071–2102.

[3] Russell, P.St.J. (2003) Photonic crystal fibers. *Science*, **299**, 358–362.

[4] Obayya, S.S.A., Rahman, B.M.A. and Grattan, K.T.V. (2005) Accurate finite element modal solution of photonic crystal fibres. *Optoelectron. IEE Proc.*, **152**, 241–246.

[5] Hameed, M.F.O., Obayya, S.S.A., Al-Begain, K. *et al.* (2009) Modal properties of an index guiding nematic liquid crystal based photonic crystal fiber. *IEEE J. Lightwave Technol.*, **27** (21), 4754–4762.

[6] Rahman, B.M.A. and Davies, J.B. (1984) Finite-element analysis of optical and microwave waveguide problems. *IEEE Trans. Microwave Theory Tech.*, **32**, 20–28.

[7] Xu, C.L., Huang, W.P. and Chaudhuri, S.K. (1993) Efficient and accurate vector mode calculations by beam propagation method. *J. Lightwave Technol.*, **11**, 1209–1215.

[8] Kumar, A., Kaul, A.N. and Ghatak, A.K. (1985) Prediction of coupling length in a rectangular-core directional coupler: an accurate analysis. *Opt. Lett.*, **10** (2), 86–88.

[9] Kumar, A., Thyagarajan, K. and Ghatak, A.K. (1983) Analysis of rectangular-core dielectric waveguides: an accurate perturbation approach. *Opt. Lett.*, **8** (1), 63–65.

[10] Kuznetsov, M. (1983) Expressions for the coupling coefficient of a rectangular waveguide directional coupler. *Opt. Lett.*, **8** (9), 499–501.

[11] Koshiba, M. and Saitoh, K. (2003) Polarization-dependent confinement losses in actual holey fibers. *IEEE Photonics Technol. Lett.*, **15**, 691–693.

[12] Finazzi, V., Monro, T.M. and Richardson, D.J. (2002) Confinement Loss in Highly Nonlinear Optical Fibers. *Optical Fiber Communications Conference, Anaheim, CA*, Paper ThS4.

[13] Saitoh, K. and Koshiba, M. (2003) Chromatic dispersion control in photonic crystal fibers: application to ultra-flattened dispersion. *Opt. Express*, **11**, 843–852.

[14] Offerhaus, H.L., Broderick, N.G.R., Richardson, D.J., Sammut, R., Calpen, J. and Dong, L. (1998) High-energy single-transverse-mode Q-switched fiber laser based on a multimode large-mode-area erbium-doped fiber. *Opt. Lett.*, **23**, 1683–1685.

[15] Birks, T.A., Knight, J.C. and Russell, P. St. J. (1997) Endlessly single-mode photonic crystal fibre. *Opt. Lett.*, **22**, 961–963.

[16] Agrawal, G.P. (1997) *Fiber-optic Communication Systems*, 2nd edn., Wiley-Blackwell, Chichester

[17] Haakestad, M.W., Alkeskjold, T.T., Nielsen, M. *et al.* (2005) Electrically tunable photonic bandgap guidance in a liquid-crystal-filled photonic crystal fiber. *IEEE Photon. Technol. Lett.*, **17** (4), 819–821.

[18] Acharya, B.R., Baldwin, K.W., Rogers, J.A. *et al.* (2002) In-fiber nematic liquid crystal optical modulator based on in-plane switching with microsecond response time. *Appl. Phys. Lett.*, **81** (27), 5243–5245.

[19] Fang, D., Yan, Q.L. and Shin, T.W. (2004) Electrically tunable liquid-crystal photonic crystal fiber. *Appl. Phys. Lett.*, **85** (12), 2181–2183.

[20] Li, J., Wu, S.T., Brugioni, S., Meucci, R. and Faetti, S. (2005) Infrared refractive indices of liquid crystals. *J. Appl. Phys.*, **97** (7), 073501-073501-5.

[21] Ren, G., Shum, P., Yu, X. *et al.* (2008) Polarization dependent guiding in liquid crystal filled photonic crystal fibers. *Opt. Commun.*, **281**, 1598–1606.

[22] Zografopoulos, D.C., Kriezis, E.E. and Tsiboukis, T.D. (2006) Photonic crystal-liquid crystal fibers for single-polarization or high-birefringence. *Opt. Express*, **14** (2), 914–925.

[23] Alkeskjold, T.T. and Bjarklev, A. (2007) Electrically controlled broadband liquid crystal photonic bandgap fiber polarimeter. *Opt. Lett.*, **32** (12), 1707–1709.

[24] Leong, J.Y.Y. (2007) Fabrication and applications of lead-silicate glass holey fiber for 1-1.5microns: nonlinearity and dispersion trade offs. Faculty of engineering, science and mathematics Optoelectronics research centre: University of Southampton.

[25] Knight, J.C., Birks, T.A., Russell, P.St.J. and Atkin, D.M. (1996) All-silica single-mode fiber with photonic crystal cladding. *Opt. Lett.*, **21**, 1547–1549.

[26] Mori, A., Shikano, K., Enbutsu, K. *et al.* (2004) 1.5 μm Band Zero-dispersion Shifted Tellurite Photonic Crystal Fibre with a Nonlinear Coefficient of 675 W−1 km−1. ECOC 2004, 30th European Conference on Optical Communication, p. Th3.3.6.

[27] Feng, X., Mairaj, A.K., Hewak, D.W. and Monro, T.M. (2004) Towards high-index glass based monomode holey fibre with large-mode-area. *Electr. Lett.*, **40**, 167–169.

[28] Petropoulos, P., Monro, T.M., Ebendorff-Heidepriem, H. *et al.* (2003) Highly nonlinear and anomalously dispersive lead silicate glass holey fibres. *Opt. Express*, **11**, 3568–3573.

[29] Kiang, K.M., Frampton, K., Monro, T.M. *et al.* (2002) Extruded single-mode nonsilica glass holey optical fibres. *Electr. Lett.*, **38**, 546–547.

[30] Kumar, V., George, A.K., Knight, J.C. and Russell, P.S.J. (2003) Tellurite photonic crystal fibre. *Opt. Express*, **11**, 2641–2645.

[31] Fujino, S., Ijiri, H., Shimizu, F. and Morinaga, K. (1998) Measurement of viscosity of multi-component glasses in the wide range for fibre drawing. *J. Jpn. Inst. Met.*, **62**, 106–110.

[32] Nielsen, K., Noordegraaf, D., Sørensen, T. *et al.* (2005) Selective filling of photonic crystal fibers. *J. Opt. A, Pure Appl. Opt.*, **7** (8), L13–L20.

[33] Xiao, L., Jin, W., Demokan, M.S. *et al.* (2005) Fabrication of selective injection microstructured optical fibers with a conventional fusion splicer. *Opt. Express*, **13** (22), 9014–9022.

[34] Huang, Y., Xu, Y. and Yariv, A. (2004) Fabrication of functional microstructured optical fibers through a selective-filling technique. *Appl. Phys. Lett.*, **85** (22), 5182–5184.

[35] Haxha, S. and Ademgil, H. (2008) Novel design of photonic crystal fibers with low confinement losses, nearly zero ultra-flatted chromatic dispersion, negative chromatic dispersion and improved effective mode area. *J. Opt. Commun.*, **281**, 278–286.

[36] Koroty, S.K., Eisenstein, G., Tucker, R.S. *et al.* (1987) Optical intensity modulation to 40 GHz using a wideguide eletrooptic switch. *Appl. Phys. Lett.*, **50** (23), 1631–1633.

[37] Schmidt, R.V. and Alferness, R.C. (1979) Directional coupler switches, modulators and filters using alternating $\Delta\beta$ techniques. *IEEE Trans. Circuits Syst.*, **26**, 1099–1108.

[38] Bulmer, C.H. and Burns, W.K. (1984) Linear interferometric modulators in Ti : LiNbO3. *J. Lightwave Technol.*, **2**, 512–521.

[39] Obayya, S.S.A., Haxha, S., Rahman, B.M.A. and Grattan, K.T.V. (2003) Optimization of optical properties of a deeply-etched semiconductor optical modulator. *J. Lightw. Technol.*, **21** (8), 1813–1819.

[40] Yariv, A. and Yeh, P. (1984) *Optical Waves in Crystals*, John Wiley & Sons, Inc., New York.

[41] Heaton, J.M., Brouke, M.M., Jones, S.B. *et al.* (1999) Optimization of deep-etched, single-mode GaAs-AlGaAs optical waveguides using controlled leakage into substrate. *J. Lightwave Technol.*, **17**, 267–281.

[42] Obayya, S.S.A., Azizur Rahman, B.M., Grattan, Kenneth T.V. and El-Mikati, H.A. (2002) Full vectorial finite-element-based imaginary distance beam propagation solution of complex modes in optical waveguides. *J. Lightwave Technol.*, **20**, 1054.

[43] Themistos, C., Rahman, B.M.A., Hadjicharalambous, A. and Grattan, K.T.V. (1995) Loss/gain characterization of optical waveguides. *J. Lightwave Technol.*, **13**, 1760–1765.

[44] Khan, M.N., Gopinath, A., Bristow, J.P.G. and Donnelly, J.P. (1993) Technique for velocity-matched travelling-wave electrooptic modulator in AlGaAs/GaAs. *IEEE Trans. Microwave Theory Tech.*, **41**, 244–249.

[45] Wongchareon, T., Rahman, B.M.A. and Grattan, K.T.V. (1997) Electro-optic directional coupler switches characterization. *J. Lightwave Technol.*, **15**, 377–382.

4

Bidirectional Beam Propagation Method

4.1 Introduction

In this chapter, the formulation of the finite-element solution for optical waveguide discontinuity problems is presented [1]. The validity and the numerical precision of the proposed method are evaluated through the analysis of three discontinuity problems, including a junction between two different waveguide sections, a laser–air interface and a laser angled facet. In addition, a numerically efficient scalar analysis of optical fibre facet problems based on the finite-element scheme is included [2] and the simulation results of dealing with various numerical uncoated and coated optical fibre facet problems will be detailed.

4.2 Optical Waveguide Discontinuity Problem

Optical waveguide discontinuities play a very important role in the design of optical communication systems, and are quite often faced in situations such as butt-coupled waveguides, waveguide ends, laser facets and antireflection coatings [3–5]. Therefore, an accurate and efficient method for the solution of optical waveguide discontinuities is highly desirable. There are a number of methods proposed in the literature for the analysis of optical waveguide discontinuity problems. These methods can be classified, according to their solution strategy, into two groups. In the first group, the total field on either side of the discontinuity is made up as a weighted summation of all guided and radiation modes. Then, the continuity of the tangential electric and magnetic fields at the discontinuity plane is enforced using, for example, the least squares boundary residual (LSBR) [6], a combination of the finite elements and method of lines (FE-MoL) [7], the finite-element method (FEM) with analytical

Computational Photonics   Salah Obayya
© 2011 John Wiley & Sons, Ltd

techniques [8] or the free-space radiation mode (FSRM) method [9], to solve for the reflection and transmission coefficients. In this class of methods, the solution accuracy is enhanced as more guided and radiation modes are included in the field summation. This would be very time-consuming, especially in the case of 3D optical waveguide problems, where approximating the radiation modes with a continuous spectrum turns out to be tremendously complicated. Alternatively, in the second group of discontinuity analyses, the Pade approximations, in the context of the finite-difference method, are used to efficiently approximate the square root of the characteristic matrix of both discontinuity sides [5, 10, 11]. In this group of analyses, although the numerically intensive modal solution stage on each side is totally avoided, the use of higher-order Pade approximations requires matrix inversion that would be very demanding in terms of computer resources, especially in dealing with 3D problems.

In this chapter, a formulation of the finite-element solution for the optical waveguide discontinuity problems [1] is proposed. By 'lumping' the mass matrix entries into the diagonal locations [12–14], this matrix renders itself diagonal, and hence, its inversion and multiplication with the original characteristic matrix will result in a modified characteristic matrix as sparse as the original one. Then, upon the application of a Taylor's series expansion to represent the square root of the modified characteristic matrix, and satisfying the interface boundary condition at the discontinuity plane, both the reflected and transmitted fields can be efficiently solved for. Therefore, the presented FEM not only avoids the modal solution stage, but also does not require any matrix inversion to approximate the square root operator of the characteristic matrix using higher-order Taylor's series expansions.

4.3   Finite-Element Analysis of Discontinuity Problems

Consider a 2D optical waveguide discontinuity problem whose schematic diagram is depicted in Figure 4.1. The transverse and propagation directions are assumed to be y and z, respectively, and the discontinuity plane is located at $z = 0$. Supposing that there is no variation in the x-direction, the following 2D Helmholtz equation for both sides 1 and 2 can be written as [15]

$$\frac{\partial}{\partial y}\left(\frac{p_i}{s_i}\frac{\partial \psi_i}{\partial y}\right) + s_i\frac{\partial}{\partial z}\left(p_i\frac{\partial \psi_i}{\partial z}\right) + k_0^2 q_i s_i \psi_i = 0 \qquad (4.1)$$

with

$$\psi_i = E_{xi}, \quad p_i = 1, \quad q_i = n_i^2 \qquad \text{for TE modes} \qquad (4.2a)$$

$$\psi_i = H_{xi}, \quad p_i = 1/n_i^2, \quad q_i = 1 \qquad \text{for TM modes} \qquad (4.2b)$$

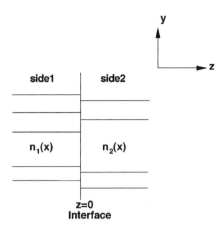

Figure 4.1 Schematic diagram of a discontinuity between two sides, each with an arbitrary refractive index distribution n_i (x). (Reproduced with permission from Obayya, S.S.A. (2004) Novel finite element analysis of optical waveguide discontinuity problems. *IEEE J. Lightwave Technol.*, **22** (5), 1420–1425. © 2004 IEEE.)

$$s_i = 1 - j \frac{3\lambda}{4\pi n_i d} \left(\frac{\rho}{d}\right)^2 \ln\left(\frac{1}{R_t}\right), \qquad \text{in PML region} \qquad (4.3a)$$

$$s_i = 1, \qquad \text{in non} - \text{PML region} \qquad (4.3b)$$

where the subscript i is 1 or 2 and corresponds to the quantities in the ith side, E_x and H_x are the x-component of the electric and magnetic fields, respectively, k_0 is the free space wavenumber, n is the refractive index distribution, λ is the operating wavelength, s is the PML parameter, ρ is the distance inside the PML region from the beginning of the PML and R_t is the theoretical reflection coefficient placed at the boundary between the PML region and the computational window.

Dividing the cross section into a number of first-order linear elements and applying the standard Galerkin's finite-element procedure results in [15]

$$[M]_i \frac{d^2\{\psi\}_i}{dz^2} + [K]_i\{\psi\}_i = \{0\} \qquad (4.4)$$

where $\{\psi\}_i$ is the nodal field vector, $\{0\}$ is the null vector, and $[M]_i$ and $[K]_i$ are the global mass and characteristic matrices whose expressions are

$$[M]_i = \sum_e \int_e p_i s_i \{N\}\{N\}^T \, dy \qquad (4.5a)$$

$$[K]_i = \sum_e \int_e \left[k_o^2 q_i s_i \{N\}\{N\}^T - \frac{p_i}{s_i} \frac{d\{N\}}{dy} \frac{d\{N\}^T}{dy} \right] dy \qquad (4.5b)$$

In Equation (4.5) $\{N\}$ represent the shape function vectors over each element, T is the transpose and the summation $\sum_e$ is performed over all the elements.

4.4   Derivation of Finite-Element Matrices

Conventionally, both the guided and radiation modes of each side are found using a modal solution technique for Equation (4.4), and then, upon matching the transverse electric and magnetic field components across the discontinuity plane, the solutions for both the reflected and transmitted fields are obtained [7]. However, the modal solution analysis is very time consuming and is complicated in dealing with radiation modes with a continuous spectrum. Alternatively, a finite-element procedure completely avoiding the modal solution stage is presented as follows. The global mass matrix of each side is lumped into a diagonal matrix using the row sum procedure [12–14]. In this procedure, the diagonal entry of the new lumped mass matrix at a particular row 'ir' is the sum of all the entries in that row of the original mass matrix as

$$\tilde{M}_{ir,ir} = \sum_{jc} M_{ir,jc} \qquad (4.6)$$

where $\tilde{M}_{ir,ir}$ is the diagonal entry of the new lumped mass matrix at the irth row, and the summation in Equation (4.6) is performed over all entries $\tilde{M}_{ir,jc}$ in that row. Having lumped the mass matrix, its inverse is merely the reciprocals of its diagonal elements. So, multiplying both sides of Equation (4.4) with the inverse of the 'lumped' mass matrix leads to

$$\frac{d^2\{\psi\}_i}{dz^2} + [A]_i\{\psi\}_i = \{0\} \qquad (4.7)$$

with

$$[A]_i = [\tilde{M}]_i^{-1}[K]_i \qquad (4.8)$$

It should be noted that the numerical advantages of lumping the mass matrix is the elimination of the need for any matrix inversion and also keeping the new characteristic matrix $[A]_i$ as sparse as the original matrix $[K]_i$. This would lead to massive savings in the use of computer resources, especially in dealing with 3D problems. The formal solution of Equation (4.7) on each side can be written as

$$\{\psi\}_i(z) = \exp\left(-j\sqrt{[A]_i}z\right)\{\psi^+\}_i + \exp\left(j\sqrt{[A]_i}z\right)\{\psi^-\}_i \qquad (4.9)$$

where $\{\psi^+\}_i$, and $\{\psi^-\}_i$ are the forward and backward propagating fields in each side at the discontinuity plane $z = 0$. Assume that the incident field is launched from side 1 and applying the interface boundary conditions of the continuity of transverse electric and magnetic fields at $z = 0$ results in

$$\left(p_1\sqrt{[A]_1} - p_2\sqrt{[A]_2}\right)\{\psi^+\}_1 = \left(p_1\sqrt{[A]_1} + p_2\sqrt{[A]_2}\right)\{\psi^-\}_1 \quad (4.10)$$

$$\left(2p_1\sqrt{[A]_1}\right)\{\psi^+\}_1 = \left(p_1\sqrt{[A]_1} + p_2\sqrt{[A]_2}\right)\{\psi^+\}_2 \quad (4.11)$$

where $\{\psi^+\}_1$, $\{\psi^-\}_1$ and $\{\psi^+\}_2$ are the incident, reflected and transmitted fields, respectively. So, upon the solution of the linear system of equations in (4.10) and (4.11), quantities such as the reflection and transmission coefficients can be determined once the incident field is known.

4.5 Application of Taylor's Series Expansion

Usually, the square root of the characteristic matrix is evaluated using a Pade approximation, which is numerically very accurate and efficient [5, 10, 11]. However, the use of higher-order Pade approximations requires matrix inversion, which may be computationally intensive, especially in dealing with 3D problems. On the other hand, the Taylor's series expansion of the square root of the characteristic matrix around a reference wavenumber β_i has been employed here, and its mth-order approximation is given as

$$\sqrt{[A]_1} = \beta_i \left([I] + \sum_{j=1}^{m} \frac{a_j}{j!}\left(\frac{[A]_i - \beta_i^2[I]}{\beta_i^2}\right)^j\right) \quad (4.12)$$

with $a_j = II_{k=0}^{j-1}(n - k), j \geq 1, n = 1/2$ and $[I]$ is the identity unity matrix.

As may be seen from Equation (4.12), the use of the Taylor's series expansion of the square root of the characteristic matrix turns out to be merely a series of matrix multiplications that can be efficiently performed by exploiting the matrix sparisty. In order to properly deal with the 'evanescent' modes generated at the interface between the two sides, the reference wavenumber has to be a complex valued [11] which can be straightforwardly implemented if a phasor term has been included so that

$$\beta_i = \beta_{oi}e^{j\Phi} \quad (4.13)$$

where β_0 is the original real-value reference wavenumber normally taken as the propagation constant of the fundamental mode, and Φ is a phase angle in the range $0° < \Phi < 90°$. This approach is equivalent to the use of the Pade approximation with

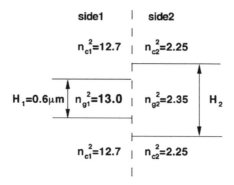

Figure 4.2 Schematic diagram of a junction between two waveguides. (Reproduced with permission from Obayya, S.S.A. (2004) Novel finite element analysis of optical waveguide discontinuity problems. *IEEE J. Lightwave Technol.*, **22** (5), 1420–1425. © 2004 IEEE.)

real coefficients and the branch cut of the square root of the characteristic matrix [10]. It will be shown in the following section that introducing Φ is mandatory to ensure the satisfaction of the power balance condition, however it has a very little effect on the accuracy of the solution.

4.6   Computation of Reflected, Transmitted and Radiation Waves

4.6.1   Junction Between Two Optical Waveguides

The first example considered [1] is a junction between two optical waveguides shown in Figure 4.2. For the range of waveguide heights considered, both waveguides are monomoded. The same structure has been considered in [6, 8]. In this example, the operating wavelength, λ, is 1.0 μm, the waveguide height of side 1, H_1, is fixed at 0.6 μm, while the waveguide height of side 2, H_2, is taken as 0.4 μm.

4.6.1.1   Effect of the Phasor Angle

For an incident mode TM_0, Figure 4.3 shows the variation of the total power, P_{tot}, normalised to the incident power, with the phasor angle, Φ. The total power is calculated as the sum of both the reflected and transmitted powers. As may be observed from Figure 4.3, the power balance condition is fulfilled for $\Phi > 1°$, either for second- or third-order Taylor's approximations. On the other hand, Figure 4.4 illustrates the effect of the phasor angle Φ on the values of the reflected $|R|^2$ and transmitted $|T|^2$ powers of the fundamental mode on sides 1 and 2, respectively. It may be noted from Figure 4.4 that for $\Phi > 1°$, both $|R|^2$ and $|T|^2$ are almost unchanged with Φ, and they converge to values of 0.16 and 0.61, respectively, which are in excellent agreement with the results obtained in [6], using the LSBR method employing an expansion of 25 modes on each side.

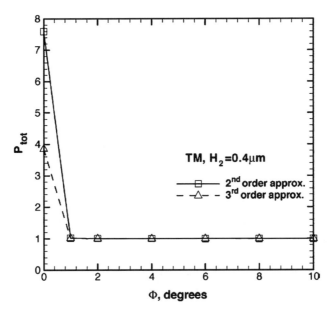

Figure 4.3 Variation of the total power, P_{tot}, with the phasor angle, Φ, for a launched TM_0 mode with $H_2 = 0.4$ μm. (Reproduced with permission from Obayya, S.S.A. (2004) Novel finite element analysis of optical waveguide discontinuity problems. *IEEE J. Lightwave Technol.*, **22** (5), 1420–1425. © 2004 IEEE.)

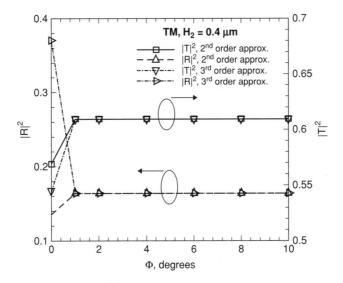

Figure 4.4 Variation of the transmitted power $|T|^2$ and the reflected power $|R|^2$ of the fundamental TM_0 modes on sides 2 and 1, respectively, with the phasor angle, Φ, for $H_2 = 0.4$ μm. (Reproduced with permission from Obayya, S.S.A. (2004) Novel finite element analysis of optical waveguide discontinuity problems. *IEEE J. Lightwave Technol.*, **22** (5), 1420–1425. © 2004 IEEE.)

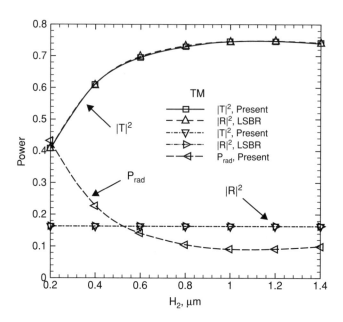

Figure 4.5 Variation of the transmitted power $|T|^2$ the reflected power $|R|^2$ of the fundamental TM_0 modes in sides 2 and 1, respectively, and the radiated power P_{rad} with the waveguide height in side 2, H_2. (Reproduced with permission from Obayya, S.S.A. (2004) Novel finite element analysis of optical waveguide discontinuity problems. *IEEE J. Lightwave Technol.*, **22** (5), 1420–1425. © 2004 IEEE.)

4.6.1.2   Effect of Waveguide Height

Next, the effect of varying the waveguide height in side 2, H_2, on the reflected, $|R|^2$ and transmitted $|T|^2$ powers of the fundamental mode in sides 1 and 2, respectively, and also the radiated power P_{rad} is shown in Figure 4.5. As may be observed from Figure 4.5, the transmitted power of the fundamental TM mode in side 2 increases with H_2 and reaches its peak value of 0.755 when H_2 is 1.2 μm. Consequently, as the transmission efficiency is enhanced with increasing H_2, the radiated power P_{rad} is continuously reducing and settles down to a value of nearly 0.1 as $H_2 > 1.0$ μm. On the other hand, the reflected power of the fundamental mode of side 1 is almost constant with the change in H_2.

The results shown in Figure 4.5 have been obtained using a third-order Taylor's expansion and they are almost indistinguishable from those obtained using the LSBR method [6]. For a TE_0 excitation from side 1, Figure 4.6 shows the variation of the reflected power $|R|^2$ and transmitted power $|T|^2$ of the fundamental modes in sides 1 and 2, respectively, and the radiation power P_{rad} with the waveguide height in side 2, H_2. As may be seen from Figure 4.6, the behaviour of $|R|^2$, $|T|^2$ and P_{rad} is quite similar to the TM excitation case, except that the maximum transmission efficiency

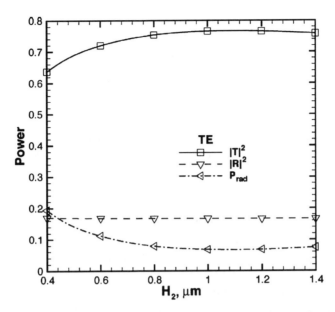

Figure 4.6 Variation of the transmitted power $|T|^2$ the reflected power $|R|^2$ of the fundamental TE$_0$ modes in sides 2 and 1, respectively, and the radiated power P_{rad} with the waveguide height in side 2, H_2. (Reproduced with permission from Obayya, S.S.A. (2004) Novel finite element analysis of optical waveguide discontinuity problems. *IEEE J. Lightwave Technol.*, **22** (5), 1420–1425. © 2004 IEEE.)

is slightly enhanced to 0.776, while the minimum radiated power drops to 0.079. In all examples considered here, the computational window is enclosed by two PML regions each of width 1.0 μm, with eight elements and a value of the theoretical reflection coefficient at the interface between the PML and the computational window R_t of 10^{-8}.

In this example, a computational window of 10.0 μm is nonuniformly represented by 200 line elements. With this discretisation, it took less than 0.5 s to calculate the reflected and transmitted power pair on a PC (Pentium IV, 2.66 GHz). On the other hand, the LSBR with the FEM method [6] employing 25 modes for each side would require 25 times the execution time and storage requirement compared to the proposed method.

4.6.2   Laser–Air Facet

The more challenging discontinuity problem of a laser–air facet is also considered [1]. The schematic diagram of the structure is shown in Figure 4.7. The refractive indexes of the core and substrate are 3.60 and 3.42, respectively, and the operating wavelength is taken as 0.86 μm. For a TE$_0$ mode launched, the variation of the reflected power

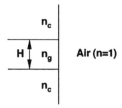

Figure 4.7 Schematic diagram of a laser–air facet. (Reproduced with permission from Obayya, S.S.A. (2004) Novel finite element analysis of optical waveguide discontinuity problems. *IEEE J. Lightwave Technol.*, **22** (5), 1420–1425. © 2004 IEEE.)

of the fundamental TE_0 mode with the waveguide height, H, is shown in Figure 4.8. It may be observed from this figure that the results of this approach converge very rapidly with those obtained using the FSRM [9].

As the index difference between the two discontinuity sides is large, it can be seen from Figure 4.8 that the first-order Taylor's expansion, which is equivalent to the paraxial approximation, is underestimating the values of the reflected power. On the other hand, the results obtained using the third-order Taylor's expansion are

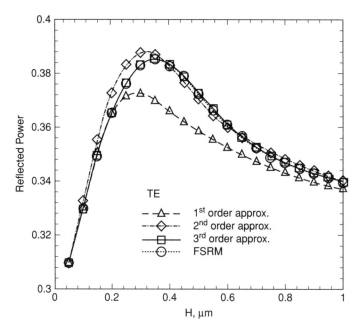

Figure 4.8 Variation of the fundamental TE_0 mode reflected power with the waveguide height, H, using the present finite-element approach with first-, second- and third-order Taylor's series expansions, and the FSRM. (Reproduced with permission from Obayya, S.S.A. (2004) Novel finite element analysis of optical waveguide discontinuity problems. *IEEE J. Lightwave Technol.*, **22** (5), 1420–1425. © 2004 IEEE.)

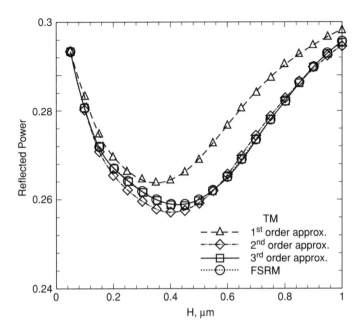

Figure 4.9 Variation of the fundamental TM_0 mode reflected power with the waveguide height, H, using the present finite-element approach with first-, second- and third-order Taylor's series expansions, and the FSRM. (Reproduced with permission from Obayya, S.S.A. (2004) Novel finite element analysis of optical waveguide discontinuity problems. *IEEE J. Lightwave Technol.*, **22** (5), 1420–1425. © 2004 IEEE.)

in excellent agreement with those obtained using the FSRM [9]. For a TM_0 mode excitation, the variation of the reflected power of the fundamental TM_0 mode with H obtained using the present finite-element approach employing the first-, second- and third-order Taylor's expansions and the FSRM [9] are shown in Figure 4.9. Again, this figure shows clearly that the use of the finite-element approach with a third-order Taylor's expansion gives reflected power results that are almost indistinguishable from those obtained using the FSRM [9]. In this example, a computational window of 6.0 μm is nonuniformly represented by 180 line elements.

4.6.3 Laser–Air Facet with a Tilt Angle

The design of a laser–air facet with a tilt angle to minimise the reflected power will also be considered [1] using the present finite-element approach. The schematic diagram of the structure is shown in Figure 4.10, where the wavelength and other parameters are the same as in the previous example. In order to apply the finite-element approach, a virtual interface perpendicular to the waveguide cross section replaces the original tilted interface, as shown in Figure 4.10. Equation (4.10) is solved for the reflected

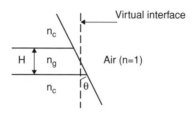

Figure 4.10 Schematic diagram of a laser–air facet with facet angle, θ. (Reproduced with permission from Obayya, S.S.A. (2004) Novel finite element analysis of optical waveguide discontinuity problems. *IEEE J. Lightwave Technol.*, **22** (5), 1420–1425. © 2004 IEEE.)

field at the normal 'virtual' interface, and then its phase front is modified to be $2\beta_{01}$ $\sin \theta(y - y_0)$, where β_{01} is the real-value propagation constant of the launched fundamental mode, θ is the facet angle and y_0 is the centre y-coordinate of the waveguide core, in order to obtain the reflected field at the real tilted interface. Then, as in the previous examples, the reflected power is calculated as the power carried by the fundamental mode normalised to the incident mode power.

For a waveguide height, H, of 0.4 μm, Figure 4.11 shows the variation of the reflected power of the fundamental mode, in decibels, with the facet angle, θ, for both

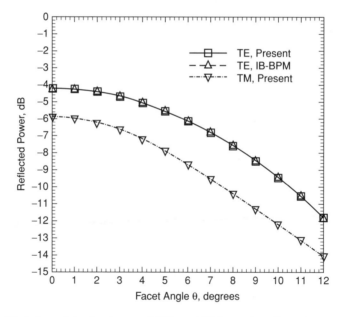

Figure 4.11 Variation of the fundamental TE_0 and TM_0 mode reflected powers, in decibels, with the facet angle θ using the present finite-element approach and the BBPM. The waveguide height, H, is 0.4 μm. (Reproduced with permission from Obayya, S.S.A. (2004) Novel finite element analysis of optical waveguide discontinuity problems. *IEEE J. Lightwave Technol.*, **22** (5), 1420–1425. © 2004 IEEE.)

TE and TM excitations. As may be noted from this figure, an increase in the facet angle has a significant effect on reducing the reflected power. In particular, for TE excitation, the reflected power is reduced from -4.1 to -11.9 dB as the facet angle θ increases from 0 to 12°. Also, for TM excitation, a reduction in the reflected power from -5.95 to -14dB occurs as θ increases from 0 to 12°. As may be observed also from Figure 4.11, the results obtained using the present finite-element approach with a third-order Taylor's expansion are in close proximity to those obtained using the bidirectional beam propagation method (BBPM) [16].

Next, the variation of the TE_0 reflected power, in decibels, with the waveguide height, H, with the facet angle, θ, as a parameter is shown in Figure 4.12. It can be noted from this figure that for $\theta > 10°$, a significant reduction in the reflected power occurs for the whole range of the waveguide heights considered. In particular, for $H = 0.7$ µm and $\theta = 15°$, reflected power as low as -30 dB can be obtained. Similarly, Figure 4.13 shows that the TM reflected power behaves in a very similar way to the TE case, for example, the TM reflected power for $H = 0.7$ µm and $\theta = 15°$ is nearly -34 dB.

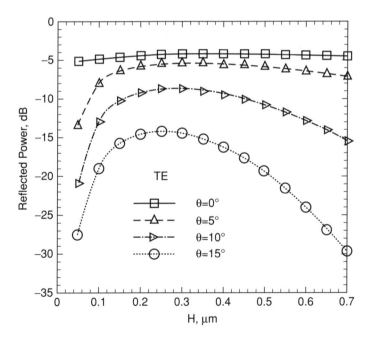

Figure 4.12 Variation of the fundamental TE_0 mode reflected powers, in decibels, with the waveguide height, H, and the facet angle, θ, as a parameter. (Reproduced with permission from Obayya, S.S.A. (2004) Novel finite element analysis of optical waveguide discontinuity problems. *IEEE J. Lightwave Technol.*, **22** (5), 1420–1425. © 2004 IEEE.)

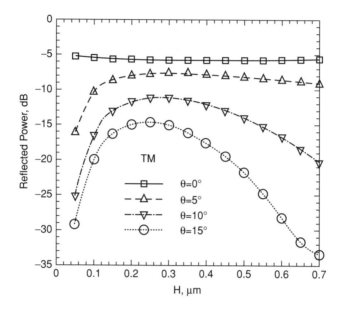

Figure 4.13 Variation of the fundamental TM_0 mode reflected powers, in decibels, with the waveguide height, H, and the facet angle, θ, as a parameter. (Reproduced with permission from Obayya, S.S.A. (2004) Novel finite element analysis of optical waveguide discontinuity problems. *IEEE J. Lightwave Technol.*, **22** (5), 1420–1425. © 2004 IEEE.)

4.7   Optical Fibre Facet Problem

The problems of light scattering at the discontinuity planes of various optical wave-guides have been considered in the literature using various numerical techniques, for example [1,4,6,11]. However, the treatment of the optical fibre facet problem has been dealt with in a relatively few publications. Based on the method of lines (MoL), a rigorous numerical approach has been suggested in [17] to account for the diffraction of the scalar LP modes of abruptly terminated optical fibers. Although accurate, the analysis has been restricted to the case of uncoated optical fibers. Further, the time-consuming modal-solution process to find both the guided and radiation modes of both discontinuity sides is inevitable prior to the calculation of the scattered fields. On the other hand, the finite-difference time domain (FDTD) has been successfully applied to study the propagation characteristics of optical fibers using an approach that adopts cylindrical coordinates [18]. However, the FDTD is known to be computationally expensive in terms of both the time and memory requirements compared to frequency-domain techniques. The semi-analytical free-space radiation mode (FSRM) method [19] has been shown to give a very accurate and yet numerically efficient full-vectorial treatment of both uncoated and coated optical fibre facet problems [19]. The approach is mainly based on the assumption that the fibre radiation modes are propagating in a uniform refractive index region. This assumption has proven valid for transverse index variation up to 10%, which is the case in most optical fibre situations.

In the next section, an alternative scalar finite-element approach [2] is introduced. Through the application of Galerkin's finite-element procedure, the characteristic matrix is generated for each side of the discontinuity plane. Conventionally, a lengthy and tedious modal-solution approach would be followed to find the guided and radiation modes by performing an eigenvalue analysis on these matrices. Instead, the mass matrix is lumped into a diagonal matrix, and its inverse is multiplied with the original characteristic matrix to produce a new characteristic matrix as sparse as the original one [1]. Then, the square-root operator of this matrix is approximated using the Taylor's series expansion, and through the enforcement of the boundary conditions at the discontinuity plane, the reflected and transmitted fields can be found numerically [1]. The numerical advantage of using the Taylor's series expansion is the avoidance of matrix inversion to represent high-order approximations as in Pade approximation analysis [11]. Although the present finite-element approach is scalar, it may be more flexible than the FRSM method [19] in dealing with situations such as coated angled fibre facets, where small angles would have a significant influence on the performance of the antireflection (AR) coating, as will be seen later from the results.

4.8   Finite-Element Analysis of Optical Fibre Facets

4.8.1   *Formulation*

The schematic diagrams of the coated optical fibre facet problem and the 3D view are shown in Figure 4.14. As may be noted from this figure, the diameter of the fibre core is D and the refractive indexes of the core and cladding are n_{core} and n_{clad}, respectively. The fibre facet can be either abruptly terminated by air or coated with an AR layer of thickness d_{AR} and refractive index n_{AR}. These parameters are to be chosen in order to minimise the reflection to the fibre from the air. First, the problem of scattering of the scalar LP_{01} mode from the uncoated fibre facet, as shown in Figure 4.15, will be considered.

As shown from Figure 4.15, cylindrical coordinates are adopted and the discontinuity plane is located at $z = 0$. For circularly symmetric structures, such as the one in hand, and starting from Maxwell's equations, the scalar wave equation governing the propagation of the LP_{0m} modes on each side of the discontinuity can be derived as [17]

$$\frac{\partial^2 E_i}{\partial z^2} + \frac{1}{\alpha_i}\frac{\partial}{\partial r}\left(\frac{1}{\alpha_i}\frac{\partial E_i}{\partial r}\right) + \frac{1}{\alpha_i^2}\frac{1}{r}\frac{\partial E_i}{\partial r} + k_o^2 n_i^2 E_i = 0 \qquad (4.14)$$

where i refers to the quantities in the ith side, n_i is the refractive-index distribution, k_0 is the free-space wavenumber and E_i is the electric field. The parameter α_i is related

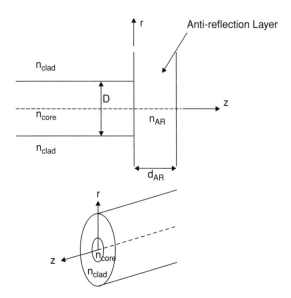

Figure 4.14 Schematic diagram of the coated optical fibre facet problem. (Reproduced with permission from Obayya, S.S.A. (2006) Scalar finite-element analysis of optical-fiber facets. *IEEE J. Lightwave Technol.*, **24** (5), 2115–2121. © 2006 IEEE.)

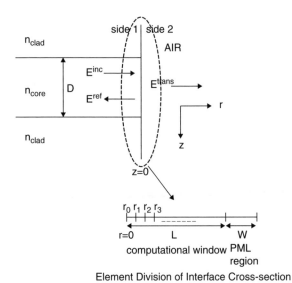

Figure 4.15 Finite-element discretisation of the discontinuity plane. (Reproduced with permission from Obayya, S.S.A. (2006) Scalar finite-element analysis of optical-fiber facets. *IEEE J. Lightwave Technol.*, **24** (5), 2115–2121. © 2006 IEEE.)

to the perfectly matched layer (PML) boundary condition, and is assumed to follow

$$\alpha = \begin{cases} 1 & \text{inside computational window} \\ 1 - j\dfrac{\sigma_{max}}{\omega \varepsilon_o \varepsilon_r} \left(\dfrac{r - L}{W} \right)^2 & \text{inside PML} \end{cases} \tag{4.15}$$

where ω is the frequency $(= 2\pi c/\lambda)$, σ_{max} is the maximum conductivity inside the PML region, W is the width of the PML region and L is the width of the computational window, as shown in Figure 4.15. Then, by dividing the radial cross section into a number of first-order line elements and applying Galerkin's procedure to Equation (4.14) leads to

$$[M]_i \frac{d^2\{E\}_i}{dz^2} + [k]_i\{E\}_i = \{0\} \tag{4.16}$$

where $\{E\}_i$ is the electric field vector, $\{0\}$ is the null vector, $[M]_i$ and $[K]_i$ are the global mass and characteristic matrices of which the expressions are given as

$$[M]_i = \sum_e \int [N][N]^T dr \tag{4.17}$$

$$[K]_i = \sum_e \int \left(k_o^2 n_i^2 [N][N]^T - \frac{1}{\alpha_i^2 r}[N][N]^T - \frac{1}{\alpha_i^2}\frac{\partial}{\partial r}[N]\frac{\partial}{\partial r}[N]^T \right) dr \tag{4.18}$$

In Equations (4.17) and (4.18), $[N]$ represent the shape function vectors over each element, T is the transpose and the summation $\sum_e$ is carried out over all the elements. Equation (4.18) gives a proper representation of the characteristic matrix as long as the singularity caused by $r = 0$ is avoided. However, as may be noted from Figure 4.15, the first element contains the singularity point, $r = 0$. In this case, to avoid the singularity, the L'Hopital rule can be used to represent the term $(1/\alpha_i^2)(1/r)(\partial E_i/\partial r)$ as $\text{Lim}_{r \to 0}(1/r)(\partial E_i/\partial r) = (\partial^2 E_i/\partial r^2)$. Therefore, for the first element only, the characteristic matrix will read

$$[K]_i = \int \left(k_0^2 n_i^2 [N][N]^T - \frac{2}{\alpha_i^2}\frac{\partial}{\partial r}[N]\frac{\partial}{\partial r}[N]^T \right) dr \tag{4.19}$$

4.8.2 Derivation of Finite-Element Matrices

In the MoL analysis [17], both guided and radiation modes of both discontinuity sides are first obtained, and then, by enforcing the continuity of the transverse electromagnetic fields at the discontinuity plane, both reflected and transmitted fields can be calculated. However, to avoid the lengthy and complicated process of performing the modal analysis, an alternative approach is adopted [1]. As explained in Section 4.4, the global mass matrix, $[M]$, is first converted into a diagonal matrix by simply using the 'lumping' rule [1, 12] in Equation (4.6) ($\tilde{M}_{ir,ir} = \sum_{jc} M_{ir,jc}$), where the diagonal entry of the new lumped mass matrix $\tilde{M}$ at a particular row 'ir' is the sum of all entries in that particular row of the original mass matrix, M. The summation in Equation (4.6) is performed over all columns 'jc' of the original matrix where nonzero entries exist. Then, the inverse of the lumped matrix would lend itself to be simply another diagonal matrix of which the entries are the reciprocals of the lumped mass matrix. Then, multiplying both sides of Equation (4.16) with the inverse of the lumped mass matrix results in

$$\frac{d^2\{E\}_i}{dz^2} + [L]_i\{E\}_i = \{0\} \text{ with } [L]_i = [\tilde{M}]_i^{-1}[K]_i \qquad (4.20)$$

It should be emphasised that the new characteristic matrix $[L]$ is as sparse as the original matrix $[K]$, thanks to the lumping of the mass matrix into a diagonal matrix. The formal solution of Equation (4.20) would be written on each side as

$$\{E\}_i(z) = \exp\left(-j\sqrt{[L]_i}z\right)\{E^+\}_i + \exp\left(j\sqrt{[L]_i}z\right)\{E^-\}_i \qquad (4.21)$$

where $\{E^+\}_i$, and $\{E^-\}_i$ are the forward and backward propagating fields on each side at the discontinuity plane $z = 0$. Assuming an LP_{0m} mode is incident from the fibre on side 1, as shown in Figure 4.15, and applying the continuity of the transverse electric and magnetic fields at $z = 0$ leads to

$$\left(\sqrt{[L]_1} + \sqrt{[L]_2}\right)E^{\text{ref}} = \left(\sqrt{[L]_1} - \sqrt{[L]_2}\right)E^{\text{inc}} \qquad (4.22)$$

$$\left(2\sqrt{[L]_1}\right)E^{\text{inc}} = \left(\sqrt{[L]_1} + \sqrt{[L]_2}\right)E^{\text{trans}} \qquad (4.23)$$

where E^{inc}, E^{ref} and E^{trans} denote the incident, reflected and transmitted fields, respectively. The solution of the linear system of equations given in (4.22) and (4.23) yields the reflected and transmitted fields once the incident field is known.

4.8.3   Application of Taylor's Series Expansion

The square-root operator of the characteristic matrix can be evaluated accurately and efficiently using Pade approximations [11]. However, the calculation of high-order Pade approximations may be computationally intensive, as it requires the inversion of the matrix. Alternatively, Taylor's series expansion is used here instead to represent the square-root operator of the characteristic matrix around a reference wavenumber, β, and the nth order approximation is given by [1]

$$\sqrt{L} = \beta \left([I] + \sum_{j=1}^{n} \frac{a_j}{j!} \left(\frac{[L] - \beta^2[I]}{\beta^2} \right)^j \right) \tag{4.24}$$

where the expansion coefficients are given by $a_j = \prod_{k=0}^{j-1}(n-k), j \geq 1, n = 1/2$ and $[I]$ is the identity matrix. As explained in Section 4.5, the reference wavenumber has to be chosen as a complex number to truly represent both the guided and evanescent modes generated at the discontinuity plane and, in turn, to satisfy the power-balance condition. Therefore, the reference wavenumber is represented as $\beta = \beta_0 e^{j\varphi}$, where β_0 is the real-value reference wavenumber and φ is taken in the range from 1 to $10°$ [1,11].

4.9 Iterative Analysis of Multiple Discontinuities

The fibre facet as schematically shown in Figure 4.14 can be either abruptly terminated by air or coated with an AR layer of thickness d_{AR} and refractive index n_{AR}. These parameters are to be chosen in order to minimise the reflection to the fibre from the air. As may be noted from Figure 4.14, the diameter of the fibre core is D and the refractive indexes of the core and cladding are n_{core} and n_{clad}, respectively.

In order to numerically simulate the wave propagation in the fibre facet terminated by AR layer, an iterative bidirectional beam-propagation algorithm has been implemented, based on the picture of the wave propagation given in Figure 4.16. The

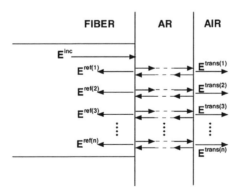

Figure 4.16 Picture depicting the wave propagation in coated optical-fibre facet, where n is the propagation trip number. (Reproduced with permission from Obayya, S.S.A. (2006) Scalar finite-element analysis of optical-fiber facets. *IEEE J. Lightwave Technol.*, **24** (5), 2115–2121. © 2006 IEEE.)

incident wave is the scalar LP_{01} mode from the optical-fibre side, and the discontinuity analysis is performed at the interface separating the fibre and the AR coating to find the reflected- and transmitted-field pair. The transmitted field is now allowed to propagate forward inside the AR coating until it reaches the other discontinuity end separating the AR coating and the air. This forward propagation is performed using the conventional paraxial BPM. Upon performing the forward propagation in the AR coating, a discontinuity analysis is carried out at the interface between the AR coating and the air. As a result, the reflected field is allowed to propagate backward in the AR coating until it reaches the other end separating the AR coating and the fibre, where the discontinuity analysis is carried out to find the total reflected and transmitted fields. On the fibre and air sides, the total reflected and transmitted fields are updated at the end of each propagation trip by appropriately adding up the contributions from the scattered fields, as shown in Figure 4.16. This process is repeated until a certain convergence criterion is met. In this work, it has been chosen to stop the propagation when the difference between the reflectivities calculated in the last two iterations is less than 0.001 dB.

4.10   Numerical Assessment

4.10.1   Uncoated Fibre Facet

The problem of scattering of the scalar LP_{01} mode from the uncoated fibre facet [2], as shown in Figure 4.15 will be investigated first. In this study, the core and cladding indexes of the optical fibre are taken as 1.4516 and 1.4473, respectively, and abruptly terminated by air. At a wavelength of 1.55 μm, the fibre is excited with the scalar LP_{01} mode of which the reflectivity against the core diameter is shown in Figure 4.17. In all fibre facet examples presented here, the fibre cross section is enclosed in a 15 μm computational window represented by a nonuniform mesh of 500 line elements and terminated by a PML layer of 1 μm width divided into five line elements with a maximum conductivity σ_{max} of 10.

   The reflectivity of the LP_{01} mode is calculated by using

$$\text{Reflectivity of } LP_{01} \text{ mode} = \frac{\int r E^{\text{inc},*} E^{\text{ref}} \, dr}{\int r \left| E^{\text{inc}} \right|^2 \, dr} \tag{4.25}$$

where $*$ denotes the complex conjugate, and E^{inc}, E^{ref} and E^{trans} denote the incident, reflected and transmitted fields, as shown in Figure 4.15. Also, in all simulations, a third-order Taylor's series expansion was used as it gives accurate results. As may be noted from Figure 4.17, the reflectivity of the scalar LP_{01} mode increases monotonically with increasing the core diameter. Also, it is interesting to note that these scalar finite-element results are in very good agreement with their scalar counterparts obtained using the FSRM and reported in [19].

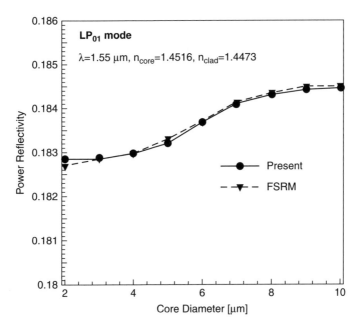

Figure 4.17 Variation of the scalar LP$_{01}$ mode reflectivity with the core diameter of an uncoated optical fibre. (Reproduced with permission from Obayya, S.S.A. (2006) Scalar finite-element analysis of optical-fiber facets. *IEEE J. Lightwave Technol.*, **24** (5), 2115–2121. © 2006 IEEE.)

Also, shown in Figure 4.18 is the variation of the total reflected and total transmitted powers, normalised to the incident power, with the core diameter. The incident, total transmitted and total reflected powers are calculated by using

$$\text{Incident power} = \text{Re}\left(\int E^{\text{inc}}\left(\sqrt{[L]_1}E^{\text{inc}}\right)^* r\,dr\right) \qquad (4.26)$$

$$\text{Reflected power} = \text{Re}\left(\int E^{\text{ref}}\left(\sqrt{[L]_1}E^{\text{ref}}\right)^* r\,dr\right) \qquad (4.27)$$

$$\text{Incident power} = \text{Re}\left(\int E^{\text{trans}}\left(\sqrt{[L]_2}E^{\text{trans}}\right)^* r\,dr\right) \qquad (4.28)$$

As may be observed from Figure 4.18, the total power (reflected + transmitted) is balanced to the incident power within less than 0.05%, which clearly shows that the power-balance condition is satisfied, thanks to the proper treatment of the evanescent modes generated at the discontinuity plane by using the concept of the complex-value reference wavenumber.

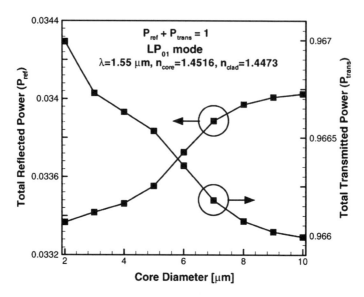

Figure 4.18 Variation of the total reflected (P_{ref}) and total transmitted (P_{trans}) powers with the core diameter of an uncoated optical fibre (power balance). (Reproduced with permission from Obayya, S.S.A. (2006) Scalar finite-element analysis of optical-fiber facets. *IEEE J. Lightwave Technol.*, **24** (5), 2115–2121. © 2006 IEEE.)

For a specific value of the core diameter of 8.7 μm, the spectral variation of the scalar LP_{01} mode reflectivity in the wavelength range from 1 to 3 μm is shown in Figure 4.19. It could be noted from this figure that the scalar LP_{01} mode reflectivity is monotonically decreasing as the wavelength is increased. These results can be explained by adopting the simple, but less accurate, Fresnel's formula for predicting the reflectivity by using

$$\text{Reflectivity} = \frac{(n_{eff} - 1)}{(n_{eff} + 1)} \qquad (4.29)$$

where n_{eff} is the effective index of the scalar LP_{01} mode. As the wavelength is increased, the fundamental LP_{01} mode becomes less confined to the core and more spread into the cladding, giving rise to lower values of the effective index n_{eff}, and hence to lower values of the reflectivity, as predicted by Equation (4.29). Once again, there is an excellent agreement between the results obtained using this finite-element approach and its scalar counterparts reported in [19] using the FSRM method.

4.10.2 Single Antireflection Layer

The next example, shown in Figure 4.14, is an optical fibre with the same refractive-index distribution and with a core diameter of 8.7 μm, and the facet is coated with a single quarter-wavelength AR coating [2] with a refractive index of $(n_{eff})^{0.5}$ and a

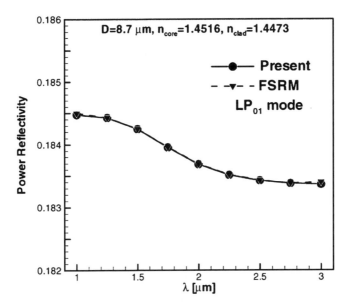

Figure 4.19 Spectral variation of the scalar LP_{01} mode reflectivity from an uncoated optical-fibre facet. (Reproduced with permission from Obayya, S.S.A. (2006) Scalar finite-element analysis of optical-fiber facets. *IEEE J. Lightwave Technol.*, **24** (5), 2115–2121. © 2006 IEEE.)

thickness d_{AR} of $1.3/(4n_{eff}^{0.5})$, where n_{eff} is the effective index of the LP_{01} mode at the central design wavelength, taken here as 1.3 μm. In order to numerically simulate the wave propagation in the fibre facet terminated by the AR layer, the iterative bidirectional beam-propagation algorithm has been used, as shown in Figure 4.16. At a wavelength of 1.3 μm, Figure 4.20 shows the convergence of the calculated reflectivity of the coated optical fibre with the number of propagation trips using the presented iterative bidirectional BPM algorithm. As maybe noticed, less than five propagation trips are enough to achieve convergence.

Figure 4.21 shows the variation of the LP_{01} mode reflectivity with the wavelength in the case of an AR coated optical fibre. As may be seen here, the reflectivity at the central design wavelength, 1.3 μm, is about −63 dB, thanks to the proper selection of a quarter-wavelength AR coating. The fact that the reflectivity at the central wavelength is quite small is also quite evident from the reflected field distribution inside the fibre, with and without the AR coating, as shown in Figure 4.22. It may also be noticed that the results obtained using the finite-element-based bidirectional BPM are in very good agreement with their scalar counterparts obtained using the FSRM method and reported in [19], as shown in Figure 4.21, and also as evident from Table 4.1. It should be mentioned that each of the simulation points for the uncoated and coated fibre facets, represented by a nonuniform 500 line elements, takes about 20 and 30 s, respectively, on a PC (Pentium IV, 2.66 GHz).

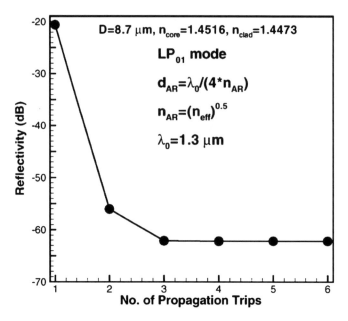

Figure 4.20 Numerical convergence of the reflectivity from a coated optical-fibre facet with the number of propagation trips in the bidirectional BPM. (Reproduced with permission from Obayya, S.S.A. (2006) Scalar finite-element analysis of optical-fiber facets. *IEEE J. Lightwave Technol.*, **24** (5), 2115–2121. © 2006 IEEE.)

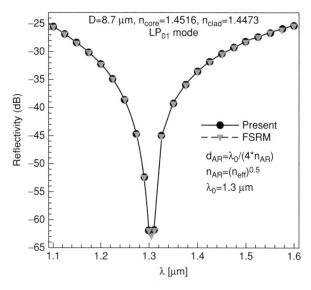

Figure 4.21 Spectral variation of the scalar LP_{01} mode reflectivity from an optical fibre facet coated with a single AR layer. (Reproduced with permission from Obayya, S.S.A. (2006) Scalar finite-element analysis of optical-fiber facets. *IEEE J. Lightwave Technol.*, **24** (5), 2115–2121. © 2006 IEEE.)

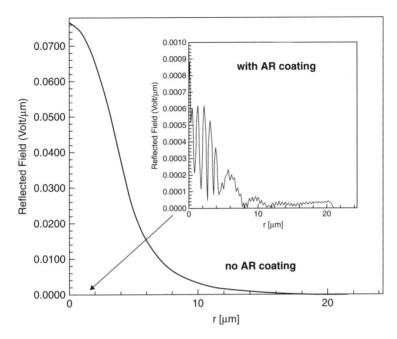

Figure 4.22 Reflected electric-field distribution inside a fibre, with and without AR coating. (Reproduced with permission from Obayya, S.S.A. (2006) Scalar finite-element analysis of optical-fiber facets. *IEEE J. Lightwave Technol.*, **24** (5), 2115–2121. © 2006 IEEE.)

Table 4.1 Values of the reflectivity of the fundamental LP_{01} mode at different wavelengths for an optical fiber with an AR coating (the same as shown in Figure 4.21), obtained using the present finite-element approach (column 2), and their scalar counterparts reported in [19] using the FSRM method (column 3). (Reproduced with permission from Obayya, S.S.A. (2006) Scalar finite-element analysis of optical-fiber facets. *IEEE J. Lightwave Technol.*, **24** (5), 2115–2121. © 2006 IEEE.)

Wavelength, λ (μm)	Reflectivity (dB), Present FE Approach	Reflectivity (dB), FRSM [19]
1.100	−25.547	−25.60
1.250	−38.562	−38.60
1.275	−44.697	−44.70
1.290	−52.385	−52.40
1.300	−62.152	−63.00
1.325	−45.003	−45.00
1.350	−39.218	−39.20
1.400	−33.549	−33.60
1.500	−28.188	−28.20

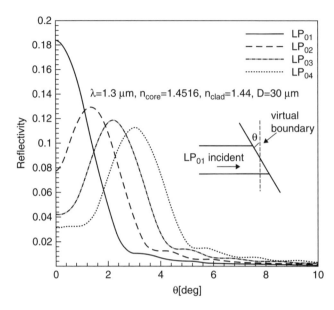

Figure 4.23   Variation of the reflectivity of the first four LP_{0m} modes with the tilt angle, θ, of the uncoated fibre facet. (Reproduced with permission from Obayya, S.S.A. (2006) Scalar finite-element analysis of optical-fiber facets. *IEEE J. Lightwave Technol.*, **24** (5), 2115–2121. © 2006 IEEE.)

4.10.3   Slanted Interfaces

The case of an uncoated multimode fibre with angled facet [2], as shown in the inset of Figure 4.23, is also considered using the present scalar finite-element approach. At a wavelength of 1.3 μm, the fibre is selected to have a core diameter of 30 μm, and core and cladding refractive indexes of 1.4516 and 1.44, respectively, to be multimoded. In order to apply the present finite-element approach, a virtual interface perpendicular to the fibre cross section replaces the original angled facet, as shown in the inset of Figure 4.23. Then, after performing the discontinuity analysis on this perpendicular 'virtual' interface, the phase front of the calculated reflected field is modified to be $2nk_o r \sin \theta$, where n is the refractive-index distribution of the fibre, and θ is the tilt angle, in order to obtain the reflected field at the real angled interface [1]. The incident field is the LP_{01} mode, and due to mode coupling, the reflected field spectrum includes both the fundamental and higher-order modes. The reflectivities of the first four LP_{0m} modes are calculated, and their variations with the tilt angle θ are shown in Figure 4.23. Those reflectivities are calculated by simply performing overlap integrals of the reflected field and the normalised electric-field profiles of the first four LP_{0m} modes using expressions similar to (4.24). As can be observed from this figure, the reflectivity of the LP_{01} mode is monotonically decreasing as the tilt angle θ increases. On the other hand, it can be observed from Figure 4.23 that the reflectivity of any

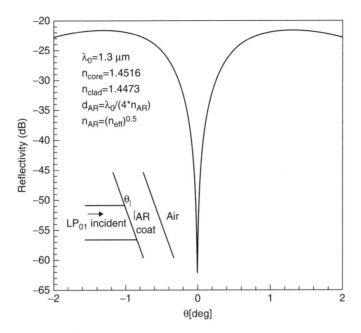

Figure 4.24 Variation of the reflectivity of the fundamental LP_{01} mode with the tilt angle θ of the angled-facet fibre coated with an AR coating. (Reproduced with permission from Obayya, S.S.A. (2006) Scalar finite-element analysis of optical-fiber facets. *IEEE J. Lightwave Technol.*, **24** (5), 2115–2121. © 2006 IEEE.)

higher-order LP_{0m} mode reaches a peak value at a particular tilt angle, and then starts to monotonically decrease as the tilt angle increases. Also, it can be noted that the peak value of the reflectivity of any higher-order LP_{0m} mode decreases as m increases, and its tilt angle tends to increase.

Finally, the case of an angled-facet fibre with an AR coating is investigated. The parameters of the structure, shown as an inset in Figure 4.24, are the same as those considered in Figure 4.21. At the central design wavelength of 1.3 μm, the variation of the reflectivity of the fundamental LP_{01} mode with the tilt angle θ is shown in Figure 4.24. It can be noticed from Figure 4.24 that the reflectivity of the LP_{01} mode has drastically changed from -63 to about -21 dB for a tilt angle of only $\pm 1°$. These results show clearly that even small tilt angles of the fibre interface can lead to a deterioration of the performance of the AR coating.

References

[1] Obayya, S.S.A. (2004) Novel finite element analysis of optical waveguide discontinuity problems. *J. Lightwave Technol.*, **22** (5), 1420–1425.

[2] Obayya, S.S.A. (2006) Scalar finite-element analysis of optical-fiber facets. *J. Lightwave Technol.*, **24** (5), 2115–2121.

[3] Ikegami, T. (1972) Reflectivity of mode at facet and oscillation mode in double-heterostructure injection lasers. *IEEE J. Quantum Electron.*, **8**, 470–476.

[4] Gerdes, B., Lunitz, B., Benish, D. and Pregla, R. (1992) Analysis of slab waveguide discontinuities including radiation and absorption effects. *Electron. Lett.*, **28** (11), 1013–1015.

[5] Chiou, Y.P. and Chang, H.C. (1997) Analysis of optical waveguide discontinuity problems. *IEEE Photon. Technol. Lett.*, **9**, 964–966.

[6] Rahman, B.M.A. and Davies, J.B. (1988) Analysis of optical waveguide discontinuities. *J. Lightwave Technol.*, **6**, 52–57.

[7] Kawano, K., Kitoh, T., Kohtoku, M. *et al.* (1998) Bidirectional finite-element method-of-line beam propagation method (FE-MOL-BPM) for analyzing optical waveguide with discontinuities. *IEEE Photon. Technol. Lett.*, **10**, 244–245.

[8] Koshiba, M., Ooishi, K., Miki, T. and Suzuki, M. (1982) Finite-element analysis of the discontinuities in a dielectric slab waveguide bounded by parallel plates. *Electron. Lett.*, **18** (1), 33–34.

[9] Reed, M., Benson, T.M., Kendall, P.C. and Sewell, P. (1996) Antireflection-coated angled facet design. *IEE Proc. Optoelectron.*, **143** (4), 214–220.

[10] El-Refaei, H., Betty, I. and Yevick, D. (2000) The application of complex Pade approximants to reflection at optical waveguide facets. *IEEE Photon Technol. Lett.*, **12**, 158–160.

[11] Rao, H., Steel, M.J., Scarmozzino, R. and Osgood, R.M. (2000) Complex propagators for evanescent waves in bidirectional beam propagation method. *J. Lightwave Technol.*, **18**, 1155–1160.

[12] Zienkiewicz, O.C. and Taylor, R.L. (2000) *The Finite Element Method*, Butterworth-Heinemann, Oxford.

[13] Yamada, T. and Tani, K. (1997) Finite element time domain method using hexahedral elements. *IEEE Trans. Magn.*, **33**, 1476–1479.

[14] Yong, Y.K. and Cho, Y. (1994) Algorithm for Eigenvalue Problems in Piezoelectric Finite Element Analysis. Proceedings of the IEEE Ultrasonic Symposium, vol. **2**, pp. 1057–1062.

[15] Tsuji, Y. and Koshiba, M. (2002) Finite element method using port trauncation by perfectly matched layer boundary conditions for optical waveguide discontinuity problems. *J. Lightwave Technol.*, **20**, 463–468.

[16] Kaczmarski, P. and Lagasse, P.E. (1988) Bidirectional beam propagation method. *Electron. Lett.*, **24** (11), 675–676.

[17] Imtaar, M. and Al-Bader, S. J. (1995) Analysis of diffraction from abruptlyterminated optical fibers by the method of lines. *J. Lightw. Technol.*, **13** (2), 137–141.

[18] Shen, G., Chen, Y. and Mittra, R. (1999) A nonuniform FDTD technique for efficient analysis of propagation characteristics of optical-fiber waveguides. *IEEE Trans. Microw. Theory Tech.*, **47** (3), 345–349.

[19] Vorgul, I., Vukovic, A., Sewell, P. and Benson, T.M. (2003) Optical fibre facets: A vector analysis. *Proc. Inst. Elect. Eng. Optoelectron.*, **150** (6), 508–512.

5

Complex-Envelope Alternating-Direction-Implicit Finite-Difference Time-Domain Method

5.1 Introduction

In this chapter, emphasis will be put on the finite-difference time-domain technique, taking into account its advantages and its main drawbacks. In particular, a detailed description of how those drawbacks have been solved with the advent of the alternating-direction implicit FDTD (ADI-FDTD) technique and the complex-envelope ADI-FDTD (CE-ADI-FDTD) technique is given. Also, a comprehensive analysis of the perfectly matched layer (PML) boundary conditions will be also given, putting more emphasis on the new formulation of the uniaxial PML (UPML) boundary conditions for the CE-ADI-FDTD. At the end of the chapter, numerous examples will be given in order to assess the effectiveness of the new formulation of the UPML boundary conditions for the CE-ADI-FDTD in the context of photonic crystal devices. The great improvement in the absorption properties of the proposed boundary conditions will be shown to be positively reflected in the excellent enhancement of the stability properties of the numerical code. In this way, larger time-step sizes can be employed with negligible impact on the numerical accuracy of the numerical code and with consequent huge savings in computational resources.

Computational Photonics   Salah Obayya
© 2011 John Wiley & Sons, Ltd

5.2   Maxwell's Equations

All electromagnetics problems can be solved by considering Maxwell's equations with appropriate boundary conditions. In their differential form, Maxwell's equations are expressed by

$$\nabla \times \bar{E} = -\frac{\partial \bar{B}}{\partial t} \tag{5.1}$$

$$\nabla \times \bar{H} = \bar{J} + \frac{\partial \bar{D}}{\partial t} \tag{5.2}$$

$$\nabla \cdot \bar{D} = \rho \tag{5.3}$$

$$\nabla \cdot \bar{B} = 0 \tag{5.4}$$

where $\bar{E}$ is the electric field, $\bar{D}$ is the electric flux density, $\bar{H}$ is the magnetic field, $\bar{B}$ is the magnetic flux density, ρ is the electric charge density and $\bar{J}$ is the current density. At first sight, it can be said that for a given electric charge distribution, ρ and a current density, $\bar{J}$, solving Equations (5.1)–(5.4) can determine the solution of any given electromagnetic problem. However, this is not the case because Equations (5.1)–(5.4) do not form a well-posed problem. It can be easily demonstrated that Equations (5.1) and (5.4) are not independent, as well as Equations (5.2) and (5.3). In order to overcome this problem two more vectorial relationships need to be introduced, and these two relationships are represented by the constitutive relations, written as

$$\bar{D} = \varepsilon \bar{E} \tag{5.5}$$

$$\bar{B} = \mu \bar{H} \tag{5.6}$$

where ε represents the permittivity and μ represents the permeability of the medium. It has to be mentioned that Equations (5.5)–(5.6) are valid for a linear, isotropic, homogeneous medium which implies ε and μ are scalar quantities.

5.3   Brief History of the Finite-Difference Time-Domain (FDTD) Method

The finite-difference time domain (FDTD) method is one of the most popular computational techniques employed in the research environment for a wide variety of applications covering many different areas. The FDTD method was first proposed by Yee in 1966 [1] and it represented a simple and accurate yet efficient way to discretise Maxwell's equations. However, computer technology was not mature enough to fully exploit the potential of the method at the time. With the increase in computational power and its decreasing costs, the FDTD method started to become more and more attractive and its popularity has grown exponentially since. Taflove was one of the first to give a boost to the FDTD method [2] with a rigorous analysis of the numerical dispersion and anisotropy errors of the method. A couple of years later, the first application of the FDTD in the context of electromagnetic analysis was presented by Holland [3]. In 1981, the first second-order and numerically stable absorbing boundary conditions (ABCs) were introduced by Mur [4]. The performance of these new ABCs, in terms of the absorption of the incoming waves at the boundaries of the computational domain, was very good when compared with the, at the time, widely used radiating conditions, with a consequent reduction in the computational domain size and the computational burden. A few years later, the FDTD method was also used for the first time for the simulation of waveguide structures by Choi [5]. In that work, the efficiency, in terms of CPU time and memory allocation compared to the transmission line matrix (TLM) method, was shown. In 1990 in work published by Kashiwa and Fukay [6], the FDTD was for the first time applied for the modelling of dispersive media, optimised by Joseph *et al.* [7] in 1991, and successively extended to the simulation of nonlinear media by Goorijan and Taflove [8]. In similar work conducted by Ziolkowsky and Judkins [9], the self-focusing phenomenon was analysed with the use of FDTD and was later extended to the analysis of active materials by Toland *et al.* [10]. In innovative work published in 1994, Berenger [11] developed a novel class of ABCs, the perfectly matched layers (PMLs), an artificial medium surrounding the computational domain which acts as absorber of the impinging waves. The approach developed by Berenger has been shown to be robust and with absorbing performances much better than any other boundary conditions established at the time. In 1995, Sacks *et al.* [12] and Gedney [13] developed the uniaxial perfectly matched layer (UPML) ABC which represented a physically realisable version of the PML. Recently, in work published by Chang and Taflove [14], the FDTD has been also extended to the simulation of quantum-mechanical effects with a model of lasing in a four-level, two-electron atomic system.

5.4 Finite-Difference Time-Domain (FDTD) Method

The FDTD method is a quite general method for the simulation of electromagnetic devices for all range of frequencies from the microwave to the optical regime. The power of the method lies in its simple formulation in which no restrictive assumptions are made in order to preserve its applicability to a wide range of problems. Considering a 3D space with no electric or magnetic current sources, but with materials that present electric and magnetic conductivity, Maxwell's equations are then written as

$$\nabla \times \bar{E} = -\mu_0 \mu_r \frac{\partial \bar{H}}{\partial t} - \sigma^* \bar{H} \tag{5.7}$$

$$\nabla \times \bar{H} = \varepsilon_0 \varepsilon_r \frac{\partial \bar{E}}{\partial t} + \sigma \bar{E} \tag{5.8}$$

where $\bar{E} = \left(E_x \bar{x} \ E_y \bar{y} \ E_z \bar{z} \right)^{\mathrm{T}}$ is the electric field, $\bar{H} = \left(H_x \bar{x} \ H_y \bar{y} \ H_z \bar{z} \right)^{\mathrm{T}}$ is the magnetic field and where T represents the matrix transpose, ε_0 and μ_0 are the permittivity and permeability of the free space, respectively, ε_r and μ_r are the relative permittivity and the relative permeability of the medium, respectively, and σ and σ^* are the electric and magnetic conductivity of the medium, respectively. Equations (5.7) and (5.8) represent Maxwell's equations written in differential form and the electric field $\bar{E}$ and the magnetic field $\bar{H}$ are vector quantities. Each of the two previous equations represents three scalar equations, one for each electromagnetic field component, so that vectorial Equations (5.7) and (5.8) can be written in the following six scalar equations:

$$\frac{\partial E_z}{\partial y} - \frac{\partial E_y}{\partial z} = -\mu_0\mu_r\frac{\partial H_x}{\partial t} - \sigma^* H_x \tag{5.9}$$

$$\frac{\partial E_x}{\partial z} - \frac{\partial E_z}{\partial x} = -\mu_0\mu_r\frac{\partial H_y}{\partial t} - \sigma^* H_y \tag{5.10}$$

$$\frac{\partial E_y}{\partial x} - \frac{\partial E_x}{\partial y} = -\mu_0\mu_r\frac{\partial H_z}{\partial t} - \sigma^* H_z \tag{5.11}$$

$$\frac{\partial H_z}{\partial y} - \frac{\partial H_y}{\partial z} = \varepsilon_0\varepsilon_r\frac{\partial E_x}{\partial t} + \sigma E_x \tag{5.12}$$

$$\frac{\partial H_x}{\partial z} - \frac{\partial H_z}{\partial x} = \varepsilon_0\varepsilon_r\frac{\partial E_y}{\partial t} + \sigma E_y \tag{5.13}$$

$$\frac{\partial H_y}{\partial x} - \frac{\partial H_x}{\partial y} = \varepsilon_0\varepsilon_r\frac{\partial E_z}{\partial t} + \sigma E_z \tag{5.14}$$

Equations (5.9)–(5.14) can be used for the solution of any electromagnetic problem once the appropriate boundary conditions are given. But for a numerical analysis Equations (5.9)–(5.14) are of no use in the present form. In order to store the equations on a computer memory, a process of discretisation in time and in space needs to be applied to switch from infinite and continuous domains to finite and discretised domains.

5.4.1 Discretisation in Space and Time: The Yee's Algorithm and the Leapfrog Scheme

The basic idea of Yee's algorithm is to position the electromagnetic field components in specified places in a unit 3D cell in order to construct a finite-difference notation of Maxwell's equations capable of calculating the time and space derivative with second-order accuracy [15]. The electric and magnetic field components are placed as shown in Figure 5.1. The position of the electromagnetic field components as shown in Figure 5.1 leads to a series of advantages:

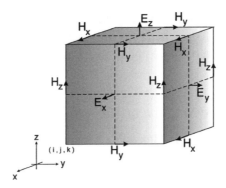

Figure 5.1 Arrangement of the six electromagnetic field components on the Yee cell for a 3D case.

• The update of the electric and magnetic field components is made considering the coupled set of Maxwell's equations instead of evaluating them considering only the electric or magnetic field as happens for the wave equation.
• The electric, $\bar{E}$, and magnetic, $\bar{H}$, components are placed in such a way that each single component of a field is surrounded by four components of the other field.
• The continuity relation of the tangential components at the interface between two different materials is held in a natural and straightforward manner if this interface is parallel to one of the directions of the unit 3D cell, so that no further relations for the electromagnetic field components have to be forced.
• The electric and magnetic field components are also placed in an alternate manner in the discretised time domain in a way that in order to update the electric field components the magnetic field components calculated in the previous step are used, and the updated electric field components obtained are then used to update the magnetic field components. This is shown in Figure 5.2, where a 1D case is considered. This update process is called *time-stepping* and the electric and magnetic field components are placed in time with a scheme called *leapfrog* [15]. This scheme permits the numerical wave to propagate in the computational domain as the simulation is running without any artificial decay due to the *time-stepping* process.

Figure 5.2 Arrangement of the electric- and magnetic-field components with the leapfrog scheme for a 1D case.

The collocation of the electromagnetic field components described by the Yee cell in space and by the leapfrog scheme in time permits second-order accuracy to be achieved in the calculation of space and time derivatives of Maxwell's equations.

5.4.2   Finite-Difference Notation of Maxwell's Equations

The finite-difference notation of Maxwell's equations is obtained by applying Yee's algorithm to Equations (5.9)–(5.14). Starting from Equation (5.9) the finite-different notation is obtained as follows. First, Equation (5.9) needs to be rewritten in such a way that the time derivative is be put on the LHS of the equation itself, while the rest of the equation should be on the RHS

$$\frac{\partial H_x}{\partial t} = -\frac{1}{\mu_0 \mu_r}\left(\frac{\partial E_z}{\partial y} - \frac{\partial E_y}{\partial z} + \sigma^* H_x\right) \tag{5.15}$$

Following the notation in [15], Equation (5.15) is then dicretised following the leapfrog scheme for the time derivative and the Yee cell to discretise the space derivatives

$$\frac{H_x|_{i-1/2,j+1,k+1}^{n+1} - H_x|_{i-1/2,j+1,k+1}^{n}}{\Delta t} = -\frac{1}{\mu_0 \mu_r|_{i-1/2,j+1,k+1}}$$
$$\left(\frac{E_z|_{i-1/2,j+3/2,k+1}^{n+1/2} - E_z|_{i-1/2,j+1/2,k+1}^{n+1/2}}{\Delta y} - \frac{E_y|_{i-1/2,j+1,k+3/2}^{n+1/2} - E_y|_{i-1/2,j+1,k+1/2}^{n+1/2}}{\Delta z}\right.$$
$$\left. + \sigma^*|_{i-1/2,j+1,k+1} H_x|_{i-1/2,j+1,k+1}^{n+1/2}\right) \tag{5.16}$$

where Δt is the time increment or *time step*, and Δy and Δz are the space increments along the y- and z-axes, respectively. As can be seen from Equation (5.16), the H_x component on the RHS of the equation is evaluated at the time step $n + 1/2$, which implies that this magnetic-field component is collocated in time with the electric-field components, in clear violation of the leapfrog scheme (see Figure 5.2). In order to solve this problem, the H_x component on the RHS is calculated using a linear approximation

$$H_x|_{i-1/2,j+1,k+1}^{n+1/2} = \frac{H_x|_{i-1/2,j+1,k+1}^{n} + H_x|_{i-1/2,j+1,k+1}^{n+1}}{2} \tag{5.17}$$

The substitution of Equation (5.17) into Equation (5.16) yields

$$\frac{H_x|_{i-1/2,j+1,k+1}^{n+1} - H_x|_{i-1/2,j+1,k+1}^{n}}{\Delta t} = -\frac{1}{\mu_0\mu_r|_{i-1/2,j+1,k+1}}$$

$$\left(\frac{E_z|_{i-1/2,j+3/2,k+1}^{n+1/2} - E_z|_{i-1/2,j+1/2,k+1}^{n+1/2}}{\Delta y} - \frac{E_y|_{i-1/2,j+1,k+3/2}^{n+1/2} - E_y|_{i-1/2,j+1,k+1/2}^{n+1/2}}{\Delta z}\right.$$

$$\left.+\sigma^*|_{i-1/2,j+1,k+1}\frac{H_x|_{i-1/2,j+1,k+1}^{n+1} - H_x|_{i-1/2,j+1,k+1}^{n}}{2}\right)$$

$$(5.18)$$

and then isolating H_x at the time step $n + 1$, the following equation is obtained

$$H_x|_{i-1/2,j+1,k+1}^{n+1}$$

$$= \left(\frac{1 - \dfrac{\sigma^*|_{i-1/2,j+1,k+1}\Delta t}{2\mu_0\mu_r|_{i-1/2,j+1,k+1}}}{1 + \dfrac{\sigma^*|_{i-1/2,j+1,k+1}\Delta t}{2\mu_0\mu_r|_{i-1/2,j+1,k+1}}}\right) H_x|_{i-1/2,j+1,k+1}^{n} - \left(\frac{\dfrac{\Delta t}{\mu_0\mu_r|_{i-1/2,j+1,k+1}}}{1 + \dfrac{\sigma^*|_{i-1/2,j+1,k+1}\Delta t}{2\mu_0\mu_r|_{i-1/2,j+1,k+1}}}\right)$$

$$\left(\frac{E_z|_{i-1/2,j+3/2,k+1}^{n+1/2} - E_z|_{i-1/2,j+1/2,k+1}^{n+1/2}}{\Delta y} - \frac{E_y|_{i-1/2,j+1,k+3/2}^{n+1/2} - E_y|_{i-1/2,j+1,k+1/2}^{n+1/2}}{\Delta z}\right)$$

$$(5.19)$$

Equation (5.19) is the finite difference form of Equation (5.15). Applying the same procedure followed for Equation (5.9) to Equations (5.10)–(5.14), the following set of discretised equations is obtained

$$H_y|_{i,j+1/2,k+1}^{n+1}$$

$$= \left(\frac{1 - \dfrac{\sigma^*|_{i,j+1/2,k+1}\Delta t}{2\mu_0\mu_r|_{i,j+1/2,k+1}}}{1 + \dfrac{\sigma^*|_{i,j+1/2,k+1}\Delta t}{2\mu_0\mu_r|_{i,j+1/2,k+1}}}\right) H_y|_{i,j+1/2,k+1}^{n} - \left(\frac{\dfrac{\Delta t}{\mu_0\mu_r|_{i,j+1/2,k+1}}}{1 + \dfrac{\sigma^*|_{i,j+1/2,k+1}\Delta t}{2\mu_0\mu_r|_{i,j+1/2,k+1}}}\right)$$

$$\left(\frac{E_x|_{i,j+1/2,k+3/2}^{n+1/2} - E_x|_{i,j+1/2,k+1/2}^{n+1/2}}{\Delta z} - \frac{E_z|_{i+1/2,j+1/2,k+1}^{n+1/2} - E_z|_{i-1/2,j+1/2,k+1}^{n+1/2}}{\Delta x}\right)$$

$$(5.20)$$

$$H_z\big|_{i,j+1,k+1/2}^{n+1}$$

$$= \left(\frac{1 - \dfrac{\sigma^*|_{i,j+1,k+1/2}\Delta t}{2\mu_0\mu_r|_{i,j+1,k+1/2}}}{1 + \dfrac{\sigma^*|_{i,j+1,k+1/2}\Delta t}{2\mu_0\mu_r|_{i,j+1,k+1/2}}}\right) H_z\big|_{i,j+1,k+1/2}^{n} - \left(\frac{\dfrac{\Delta t}{\mu_0\mu_r|_{i,j+1,k+1/2}}}{1 + \dfrac{\sigma^*|_{i,j+1,k+1/2}\Delta t}{2\mu_0\mu_r|_{i,j+1,k+1/2}}}\right)$$

$$\left(\frac{E_y\big|_{i+1/2,j+1,k+1/2}^{n+1/2} - E_y\big|_{i-1/2,j+1,k+1/2}^{n+1/2}}{\Delta x} - \frac{E_x\big|_{i,j+3/2,k+1/2}^{n+1/2} - E_x\big|_{i,j+1/2,k+1/2}^{n+1/2}}{\Delta y}\right)$$

$$(5.21)$$

$$E_x\big|_{i,j+1/2,k+1/2}^{n+1/2}$$

$$= \left(\frac{1 - \dfrac{\sigma|_{i,j+1/2,k+1/2}\Delta t}{2\varepsilon_0\varepsilon_r|_{i,j+1/2,k+1/2}}}{1 + \dfrac{\sigma|_{i,j+1/2,k+1/2}\Delta t}{2\varepsilon_0\varepsilon_r|_{i,j+1/2,k+1/2}}}\right) E_x\big|_{i,j+1/2,k+1/2}^{n-1/2} - \left(\frac{\dfrac{\Delta t}{\varepsilon_0\varepsilon_r|_{i,j+1/2,k+1/2}}}{1 + \dfrac{\sigma|_{i,j+1/2,k+1/2}\Delta t}{2\mu_0\mu_r|_{i,j+1/2,k+1/2}}}\right)$$

$$\left(\frac{H_z\big|_{i,j+1,k+1/2}^{n} - H_z\big|_{i,j,k+1/2}^{n}}{\Delta y} - \frac{H_y\big|_{i,j+1/2,k+1}^{n} - H_y\big|_{i,j+1/2,k}^{n}}{\Delta z}\right)$$

$$(5.22)$$

$$E_y\big|_{i-1/2,j+1,k+1/2}^{n+1/2}$$

$$= \left(\frac{1 - \dfrac{\sigma|_{i-1/2,j+1,k+1/2}\Delta t}{2\varepsilon_0\varepsilon_r|_{i-1/2,j+1,k+1/2}}}{1 + \dfrac{\sigma|_{i-1/2,j+1,k+1/2}\Delta t}{2\varepsilon_0\varepsilon_r|_{i-1/2,j+1,k+1/2}}}\right) E_y\big|_{i-1/2,j+1,k+1/2}^{n-1/2} - \left(\frac{\dfrac{\Delta t}{\varepsilon_0\varepsilon_r|_{i-1/2,j+1,k+1/2}}}{1 + \dfrac{\sigma|_{i-1/2,j+1,k+1/2}\Delta t}{2\varepsilon_0\varepsilon_r|_{i-1/2,j+1,k+1/2}}}\right)$$

$$\left(\frac{H_x\big|_{i-1/2,j+1,k+1}^{n} - H_x\big|_{i-1/2,j+1,k}^{n}}{\Delta z} - \frac{H_z\big|_{i,j+1,k+1/2}^{n} - H_z\big|_{i-1,j+1,k+1/2}^{n}}{\Delta x}\right)$$

$$(5.23)$$

$$E_z\big|_{i-1/2,j+1/2,k+1}^{n+1/2}$$

$$= \left(\frac{1 - \dfrac{\sigma|_{i-1/2,j+1/2,k+1}\Delta t}{2\varepsilon_0\varepsilon_r|_{i-1/2,j+1/2,k+1}}}{1 + \dfrac{\sigma|_{i-1/2,j+1/2,k+1}\Delta t}{2\varepsilon_0\varepsilon_r|_{i-1/2,j+1/2,k+1}}}\right) E_z\big|_{i-1/2,j+1/2,k+1}^{n-1/2} - \left(\frac{\dfrac{\Delta t}{\varepsilon_0\varepsilon_r|_{i-1/2,j+1/2,k+1}}}{1 + \dfrac{\sigma|_{i-1/2,j+1/2,k+1}\Delta t}{2\varepsilon_0\varepsilon_r|_{i-1/2,j+1/2,k+1}}}\right)$$

$$\left(\frac{H_y\big|_{i,j+1/2,k+1}^{n} - H_y\big|_{i,j+1/2,k+1}^{n}}{\Delta x} - \frac{H_x\big|_{i-1/2,j+1,k+1}^{n} - H_x\big|_{i-1/2,j,k+1}^{n}}{\Delta y}\right)$$

$$(5.24)$$

The updating process of Equations (5.19)–(5.24) starts with the calculation of the electric-field components at the time step $n + 1/2$ using the electric-field components at the time step $n - 1/2$ and the magnetic-field components at the time step n. Following, the update of the magnetic field components at the time step $n + 1$ is performed using the magnetic-field components at the time step n and the electric-field components just calculated at the time step $n + 1/2$. This procedure is then repeated for each time step until a steady state is reached or the fixed maximum number of time steps is reached.

5.4.3 Numerical Stability

The choice of the time increment and the space increments with which the computational domain is discretised can affect the velocity of propagation of the numerical waves in the computational domain with consequences for the numerical accuracy of the FDTD scheme [15]. Particular attention needs to be paid in the choice of the time step Δt in order to avoid the accumulation of numerical error during the process of time-stepping, which can increase without limit leading to instability of the FDTD scheme. The analysis of the numerical stability can be performed in two different ways:

- With the Courant, Friedrich, Levy and Von Neumann criterion which basically studies the stability relative to the spatial derivatives of the wave equation and the stability relative to the time derivative of the same equation separately. This implies a split of the stability analysis into two relatively simple problems to be solved. As a necessary condition, it is required that the eigenspace of the spatial derivatives must be a subset of the eigenspace of the stable solutions of the time derivative [15].
- With complex-frequency analysis, in which complex solutions of the dispersion relation for the FDTD grid are considered [15, 16].

Both analyses lead to a relationship which links the time increment, the *time step*, to the space increments, the *space steps*, which also guarantees the numerical stability of the FDTD scheme. In order to guarantee the stability of the FDTD scheme, it is sufficient that

$$2\sqrt{\frac{1}{\Delta x^2} + \frac{1}{\Delta y^2} + \frac{1}{\Delta z^2}} \leq \frac{2}{\Delta t} \tag{5.25}$$

where Δx, Δy and Δz are the space steps along the x-, y- and z-axes, respectively, and Δt is the time step. It needs to be mentioned that the previous relationship has been normalised to the speed of light, c, in the medium considered in the computational domain. With simple algebraic manipulation and considering a generic medium inside the computational domain, the relationship is finally obtained that bounds the time step of the FDTD scheme in order to ensure the stability of the numerical method

$$\Delta t \leq \frac{1}{c\sqrt{\dfrac{1}{\Delta x^2} + \dfrac{1}{\Delta y^2} + \dfrac{1}{\Delta z^2}}} \qquad (5.26)$$

Equation (5.26) is also known as the Courant–Friedrich–Levy stability criterion. As an example of the application of Equation (5.26), if the Yee cell considered for the discretisation of the computational domain is cubic with $\Delta x = \Delta y = \Delta z = \Delta$, from Equation (5.26)

$$\Delta t \leq \frac{1}{c\sqrt{\dfrac{1}{\Delta^2} + \dfrac{1}{\Delta^2} + \dfrac{1}{\Delta^2}}} = \frac{1}{c\sqrt{\dfrac{3}{\Delta^2}}} = \frac{\Delta}{c\sqrt{3}} \qquad (5.27)$$

It needs to be mentioned that the previous relationships have been calculated by considering a computational domain filled with a homogeneous medium. Nevertheless, the validity of Equation (5.26) still holds, even if the computational domain is filled with a nonhomogeneous medium, because it simply represents the worst-case scenario in the choice of the time step, Δt. For this reason, the validity of Equation (5.26) is sufficient to ensure the stability of the numerical scheme for an indefinite number of time steps. On the other hand, it has to be noted that the stability of the FDTD scheme is not only affected by the validity of Equation (5.26). Other factors, such as the boundary conditions (BCs) applied to the computational domain, the employment of nonuniform meshes for the discretisation of the computational domain, dispersive media, nonlinear media and media with loss can affect the stability of the FDTD method. Nevertheless, huge numbers of simulations run for these cases in the research environment have shown that the FDTD method can be usefully applied because the stability of the scheme, even though not for an indefinite number of time steps, is ensured for thousands of time steps, or at least for the number of time steps necessary to extract all the essential information from the simulation [15].

5.4.4   Numerical Dispersion

In simulations carried out with the FDTD method it is possible to note that the phase velocity of the propagating waves involved in the updating of the numerical scheme can travel in the computational domain with different velocities for each wavelength. This phenomenon is known as numerical dispersion and it is due to different factors, such as the wavelength of the travelling wave, the direction of propagation of the wave inside the computational domain and the resolution of the mesh chosen for the discretisation of the computational domain. This is a nonphysical phenomenon and it is undesirable because it can add delays or phase errors to the propagating wave, which can lead, as final results, to pseudo-refraction, nonphysical anisotropy and increased pulse width [17,18]. Unfortunately, this phenomenon is due to the discretised nature of the scheme, it is intrinsically related to the Yee cell and it cannot be totally eliminated. It is important then to understand how this phenomenon can affect the accuracy of the results obtained with the FDTD method and how it is possible to minimise the effects of this source of error. The numerical dispersion of the FDTD method is analysed by comparing the equation for the numerical dispersion obtained for the Yee cell with the equation for the numerical dispersion for the continuous case [15]. From this comparison it is possible to get information on how to set the discretisation steps in space in order to minimise the effects of the numerical dispersion. The equation for the numerical dispersion for a discretised computational domain is [15]

$$
\left[\frac{1}{\Delta x}\sin\left(\frac{k_x\Delta x}{2}\right)\right]^2 + \left[\frac{1}{\Delta y}\sin\left(\frac{k_y\Delta y}{2}\right)\right]^2 + \left[\frac{1}{\Delta z}\sin\left(\frac{k_z\Delta z}{2}\right)\right]^2
$$
$$
= \left[\frac{1}{c\Delta t}\sin\left(\frac{\omega\Delta t}{2}\right)\right]^2 \tag{5.28}
$$

where Δx, Δy and Δz are the space steps along the x-, y- and z-axes, respectively, Δt is the time step, c is the velocity of light, ω is the angular frequency, and k_x, k_y and k_z are the wave-vector components. For a travelling plane wave in a 3D lossless medium, the dispersion equation is

$$
\frac{\omega^2}{c^2} = k_x^2 + k_y^2 + k_z^2 \tag{5.29}
$$

Taking into account the following well-known limit

$$
\lim_{x\to 0}\frac{\sin x}{x} = 1 \tag{5.30}
$$

it is straightforward to see that Equation (5.28) is equal to Equation (5.29), when at the same time $\Delta x \to 0$, $\Delta y \to 0$, $\Delta z \to 0$ and $\Delta t \to 0$. This means that the only way to reduce the effects of the numerical dispersion is to employ very fine meshes and a very small time step at the same time. But in order to have a quantitative analysis on how small the space increments have to be in order to limit the dispersion error on the results obtained by means of the FDTD method, a 2D example will be analysed. Considering a 2D computational domain discretised using a square Yee cell ($\Delta x = \Delta y = \Delta$), and supposing a plane wave propagating in this 2D space forms an angle, α, with the positive direction of the x-axis ($k_x = k\cos(\alpha)$, and $k_y = k\sin(\alpha)$), the

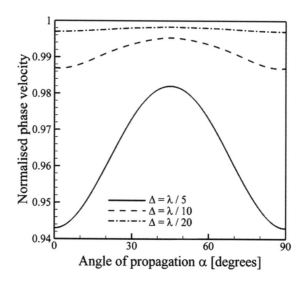

Figure 5.3 Variation of the normalised phase velocity of a plane wave propagating in a 2D computational domain with the angle of propagation for three different mesh resolutions.

equation for the numerical dispersion can be rewritten as

$$\sin^2\left(\frac{k\cos(\alpha)\,\Delta}{2}\right) + \sin^2\left(\frac{k\sin(\alpha)\,\Delta}{2}\right) = \left(\frac{\Delta}{c\Delta t}\right)\sin^2\left(\frac{\omega\Delta t}{2}\right) \tag{5.31}$$

From Equation (5.31) it is possible to evaluate the wave vector, k, for each direction of propagation, α, using, for instance, an iterative method [15]. The result of the procedure is shown in Figure 5.3 for three different resolutions of the mesh used to discretise the computational domain. From Figure 5.3 it is possible to note that the phase velocity of the propagating wave is always less than the velocity of the light and it has two minima for the directions of propagation $\alpha = 0°$ and $\alpha = 90°$ while it possesses a maximum for $\alpha = 45°$. This variation of the phase velocity with the angle of propagation of the wave doesn't depend on the grid resolution, which confirms that this numerical dispersion is intrinsic to the Yee lattice. Even though this variation cannot be eliminated, it is possible to reduce it by simply choosing an appropriate grid resolution for the computational domain. From Figure 5.3 it can be seen that with a grid resolution of $\Delta = \lambda\,/\,20$, the variation of the normalised phase velocity is reduced to 0.3%, which can be considered as a lower bound in order to get accurate results.

Different solutions to the numerical dispersion problem of the FDTD scheme have been proposed in the literature. In a work proposed by K. Suzuki *et al.* [19] a further reduction of the numerical dispersion has been reached by employing an anisotropic velocity of light which compensates the anisotropy introduced by the Yee lattice. J.S. Juntunen *et al.* [18] have proposed the compensation of the anisotropy of the phase velocity by acting directly in the dielectric properties of the simulated material inside the computational domain. Both solutions reach the goal of a further reduction in the anisotropy of the phase velocity introduced by the Yee cell without increasing the need for the computational resources of the FDTD method. But the main disadvantage of both solutions is that this reduction is effectively reached for a single frequency signal, while for signals with a broad spectrum of frequencies their effectiveness is slightly compromised.

To overcome this problem, high-order schemes have been applied to the FDTD method [17, 20, 21]. These high-order schemes also effectively reduce the numerical dispersion for coarser meshes and furthermore this reduction, even though the analysis has been formulated for a single frequency signal, has been proven to be effective for signals with a broader spectrum of frequencies. The main disadvantage of these high-order schemes is their need for computational resources, which makes them suitable for problems that require the analysis of relatively big structures, such as electrically elongated domains for waveguide problems [21].

5.5 Alternating-Direction-Implicit FDTD (ADI-FDTD): Beyond the Courant Limit

The simple formulation of the FDTD method and its ability to simulate a wide variety of devices for a large range of frequencies, from the microwave to the optical regime, have made this method one of the most popular in the research environment. But for classes of problems such as resonant cavities with very high quality factors, Q, and structures with geometrical features very small compared to the shortest wavelength involved, the requirement for computational resources of the FDTD method can be prohibitive. This is quite intuitive, taking into account that the time steps employed in FDTD simulations are bounded by Equation (5.26). For this reason, research effort has been put into finding a way to make the FDTD more efficient for such classes of problems. These efforts have given birth to the alternating-direction-implicit finite-difference time-domain (ADI-FDTD) method [22–26]. The main advantage of the ADI-FDTD method over conventional FDTD is that the time-step size that can be employed is not bounded by the CFL criterion expressed by Equation (5.26). This property allows a great reduction in the number of time steps necessary to complete a single simulation with a direct impact on computational resources. However as a drawback, the time-step size also has effects on the numerical accuracy of the ADI-FDTD method which, as a matter of fact, puts a limit on larger time-step sizes [23, 24].

In this method, the electromagnetic-field components are still placed in space following the Yee cell arrangement, as shown in Figure 5.1, but in time do not follow the leapfrog arrangement. In particular, a single time step is divided in two halves in which the electromagnetic-field components are collocated and not staggered, as in the conventional FDTD method. Following a similar procedure for the discretisation process of the Maxwell's equations, the following set of equations is obtained from Equations (5.9)–(5.14) for the first half time step:

$$
E_x\big|_{i+1/2,j,k}^{n+1/2} = E_x\big|_{i+1/2,j,k}^{n} + \frac{\Delta t}{2\varepsilon\,\Delta y}\left(H_z\big|_{i+1/2,j+1/2,k}^{n+1/2} - H_z\big|_{i+1/2,j-1/2,k}^{n+1/2}\right)
$$
$$
- \frac{\Delta t}{2\varepsilon\,\Delta z}\left(H_y\big|_{i+1/2,j,k+1/2}^{n} - H_y\big|_{i+1/2,j,k-1/2}^{n}\right) \tag{5.32}
$$

$$
E_y\big|_{i,j+1/2,k}^{n+1/2} = E_y\big|_{i,j+1/2,k}^{n} + \frac{\Delta t}{2\varepsilon\,\Delta z}\left(H_x\big|_{i,j+1/2,k+1/2}^{n+1/2} - H_x\big|_{i,j+1/2,k-1/2}^{n+1/2}\right)
$$
$$
- \frac{\Delta t}{2\varepsilon\,\Delta x}\left(H_z\big|_{i+1/2,j+1/2,k}^{n} - H_z\big|_{i-1/2,j+1/2,k}^{n}\right) \tag{5.33}
$$

$$
E_z\big|_{i,j,k+1/2}^{n+1/2} = E_z\big|_{i,j,k+1/2}^{n} + \frac{\Delta t}{2\varepsilon\,\Delta x}\left(H_y\big|_{i+1/2,j,k+1/2}^{n+1/2} - H_y\big|_{i-1/2,j,k+1/2}^{n+1/2}\right)
$$
$$
- \frac{\Delta t}{2\varepsilon\,\Delta y}\left(H_x\big|_{i,j+1/2,k+1/2}^{n} - H_x\big|_{i,j-1/2,k+1/2}^{n}\right) \tag{5.34}
$$

$$
H_x\big|_{i,j+1/2,k+1/2}^{n+1/2} = H_x\big|_{i,j+1/2,k+1/2}^{n} + \frac{\Delta t}{2\mu\,\Delta z}\left(E_y\big|_{i,j+1/2,k+1}^{n+1/2} - E_y\big|_{i,j+1/2,k}^{n+1/2}\right)
$$
$$
- \frac{\Delta t}{2\varepsilon\,\Delta y}\left(E_z\big|_{i,j+1,k+1/2}^{n} - E_z\big|_{i,j,k+1/2}^{n}\right) \tag{5.35}
$$

$$
H_y\big|_{i+1/2,j,k+1/2}^{n+1/2} = H_x\big|_{i+1/2,j,k+1/2}^{n} + \frac{\Delta t}{2\mu\,\Delta x}\left(E_z\big|_{i+1,j,k+1/2}^{n+1/2} - E_z\big|_{i,j,k+1/2}^{n+1/2}\right)
$$
$$
- \frac{\Delta t}{2\varepsilon\,\Delta z}\left(E_x\big|_{i+1/2,j,k+1}^{n} - E_x\big|_{i+1/2,j,k}^{n}\right) \tag{5.36}
$$

$$
H_z\big|_{i+1/2,j+1/2,k}^{n+1/2} = H_x\big|_{i+1/2,j+1/2,k}^{n} + \frac{\Delta t}{2\mu\,\Delta y}\left(E_x\big|_{i+1/2,j+1,k}^{n+1/2} - E_x\big|_{i+1/2,j,k}^{n+1/2}\right)
$$
$$
- \frac{\Delta t}{2\varepsilon\,\Delta x}\left(E_y\big|_{i+1,j+1/2,k}^{n} - E_y\big|_{i,j+1/2,k}^{n}\right) \tag{5.37}
$$

As can be seen from the previous set of equations, some electric- and magnetic-field components are collocated in time and for this reason Equations (5.32)–(5.37) cannot be explicitly updated. After some algebraic manipulations, for each electric- or magnetic-field component, a tri-diagonal system of equations is obtained, which can be simply solved with a small overhead of computational resources. After the tri-diagonal system being solved, the remaining magnetic- or electric-field components are explicitly updated. For the second half time step, the equations that need to be solved are the following

$$E_x\big|_{i+1/2,j,k}^{n+1} = E_x\big|_{i+1/2,j,k}^{n+1/2} + \frac{\Delta t}{2\varepsilon\,\Delta y}\left(H_z\big|_{i+1/2,j+1/2,k}^{n+1/2} - H_z\big|_{i+1/2,j-1/2,k}^{n+1/2}\right)$$
$$- \frac{\Delta t}{2\varepsilon\,\Delta y}\left(H_y\big|_{i+1/2,j,k+1/2}^{n+1} - H_y\big|_{i+1/2,j,k-1/2}^{n+1}\right) \tag{5.38}$$

$$E_y\big|_{i,j+1/2,k}^{n+1} = E_y\big|_{i,j+1/2,k}^{n+1/2} + \frac{\Delta t}{2\varepsilon\,\Delta z}\left(H_x\big|_{i,j+1/2,k+1/2}^{n+1/2} - H_x\big|_{i,j+1/2,k-1/2}^{n+1/2}\right)$$
$$- \frac{\Delta t}{2\varepsilon\,\Delta x}\left(H_z\big|_{i+1/2,j+1/2,k}^{n+1} - H_z\big|_{i-1/2,j+1/2,k}^{n+1}\right) \tag{5.39}$$

$$E_z\big|_{i,j,k+1/2}^{n+1} = E_z\big|_{i,j,k+1/2}^{n+1/2} + \frac{\Delta t}{2\varepsilon\,\Delta x}\left(H_y\big|_{i+1/2,j,k+1/2}^{n+1/2} - H_y\big|_{i-1/2,j,k+1/2}^{n+1/2}\right)$$
$$- \frac{\Delta t}{2\varepsilon\,\Delta y}\left(H_x\big|_{i,j+1/2,k+1/2}^{n+1} - H_x\big|_{i,j-1/2,k+1/2}^{n+1}\right) \tag{5.40}$$

$$H_x\big|_{i,j+1/2,k+1/2}^{n+1} = H_x\big|_{i,j+1/2,k+1/2}^{n+1/2} + \frac{\Delta t}{2\mu\,\Delta z}\left(E_y\big|_{i,j+1/2,k+1}^{n+1/2} - E_y\big|_{i,j+1/2,k}^{n+1/2}\right)$$
$$- \frac{\Delta t}{2\varepsilon\,\Delta y}\left(E_z\big|_{i,j+1,k+1/2}^{n+1} - E_z\big|_{i,j,k+1/2}^{n+1}\right) \tag{5.41}$$

$$H_y\big|_{i+1/2,j,k+1/2}^{n+1} = H_x\big|_{i+1/2,j,k+1/2}^{n+1/2} + \frac{\Delta t}{2\mu\,\Delta x}\left(E_z\big|_{i+1,j,k+1/2}^{n+1/2} - E_z\big|_{i,j,k+1/2}^{n+1/2}\right)$$
$$- \frac{\Delta t}{2\varepsilon\,\Delta z}\left(E_x\big|_{i+1/2,j,k+1}^{n+1} - E_x\big|_{i+1/2,j,k}^{n+1}\right) \tag{5.42}$$

$$H_z\big|_{i+1/2,j+1/2,k}^{n+1} = H_x\big|_{i+1/2,j+1/2,k}^{n+1/2} + \frac{\Delta t}{2\mu\,\Delta y}\left(E_x\big|_{i+1/2,j+1,k}^{n+1/2} - E_x\big|_{i+1/2,j,k}^{n+1/2}\right)$$
$$- \frac{\Delta t}{2\varepsilon\,\Delta x}\left(E_y\big|_{i+1,j+1/2,k}^{n+1} - E_y\big|_{i,j+1/2,k}^{n+1}\right) \tag{5.43}$$

For this set of equations, a similar procedure to that employed for the first half time step is followed. The time-stepping for the ADI-FDTD is given by the repetition of the two previous procedures in time.

The semi-explicit nature of the ADI-FDTD algorithm is the key factor in its unconditional stability [22, 25, 26], which makes it possible to employ time-step sizes larger than the limit imposed by Equation (5.26) for the FDTD method. However, the main drawback of this method is that the numerical accuracy of the results is seriously affected by the time-step size: the larger the time-step size employed, the larger the numerical dispersion [23, 24]. This drawback puts a limit on the time-step size that it is possible to employ in this method, which is finally dictated by the degree of the modelling accuracy of the algorithm. Nevertheless, with the ADI-FDTD it is still possible to reduce the computational resources needed by using a time-step size up to eight times larger than the Courant criterion with a good level of accuracy of the results

[23, 24]. For applications regarding resonant structure in which the frequency band-width of the signal involved is quite narrow, this limit can be extended up to 400 times the Courant criterion, with a huge saving in terms of computational resources [27].

5.6 Complex-Envelope ADI-FDTD (CE-ADI-FDTD)

With the ADI-FDTD the capabilities of the FDTD method have been extended beyond the Courant limit with consequent savings in computational resources, making it suitable for classes of problems involving signals with a relatively narrow band-width, which, with the conventional FDTD, require quite a large number of time steps and hence a huge computational burden. However, the huge potential of the ADI-FDTD is restricted by the large numerical errors obtained when large time-step sizes are used. For this reason, some research efforts have been spent in order to circumvent the restriction posed by the numerical dispersion of the ADI-FDTD. One solution that has proven to be very attractive, especially for the simulation of photonic structure, has been proposed in [28] in the context of optical device analysis. In this work, a technique called the complex-envelope alternating-direction-implicit finite-difference time-domain (CE-ADI-FDTD) method has been proposed to greatly reduce the numerical dispersion, even for large time-step sizes. The simple, yet ingenuous, idea behind this technique is to split the total electromagnetic field into fast and the slow temporal variation components, as shown in Figure 5.4.

The fast temporal variation component is then absorbed into the equations so as to consider only the solution of the slow temporal variation. Applying this concept to all six electromagnetic field components yields

$$E_x(x, y, z, t) = E_{xa}(x, y, z, t) e^{j w_c t} \tag{5.44}$$

$$E_y(x, y, z, t) = E_{ya}(x, y, z, t) e^{j w_c t} \tag{5.45}$$

$$E_z(x, y, z, t) = E_{za}(x, y, z, t) e^{j w_c t} \tag{5.46}$$

$$H_x(x, y, z, t) = H_{xa}(x, y, z, t) e^{j w_c t} \tag{5.47}$$

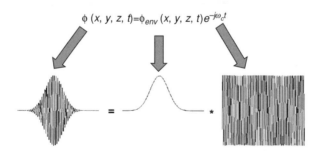

Figure 5.4 Example of the split of the total temporal variation field in its slow and fast temporal variations.

$$H_y(x, y, z, t) = H_{ya}(x, y, z, t) e^{jw_c t} \qquad (5.48)$$

$$H_z(x, y, z, t) = H_{za}(x, y, z, t) e^{jw_c t} \qquad (5.49)$$

where $\Phi_k(x, y, z, t)$ is the electromagnetic field component, $\Phi_{ka}(x, y, z, t)$ is the envelope of the electromagnetic field component, with $\Phi = E, H$ and $k = x, y, z$, and $e^{jw_c t}$ is the fast temporal variation. Substituting Equation (5.44)–(5.49) into Maxwell's equations, and after some simple algebraic manipulation, the following set of equations is obtained

$$\frac{\partial H_{xa}}{\partial t} + jw_c H_{xa} = -\frac{1}{\mu_0 \mu_r}\left(\frac{\partial E_{za}}{\partial y} - \frac{\partial E_{ya}}{\partial z}\right) \qquad (5.50)$$

$$\frac{\partial H_{ya}}{\partial t} + jw_c H_{ya} = -\frac{1}{\mu_0 \mu_r}\left(\frac{\partial E_{xa}}{\partial z} - \frac{\partial E_{za}}{\partial x}\right) \qquad (5.51)$$

$$\frac{\partial H_{za}}{\partial t} + jw_c H_{za} = -\frac{1}{\mu_0 \mu_r}\left(\frac{\partial E_{ya}}{\partial x} - \frac{\partial E_{xa}}{\partial y}\right) \qquad (5.52)$$

$$\frac{\partial E_{xa}}{\partial t} + jw_c E_{xa} = \frac{1}{\varepsilon_0 \varepsilon_r}\left(\frac{\partial H_{za}}{\partial y} - \frac{\partial H_{ya}}{\partial z}\right) \qquad (5.53)$$

$$\frac{\partial E_{ya}}{\partial t} + jw_c E_{ya} = \frac{1}{\varepsilon_0 \varepsilon_r}\left(\frac{\partial H_{xa}}{\partial z} - \frac{\partial H_{za}}{\partial x}\right) \qquad (5.54)$$

$$\frac{\partial E_{za}}{\partial t} + jw_c E_{za} = \frac{1}{\varepsilon_0 \varepsilon_r}\left(\frac{\partial H_{ya}}{\partial x} - \frac{\partial H_{xa}}{\partial y}\right) \qquad (5.55)$$

Applying the ADI-FDTD discretisation scheme to Equations (5.50)–(5.55) for the first half time step yields

$$H_{xa}\big|_{i,j+1/2,k+1/2}^{n+1/2} = \frac{4 - jw_c \Delta t}{4 + jw_c \Delta t} H_{xa}\big|_{i,j+1/2,k+1/2}^{n} - \frac{2\Delta t}{(4 + jw_c \Delta t)\mu_0 \mu_r}$$

$$\left(\frac{E_{za}\big|_{i,j+1,k+1/2}^{n+1/2} - E_{za}\big|_{i,j,k+1/2}^{n+1/2}}{\Delta y} - \frac{E_{ya}\big|_{i,j+1/2,k+1}^{n} - E_{ya}\big|_{i,j+1/2,k}^{n}}{\Delta z}\right) \qquad (5.56)$$

$$H_{ya}\big|_{i+1/2,j,k+1/2}^{n+1/2} = \frac{4 - jw_c \Delta t}{4 + jw_c \Delta t} H_{ya}\big|_{i+1/2,j,k+1/2}^{n} - \frac{2\Delta t}{(4 + jw_c \Delta t)\mu_0 \mu_r}$$

$$\left(\frac{E_{xa}\big|_{i+1/2,j,k+1}^{n+1/2} - E_{xa}\big|_{i+1/2,j,k}^{n+1/2}}{\Delta z} - \frac{E_{za}\big|_{i+1,j,k+1/2}^{n} - E_{za}\big|_{i,j,k+1/2}^{n}}{\Delta x}\right) \qquad (5.57)$$

$$H_{za}\big|_{i+1/2,j+1/2,k}^{n+1/2} = \frac{4 - jw_c \Delta t}{4 + jw_c \Delta t} H_{za}\big|_{i+1/2,j+1/2,k}^{n} - \frac{2\Delta t}{(4 + jw_c \Delta t)\mu_0 \mu_r}$$

$$\left(\frac{E_{ya}\big|_{i+1,j+1/2,k}^{n+1/2} - E_{ya}\big|_{i,j+1/2,k}^{n+1/2}}{\Delta x} - \frac{E_{ya}\big|_{i+1/2,j+1,k}^{n} - E_{ya}\big|_{i+1/2,j,k}^{n}}{\Delta y}\right) \qquad (5.58)$$

$$
E_{xa}\Big|_{i+1/2,j,k}^{n+1/2} = \frac{4 - j\omega_c\Delta t}{4 + j\omega_c\Delta t}\, E_{xa}\Big|_{i+1/2,j,k}^{n} - \frac{2\Delta t}{(4 + j\omega_c\Delta t)\,\varepsilon_0\varepsilon_r}
$$

$$
\left(\frac{H_{za}\Big|_{i+1/2,j+1/2,k}^{n+1/2} - H_{za}\Big|_{i+1/2,j-1/2,k}^{n+1/2}}{\Delta y} - \frac{H_{ya}\Big|_{i+1/2,j,k+1/2}^{n} - H_{ya}\Big|_{i+1/2,j,k-1/2}^{n}}{\Delta z} \right)
$$

$$(5.59)$$

$$
E_{ya}\Big|_{i,j+1/2,k}^{n+1/2} = \frac{4 - j\omega_c\Delta t}{4 + j\omega_c\Delta t}\, E_{ya}\Big|_{i,j+1/2,k}^{n} - \frac{2\Delta t}{(4 + j\omega_c\Delta t)\,\varepsilon_0\varepsilon_r}
$$

$$
\left(\frac{H_{xa}\Big|_{i,j+1/2,k+1/2}^{n+1/2} - H_{xa}\Big|_{i,j+1/2,k-1/2}^{n+1/2}}{\Delta z} - \frac{H_{za}\Big|_{i+1/2,j+1/2,k}^{n} - H_{ya}\Big|_{i-1/2,j+1/2,k}^{n}}{\Delta x} \right)
$$

$$(5.60)$$

$$
E_{za}\Big|_{i,j,k+1/2}^{n+1/2} = \frac{4 - j\omega_c\Delta t}{4 + j\omega_c\Delta t}\, E_{za}\Big|_{i,j,k+1/2}^{n} - \frac{2\Delta t}{(4 + j\omega_c\Delta t)\,\varepsilon_0\varepsilon_r}
$$

$$
\left(\frac{H_{ya}\Big|_{i+1/2,j,k+1/2}^{n+1/2} - H_{ya}\Big|_{i-1/2,j,k+1/2}^{n+1/2}}{\Delta x} - \frac{H_{xa}\Big|_{i,j+1/2,k+1/2}^{n} - H_{xa}\Big|_{i,j-1/2,k+1/2}^{n}}{\Delta y} \right)
$$

$$(5.61)$$

As can be seen from the previous set of equations, the electric- and magnetic-field components cannot be explicitly updated. After some algebraic manipulations, for each electric- or magnetic-field component a tri-diagonal system of equations is obtained which can be simply solved with a small overhead of computational resources. After the tri-diagonal has been solved, the remaining magnetic- or electric-field components are explicitly updated. For the second half time step similar equations are obtained and are solved with a similar procedure. In order to better show how the tri-diagonal system is set, a 2D formulation of the CE-ADI-FDTD method is fully derived here for a transverse electric polarisation case with the z-axis as the normal axis (TE$_z$). With respect to this coordinate system and under the scalar approximation, the following 2D equations can be derived from Maxwell's equations considered for a linear, isotropic, lossless medium

$$
\frac{\partial H_x}{\partial t} = \frac{1}{\mu_r\mu_0}\left(-\frac{\partial E_z}{\partial y} \right)
$$

$$(5.62)$$

$$
\frac{\partial H_y}{\partial t} = \frac{1}{\mu_r\mu_0}\left(\frac{\partial E_z}{\partial x} \right)
$$

$$(5.63)$$

$$
\frac{\partial E_z}{\partial t} = \frac{1}{\varepsilon_r\varepsilon_0}\left(\frac{\partial H_y}{\partial x} - \frac{\partial H_x}{\partial y} \right)
$$

$$(5.64)$$

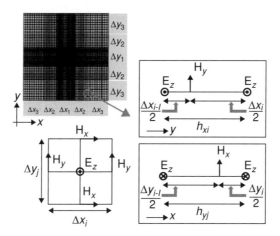

Figure 5.5 Example of a nonuniform 2D grid and, enlarged, electric- and magnetic-field components placed in a 2D nonuniform Yee unit cell.

The substitution of Equations (5.46)–(5.48) into Equations (5.62)–(5.64) yields [28]

$$\frac{\partial H_{xa}}{\partial t} + j\omega_c H_{xa} = \frac{1}{\mu_r\mu_0}\left(-\frac{\partial E_{za}}{\partial y}\right) \tag{5.65}$$

$$\frac{\partial H_{ya}}{\partial t} + j\omega_c H_{ya} = \frac{1}{\mu_r\mu_0}\left(\frac{\partial E_{za}}{\partial x}\right) \tag{5.66}$$

$$\frac{\partial E_{za}}{\partial t} + j\omega_c E_{za} = \frac{1}{\varepsilon_r\varepsilon_0}\left(\frac{\partial H_{ya}}{\partial x} - \frac{\partial H_{xa}}{\partial y}\right) \tag{5.67}$$

The discretisation in space is based on the unit cell of the Yee space lattice allowing the grid to be nonuniform, as shown in Figure 5.5. Discretisation in time is obtained by following the ADI scheme. The space cells are nonuniform so as to allow a more accurate and flexible representation of the photonic device to be simulated.

Applying the ADI scheme to Equations (5.65)–(5.67), the following set of equations for the first half time step are derived

$$H_{xa}\big|_{i,j+1/2}^{n+1/2} = \alpha_{xh}\big|_{i,j+1/2} H_{xa}\big|_{i,j+1/2}^{n} - \beta_{xh}\big|_{i,j+1/2}\left(E_{za}\big|_{i,j+1}^{n} - E_{za}\big|_{i,j}^{n}\right) \tag{5.68}$$

$$H_{ya}\big|_{i+1/2,j}^{n+1/2} = \alpha_{yh}\big|_{i+1/2,j} H_{ya}\big|_{i+1/2,j}^{n} + \beta_{yh}\big|_{i+1/2,j}\left(E_{za}\big|_{i+1,j}^{n+1/2} - E_{za}\big|_{i,j}^{n+1/2}\right) \tag{5.69}$$

$$E_{za}\big|_{i,j}^{n+1/2} = \alpha_e\big|_{i,j} E_{za}\big|_{i,j}^{n} + \beta_{xe}\big|_{i,j}\left(H_{ya}\big|_{i+1/2,j}^{n+1/2} - H_{ya}\big|_{i-1/2,j}^{n+1/2}\right) +$$
$$- \beta_{ye}\big|_{i,j}\left(H_{xa}\big|_{i,j+1/2}^{n} - H_{xa}\big|_{i,j-1/2}^{n}\right) \tag{5.70}$$

while for the second half time step, the following set of equations are obtained

$$H_{xa}|_{i,j+1/2}^{n+1} = \alpha_{xh}|_{i,j+1/2} H_{xa}|_{i,j+1/2}^{n+1/2} - \beta_{xh}|_{i,j+1/2} \left(E_{za}|_{i,j+1}^{n+1} - E_{za}|_{i,j}^{n+1} \right) \quad (5.71)$$

$$H_{ya}|_{i+1/2,j}^{n+1} = \alpha_{yh}|_{i+1/2,j} H_{ya}|_{i+1/2,j}^{n+1/2} + \beta_{yh}|_{i+1/2,j} \left(E_{za}|_{i+1,j}^{n+1/2} - E_{za}|_{i,j}^{n+1/2} \right) \quad (5.72)$$

$$E_{za}|_{i,j}^{n+1} = \alpha_e|_{i,j} E_{za}|_{i,j}^{n+1/2} + \beta_{xe}|_{i,j} \left(H_{ya}|_{i+1/2,j}^{n+1/2} - H_{ya}|_{i-1/2,j}^{n+1/2} \right) +$$
$$- \beta_{ye}|_{i,j} \left(H_{xa}|_{i,j+1/2}^{n+1} - H_{xa}|_{i,j-1/2}^{n+1} \right) \quad (5.73)$$

where the coefficients of Equations (5.68)–(5.73) are expressed as

$$\alpha_{xh}|_{i,j+1/2} = \alpha_{yh}|_{i+1/2,j} = \alpha_e|_{i,j} = \frac{4 - j\omega_c \Delta t}{4 + j\omega_c \Delta t} \quad (5.74a)$$

$$\beta_{xh}|_{i,j+1/2} = \frac{2\Delta t}{(4 + j\omega_c \Delta t)\,\mu_r \mu_0 \Delta y_j} \quad (5.74b)$$

$$\beta_{yh}|_{i+1/2,j} = \frac{2\Delta t}{(4 + j\omega_c \Delta t)\,\mu_r \mu_0 \Delta x_i} \quad (5.74c)$$

$$\beta_{xe}|_{i,j} = \frac{2\Delta t}{(4 + j\omega_c \Delta t)\,\varepsilon_r \varepsilon_0 h_{xi}} \quad (5.74d)$$

$$\beta_{ye}|_{i,j} = \frac{2\Delta t}{(4 + j\omega_c \Delta t)\,\varepsilon_r \varepsilon_0 h_{yj}} \quad (5.74e)$$

where Δt is the time step, Δx_i and Δy_j are the discretisation steps along the x- and y-directions, respectively, and h_{xi} and, h_{yj}, as also shown in Figure 5.5, are defined as

$$h_{xi} = \frac{\Delta x_i + \Delta x_{i-1}}{2} \qquad i = 2, 3, \ldots, N_x \quad (5.75a)$$

$$h_{yj} = \frac{\Delta y_j + \Delta y_{j-1}}{2} \qquad j = 2, 3, \ldots, N_y \quad (5.75b)$$

where N_x and N_y are the total number of cells of the computational domain along the x- and y-directions, respectively. The updating process of the first half time step starts with the explicit update of Equation (5.68) in order to obtain the new values of the magnetic-field component H_{xa}. Equation (5.69) cannot be explicitly solved because the electric-field component E_{za} is collocated in time with the magnetic-field component H_{ya}, hence still unknown. Substituting Equation (5.70) into Equation (5.69) and solving the derived equation for H_{ya}, the following equation is obtained

$$-\beta_{xe}|_{i+1,j} H_{ya}|_{i+3/2,j}^{n+1/2} + \left(\frac{1}{\beta_{yh}|_{i+1/2,j}} + \beta_{xe}|_{i+1,j} + \beta_{xe}|_{i,j} \right) H_{ya}|_{i+1/2,j}^{n+1/2}$$

$$-\beta_{xe}|_{i,j}H_{ya}|_{i-1/2,j}^{n+1/2} = \frac{\alpha_{yh}|_{i+1/2,j}}{\beta_{yh}|_{i+1/2,j}}H_{ya}|_{i+1/2,j}^{n} + \alpha_{e}|_{i+1,j}E_{za}|_{i+1,j}^{n} - \beta_{ye}|_{i+1,j}$$

$$\left(H_{xa}|_{i+1,j+1/2}^{n} - H_{xa}|_{i+1,j-1/2}^{n}\right) - \alpha_{e}|_{i,j}E_{za}|_{i,j}^{n} + \beta_{ye}|_{i,j}\left(H_{xa}|_{i,j+1/2}^{n} - H_{xa}|_{i,j-1/2}^{n}\right)$$

$$(5.76)$$

As can be seen from the LHS of Equation (5.76), three different values of the component H_{ya} in three different positions of the computational domain need to be calculated, and this calculation needs to be done for every value of j. In this way, a system of equations is derived whose coefficients form a tri-diagonal matrix, which can be efficiently solved in order to obtain the new values of the magnetic-field component H_{ya} inside the computational domain. Once the magnetic-field component H_{ya} has been calculated, the electric-field component E_{za} can be explicitly updated using Equation (5.70).

A similar procedure needs to be followed for the second half time step. Substituting Equation (5.73) into Equation (5.71), the following equation is obtained

$$-\beta_{ye}|_{i,j+1}H_{xa}|_{i,j+3/2}^{n+1} + \left(\frac{1}{\beta_{xh}|_{i,j+1/2}} + \beta_{ye}|_{i,j+1} + \beta_{ye}|_{i,j}\right)H_{xa}|_{i,j+1/2}^{n+1}$$

$$-\beta_{ye}|_{i,j}H_{xa}|_{i,j-1/2}^{n+1} = \frac{\alpha_{xh}|_{i,j+1/2}}{\beta_{xh}|_{i,j+1/2}}H_{xa}|_{i,j+1/2}^{n+1} - \alpha_{e}|_{i,j+1}E_{za}|_{i,j+1}^{n+1/2} - \beta_{xe}|_{i,j+1}$$

$$\left(H_{ya}|_{i+1/2,j+1}^{n+1/2} - H_{ya}|_{i-1/2,j+1}^{n+1/2}\right) + \alpha_{e}|_{i,j}E_{za}|_{i,j}^{n+1/2} + \beta_{xe}|_{i,j}$$

$$\left(H_{ya}|_{i+1/2,j}^{n+1/2} - H_{ya}|_{i-1/2,j}^{n+1/2}\right)$$

$$(5.77)$$

As can be seen from the LHS of Equation (5.77), three different values of the component H_{xa} in three different positions of the computational domain need to be calculated, and this calculation needs to be done for every value of i. In this way, a system of equations is derived whose coefficients form a tri-diagonal matrix, which can be efficiently solved in order to obtain the new values of the magnetic-field component H_{xa} inside the computational domain. Once the magnetic-field component H_{xa} has been calculated, the magnetic-field component H_{ya}, and the electric-field component E_{za} can be explicitly updated using Equations (5.72) and (5.73), respectively.

5.7   Perfectly Matched Layer (PML) Boundary Conditions

The analysis of scattering problems of electromagnetic waves propagating in optical waveguides is a problem usually studied in infinitely extended regions. Because the FDTD method relies in the finite-difference expression of Maxwell's equations, it can operate only in a finite number of points which represent the computational domain. This introduces a problem for allocation of computational resources for simulation of scattering problems because only a finite amount of points of the computational domain can be stored in a computer memory. Much effort has been diverted in the research environment to searching for a way to simulate an infinitely extended space in a computer memory, and all these efforts have led to the determination of different types of boundary conditions (BCs). These BCs can be grouped into two distinct types, analytical boundary conditions [29–33] and absorbing boundary conditions [11]. Amongst the analytical boundary conditions, the scheme proposed by Mur [4] has been mainly utilised because of its accuracy in the simulation of the propagation of outgoing waves from the computational domain. Amongst the absorbing boundary conditions, the scheme proposed by Berenger [11] is the most employed [34–38].

The innovative idea proposed by Berenger is the introduction of a nonphysical absorbing layer to terminate the FDTD computational domain which absorbs all electromagnetic waves impinging on it. This nonphysical medium is capable of absorbing all electromagnetic waves on a wide range of frequencies, with any polarisation and for all angles of incidence. For these matching properties this medium is called a 'perfectly matched layer' (PML). The formulation derived by Berenger for the PML relies on the splitting of the electromagnetic-field components of Maxwell's equations in such a way that each component is split into two sub-components that are orthogonal to each other. In this way it is possible to assign an appropriate loss parameter to each of these components. A brief mathematical treatment of the derivation of the properties of this nonphysical medium is discussed here. In order to do so, a 2D space is considered whose schematic is represented in Figure 5.6.

This 2D space is divided in two half-spaces: free space for $x < 0$, indicated as medium 1 and a medium with losses for $x > 0$ indicated as medium 2. A uniform plane wave with TE$_z$ polarisation is considered and it is propagating from medium 1 (free space) towards medium 2 (lossy medium), as shown also in Figure 5.6. The electric permittivity and magnetic permeability of medium 1 are $\varepsilon_1 = \varepsilon_0$ and $\mu_1 = \mu_0$, respectively, while the electric permittivity, magnetic permeability, electric and

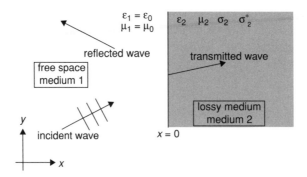

Figure 5.6 Schematic diagram of a 2D TE$_z$ polarised plane wave impinging on a medium with loss.

magnetic conductivity for the medium 2 are ε_2, μ_2, σ and σ^*, respectively. With respect to the geometry shown in Figure 5.6, the incident TE_z polarised plane wave is described as

$$\vec{H}_{inc} = H_0 e^{-(j\beta_x^i x + j\beta_y^i y)} \vec{u}_z \qquad (5.78)$$

where $\vec{H}_{inc}$ is the incident magnetic field, H_0 is the amplitude of the magnetic field, β_x^i and β_y^i are the propagation constant components of the plane wave along the x- and y-directions, respectively, and $\vec{u}_z$ is the unit normal vector of the z-axis. Considering that at the interface between the two media the plane wave is partially reflected back to region 1 and partially transmitted in region 2, the total electromagnetic field in region 1 ($x < 0$) is

$$\vec{H}_1 = H_0 \left(1 + \Gamma e^{j2\beta_x^i} \right) e^{-(j\beta_x^i x + j\beta_y^i y)} \vec{u}_z \qquad (5.79)$$

$$\vec{E}_1 = \left[-\frac{\beta_y^i}{\omega\varepsilon_1} \left(1 + \Gamma e^{j2\beta_x^i} \right) \vec{u}_x + \frac{\beta_x^i}{\omega\varepsilon_1} \left(1 - \Gamma e^{j2\beta_x^i} \right) \vec{u}_y \right] H_0 e^{-(j\beta_x^i x + j\beta_y^i y)} \quad (5.80)$$

where $\vec{E}_1$ is the electric field, ω is the frequency of the plane wave, Γ is the reflection coefficient, and $\vec{u}_x$ and $\vec{u}_y$ are the unit normal vectors of the x- and y-axes, respectively. Maxwell's equations in a sinusoidal regime for a TE_z polarised plane wave propagating in medium 2 can be expressed as

$$j\omega\varepsilon_2 \left(1 + \frac{\sigma}{j\omega\varepsilon_0} \right) E_x = \frac{\partial H_z}{\partial y} \qquad (5.81)$$

$$j\omega\varepsilon_2 \left(1 + \frac{\sigma}{j\omega\varepsilon_0} \right) E_y = -\frac{\partial H_z}{\partial x} \qquad (5.82)$$

$$j\omega\mu_2 \left(1 + \frac{\sigma^*}{j\omega\mu_0} \right) H_z = \frac{\partial E_x}{\partial y} - \frac{\partial E_y}{\partial x} \qquad (5.83)$$

where the electric and magnetic conductivities, σ and σ^*, are normalised with respect to the electric permittivity ε_0 and the magnetic permeability μ_0, respectively. Equations (5.81)–(5.83) are then rewritten considering the splitting of the magnetic-field component, H_z

$$H_z = H_{zx} + H_{zy} \qquad (5.84)$$

and with the use of the following variables

$$s_i = 1 + \frac{\sigma_i}{j\omega\varepsilon_0} \qquad (5.85a)$$

$$s_i^* = 1 + \frac{\sigma_i^*}{j\omega\mu_0} \qquad (5.85b)$$

with $i = x, y$ obtaining

$$j\omega\varepsilon_2 s_y E_x = \frac{\partial \left(H_{zx} + H_{zy} \right)}{\partial y} \qquad (5.86)$$

$$j\omega\varepsilon_2 s_x E_y = -\frac{\partial \left(H_{zx} + H_{zy} \right)}{\partial x} \qquad (5.87)$$

$$j\omega\mu_2 s_x^* H_{zx} = -\frac{\partial E_y}{\partial x} \qquad (5.88)$$

$$j\omega\mu_2 s_y^* H_{zy} = \frac{\partial E_x}{\partial y} \qquad (5.89)$$

The next step is to determine a solution for Maxwell's equations for the electromagnetic wave propagating inside the medium with loss. Differentiating Equation (5.86) with respect to y and Equation (5.87) with respect to x and substituting into Equations (5.88) and (5.89), respectively, the following two equations are obtained

$$-\omega^2 \mu_2 \varepsilon_2 H_{zx} = -\frac{1}{s_x^*} \frac{\partial}{\partial x} \frac{1}{s_x} \frac{\partial}{\partial x} \left(H_{zx} + H_{zy} \right) \qquad (5.90)$$

$$-\omega^2 \mu_2 \varepsilon_2 H_{zy} = -\frac{1}{s_y^*} \frac{\partial}{\partial y} \frac{1}{s_y} \frac{\partial}{\partial y} \left(H_{zx} + H_{zy} \right) \qquad (5.91)$$

The summation of Equations (5.90)–(5.91) leads to the following wave equation

$$\frac{1}{s_x^*} \frac{\partial}{\partial x} \frac{1}{s_x} \frac{\partial}{\partial x} H_z + \frac{1}{s_y^*} \frac{\partial}{\partial y} \frac{1}{s_y} \frac{\partial}{\partial y} H_z + \omega^2 \mu_2 \varepsilon_2 H_z = 0 \qquad (5.92)$$

Equation (5.92) has the following solution

$$H_z = H_0 T e^{-\left(j\sqrt{s_x s_x^*} \beta_x x + j\sqrt{s_y s_y^*} \beta_y y \right)} \qquad (5.93)$$

with the dispersion relation given by $(\beta_x)^2 + (\beta_y)^2 = (k_2)^2$. The substitution of Equation (5.93) into Equations (5.86)–(5.87) leads to the following equations for the electric-field components

$$E_x = H_0 T \frac{\beta_y}{\omega\varepsilon_2} \sqrt{\frac{s_y^*}{s_y}} e^{-(j\sqrt{s_x s_x^*}\beta_x x + j\sqrt{s_y s_y^*}\beta_y y)} \tag{5.94}$$

$$E_y = H_0 T \frac{\beta_x}{\omega\varepsilon_2} \sqrt{\frac{s_x^*}{s_x}} e^{-(j\sqrt{s_x s_x^*}\beta_x x + j\sqrt{s_y s_y^*}\beta_y y)} \tag{5.95}$$

The reflection and transmission coefficients can be obtained by imposing the continuity of the tangential components of the electromagnetic field at the interface

$$\Gamma = \frac{\dfrac{\beta_x^i}{\omega\varepsilon_1} - \dfrac{\beta_x}{\omega\varepsilon_2}\sqrt{\dfrac{s_x^*}{s_x}}}{\dfrac{\beta_x^i}{\omega\varepsilon_1} + \dfrac{\beta_x}{\omega\varepsilon_2}\sqrt{\dfrac{s_x^*}{s_x}}} \tag{5.96}$$

with $\beta_y = \beta_y^i = k_1 \sin\theta_i$, and $s_y = s_y^* = 1$. The aim is to determine the characteristics of s_x, s_x^*, s_y and s_y^* in order to eliminate all reflection at the interface of the absorbing medium. From Equation (5.96) it can be noted that if $\varepsilon_1 = \varepsilon_2$ and $s_x = s_x^*$ then $\beta_x = \beta_x^i$, which implies that the reflection coefficient $\Gamma = 0$ and, consequently, the transmission coefficient $T = 1$, and these relations stand for all angles of incidence θ_i. It should be noted from the definition of the variables s_x and s_x^* that the relationship $s_x = s_x^*$ implies $\sigma_x/\varepsilon_0 = \sigma_x^*/\mu_0$. The transmitted field in medium 2 is then

$$H_z = H_0 e^{-(js_x\beta_x^i x + js\beta_y^i y)} = H_0 e^{-(j\beta_x^i x + js\beta_y^i y)} e^{-\sigma\eta_1\varepsilon_1\cos(\theta_i)x} \tag{5.97}$$

$$E_x = H_0\eta_1 \sin(\theta_i) e^{-(j\beta_x^i x + js\beta_y^i y)} e^{-\sigma\eta_1\varepsilon_1\cos(\theta_i)x} \tag{5.98}$$

$$E_y = H_0\eta_1 \cos(\theta_i) e^{-(j\beta_x^i x + js\beta_y^i y)} e^{-\sigma\eta_1\varepsilon_1\cos(\theta_i)x} \tag{5.99}$$

From Equations (5.97)–(5.99) it is possible to see that the transmitted field is propagating inside the medium with loss with the same phase velocity of the incident field, while it is attenuating along the normal direction with a factor of $\sigma\eta_1\cos(\theta_i)$ and these properties are satisfied for every angle of incidence θ_i.

A similar procedure can be repeated for TM$_z$ polarised plane waves. In this case the *splitting* of the electromagnetic field components is applied to the electric field $E_z = E_{zx} + E_{zy}$. In this way, applying the matching condition $\mu_1 = \mu_2$ and $s_x = s_x^*$, which implies $\sigma_x/\varepsilon_0 = \sigma_x^*/\mu_0$, the condition $\beta_x = \beta_x^i$ is satisfied and consequently $\Gamma = 0$ and $T = 1$. It should be noted that all these relations still stand for all angles of incidence θ_i.

In conclusion, a medium with loss that satisfies the properties previously described and $\sigma_x/\varepsilon_0 = \sigma_x^*/\mu_0$, with the conditions $\varepsilon_1 = \varepsilon_2$ for TE$_z$ polarised waves and $\mu_1 =$

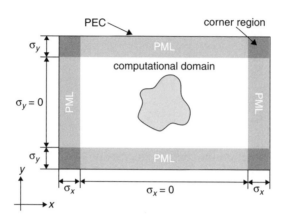

Figure 5.7 Schematic diagram of a 2D computational domain surrounded by four layers of PML boundary conditions.

μ_2 for TM$_z$ polarised waves, is a medium perfectly matched and perfectly absorbing for all electromagnetic waves impinging in its interface, in a way that no reflections can be generated at its interface. Furthermore, from Equations (5.97)–(5.99) it can be noted that the transmitted wave in the PML propagates with same speed of the incident wave while attenuating during propagation along the normal direction (the x-axis in the geometry considered in Figure 5.6). With these considerations, the FDTD grid can be surrounded by layers of this nonphysical and absorbing medium (PML) with the absorbing properties previously described and perfect electrical conductor (PEC) walls to terminate the whole computational domain, as shown in Figure 5.7 for a 2D case. In a general 3D case, six layers of PML are needed to truncate the computational domain, and all the electromagnetic-field components need to be treated with the *splitting* technique obtaining a set of 12 equations

$$\left(\varepsilon_0\varepsilon_r\frac{\partial}{\partial t} + \sigma_y\varepsilon_r\right)E_{xy} = \frac{\partial}{\partial y}\left(H_{zx} + H_{zy}\right) \tag{5.100}$$

$$\left(\varepsilon_0\varepsilon_r\frac{\partial}{\partial t} + \sigma_z\varepsilon_r\right)E_{xz} = -\frac{\partial}{\partial z}\left(H_{yx} + H_{yz}\right) \tag{5.101}$$

$$\left(\varepsilon_0\varepsilon_r\frac{\partial}{\partial t} + \sigma_z\varepsilon_r\right)E_{yz} = \frac{\partial}{\partial z}\left(H_{xy} + H_{xz}\right) \tag{5.102}$$

$$\left(\varepsilon_0\varepsilon_r\frac{\partial}{\partial t} + \sigma_x\varepsilon_r\right)E_{yx} = -\frac{\partial}{\partial x}\left(H_{zx} + H_{zy}\right) \tag{5.103}$$

$$\left(\varepsilon_0\varepsilon_r\frac{\partial}{\partial t} + \sigma_x\varepsilon_r\right)E_{zx} = \frac{\partial}{\partial x}\left(H_{yx} + H_{yz}\right) \tag{5.104}$$

$$\left(\varepsilon_0\varepsilon_r\frac{\partial}{\partial t} + \sigma_y\varepsilon_r\right)E_{zy} = -\frac{\partial}{\partial y}\left(H_{xy} + H_{xz}\right) \tag{5.105}$$

$$\left(\mu_0\mu_r\frac{\partial}{\partial t} + \sigma_y^*\mu_r\right)H_{xy} = -\frac{\partial}{\partial y}\left(E_{zx} + E_{zy}\right) \tag{5.106}$$

$$\left(\mu_0\mu_r\frac{\partial}{\partial t} + \sigma_z^*\mu_r\right)H_{xz} = \frac{\partial}{\partial z}\left(E_{yx} + E_{yz}\right) \tag{5.107}$$

$$\left(\mu_0\mu_r\frac{\partial}{\partial t} + \sigma_z^*\mu_r\right)H_{yz} = -\frac{\partial}{\partial z}\left(E_{xy} + E_{xz}\right) \tag{5.108}$$

$$\left(\mu_0\mu_r\frac{\partial}{\partial t} + \sigma_x^*\mu_r\right)H_{yx} = \frac{\partial}{\partial x}\left(E_{zx} + E_{zy}\right) \tag{5.109}$$

$$\left(\mu_0\mu_r\frac{\partial}{\partial t} + \sigma_x^*\mu_r\right)H_{zx} = -\frac{\partial}{\partial x}\left(E_{yx} + E_{yz}\right) \tag{5.110}$$

$$\left(\mu_0\mu_r\frac{\partial}{\partial t} + \sigma_y^*\mu_r\right)H_{zy} = \frac{\partial}{\partial y}\left(E_{xy} + E_{xz}\right) \tag{5.111}$$

The matching conditions are similar to those previously derived and in particular $\sigma_i/\varepsilon_0 = \sigma_i^*/\mu_0$, with $i = x, y, z$, $\varepsilon_1 = \varepsilon_2$ for TE$_z$ polarised waves (where the subscript 2 stands for the PML and the subscript 1 stands for the medium adjacent the PML) and $\mu_1 = \mu_2$ for TM$_z$ polarised waves.

5.8 Uniaxial Perfectly Matched Layer (UPML) Absorbing Boundary Condition

The formulation introduced by Berenger employing the *splitting* technique for the electromagnetic-field component permits the definition of a mathematical model for a nonphysical medium with well-defined electric and magnetic properties that is perfectly matched for all waves impinging in its interface. The way in which its loss terms are defined makes this medium an anisotropic medium, if such a medium could physically exist. Based on this consideration, it is then possible to define a medium that is anisotropic and uniaxial which is characterised by well-defined tensors for the electric permittivity and magnetic permeability. This new approach leads to a different formulation for an absorbing layer to be used with FDTD grids for simulations in open regions. The absorbing layer obtained with this approach is called a uniaxial perfectly matched layer (UPML). The formulation for the UPML is different from the formulation for the PML, even if it relies on a basic concept similar to the concept used for the PML formulation. The main advantage of the UPML formulation is that it doesn't need any *splitting* of the electromagnetic-field component, even though the absorbing properties remain unchanged. Referring to the geometry of Figure 5.6, a TE$_z$ polarised plane wave is propagating in free space towards an uniaxial medium whose interface is at $x = 0$ and whose electric and magnetic tensors are given by [15] $\bar{\bar{\varepsilon}} = \varepsilon_2 \bar{\bar{s}}, \bar{\bar{\mu}} = \mu_2 \bar{\bar{s}}$ with

$$\bar{\bar{s}} = \begin{bmatrix} s_x^{-1} & 0 & 0 \\ 0 & s_x & 0 \\ 0 & 0 & s_x \end{bmatrix} \quad (5.112)$$

where s_x is defined as in Equation (5.85), and ε_2 and μ_2 are the electric permittivity and the magnetic permeability, respectively, of the uniaxial medium. The plane wave is completely transmitted in the uniaxial medium without any reflection generated at the interface, regardless its frequency and its angle of incidence. Such a medium is basically identical to the PML medium introduced by Berenger and it is defined as a UPML because of its uniaxial anisotropy. For the TE_z polarised plane wave considered here, the reflection coefficient is $\Gamma = 0$ and the transmission coefficient is $T = 1$ so that the expressions for the transmitted electromagnetic-field components are

$$H_z = H_0 e^{-\left(js_x\beta_x^i x + js\beta_y^i y\right)} = H_0 e^{-\left(j\beta_x^i x + j\beta_y^i y\right)} e^{-\sigma\eta_1\varepsilon_1\cos(\theta_i)x} \quad (5.113)$$

$$E_x = -H_0 s_x \eta_1 \sin(\theta_i) e^{-\left(j\beta_x^i x + js\beta_y^i y\right)} e^{-\sigma\eta_1\varepsilon_1\cos(\theta_i)x} \quad (5.114)$$

$$E_y = H_0 \eta_1 \cos(\theta_i) e^{-\left(j\beta_x^i x + js\beta_y^i y\right)} e^{-\sigma\eta_1\varepsilon_1\cos(\theta_i)x} \quad (5.115)$$

where θ_i is the angle of incidence relative to the x-axis. From Equations (5.113)–(5.115), it can be noted that the transmitted field is propagating with the same phase velocity inside the absorbing medium, while attenuating along the direction normal to the interface with a factor of $\sigma\varepsilon_1\cos(\theta_i)$. It can be seen that this characteristic is similar to that described for the PML. By comparison of Equations (5.113)–(5.115) with Equations (5.97)–(5.99), it can be seen that the propagation characteristics of the two electromagnetic fields are identical. In particular, it can be noted that the tangential electric- and magnetic-field components, H_z and E_y, are identical in the two approaches while the normal components, E_x, differ by a factor s_x. Comparing the normal components of the transmitted field obtained from the two different approaches with the respective incident field components, it can be seen that in the PML formulation, the E_x component is continuous at the interface $x = 0$, while in the UPML formulation E_x is discontinuous with $D_x = \varepsilon s_x^{-1} E_x$ continuous. This can be explained by considering that the two formulations use two different formulations of the divergence theorem. Although different, these two formulations guarantee the same matching and absorbing properties for the transmitted waves.

Each side of the FDTD grid can be bounded with a layer of UPML. But there are regions in which the UPML itself is not uniaxial in the strict sense of the definition. These regions are the corner regions, as shown in Figure 5.7 for a 2D case and in Figure 5.8 for a 3D case. From this figure, it can be seen that in the corner regions there is superposition of different UPML layers. In this case the expression of the tensor

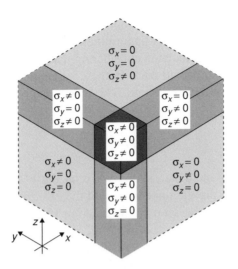

Figure 5.8 Detail of the PML boundary conditions applied for a 3D computational domain.

that multiplies the electric permittivity and magnetic permeability for this medium is given by

$$\bar{\bar{s}} = \begin{bmatrix} s_x^{-1} & 0 & 0 \\ 0 & s_x & 0 \\ 0 & 0 & s_x \end{bmatrix} \cdot \begin{bmatrix} s_y & 0 & 0 \\ 0 & s_y^{-1} & 0 \\ 0 & 0 & s_y \end{bmatrix} = \begin{bmatrix} s_x^{-1} s_y & 0 & 0 \\ 0 & s_x s_y^{-1} & 0 \\ 0 & 0 & s_x s_y \end{bmatrix} \quad (5.116)$$

with s_x and s_y defined as in Equation (5.85), while for the 3D case, the tensor at the corner region is defined as

$$\bar{\bar{s}} = \begin{bmatrix} s_x^{-1} & 0 & 0 \\ 0 & s_x & 0 \\ 0 & 0 & s_x \end{bmatrix} \cdot \begin{bmatrix} s_y & 0 & 0 \\ 0 & s_y^{-1} & 0 \\ 0 & 0 & s_y \end{bmatrix} \cdot \begin{bmatrix} s_z & 0 & 0 \\ 0 & s_z & 0 \\ 0 & 0 & s_z^{-1} \end{bmatrix} = \begin{bmatrix} \dfrac{s_y s_z}{s_x} & 0 & 0 \\ 0 & \dfrac{s_x s_z}{s_y} & 0 \\ 0 & 0 & \dfrac{s_x s_y}{s_z} \end{bmatrix}$$

$$(5.117)$$

5.9 PML Parameters

The performance of the PML absorbing boundary conditions can be influenced by the parameters which fix the absorption rate of the wave that is propagating inside the PML itself. These parameters are identified with the electric and magnetic conductivity σ and σ^*, respectively. The performance of the PML has also been derived by Berenger [11], applying the transmission-line concept to the PML formulation. In this formulation, each layer of PML is terminated by a layer of PEC and the reflection coefficient R has been calculated considering that, upon reaching the PEC layer, the propagating wave is reflected back to the computational domain, affecting the accuracy of the results of the simulation. The reflection coefficient has been derived as

$$R(\theta) = e^{-2\sigma\eta\varepsilon_r d \cos(\theta)} \tag{5.118}$$

where θ is the angle of incidence of the wave on the PML, d is the width of the PML layer, and η and σ are the impedance and the electric conductivity of the PML layer, respectively. As can be seen from Equation (5.118), the reflection from the PML is exponentially reduced as the width, d, and the conductivity, σ, increase. It is obvious that for increasing width, d, of the PML layer the computational burden increases and for this reason it is clear that the choice of the conductivity, σ, is crucial in order to obtain good performance of the PML in terms of low reflection. For this reason, choosing really high values for the electric and magnetic conductivities σ and σ^* inside the PML layer seems to be a good choice in terms of good absorption. Nevertheless, it has to be taken into account that high values of the electric and magnetic conductivity for the PML layer introduce a high discontinuity at the interface between the computational domain and the PML layer. This discontinuity is indeed a source of reflections which affect the accuracy of the simulation results. A solution to this problem has been proposed in [11, 15] in which a variation of the electric and magnetic conductivity profiles along the transverse direction has been adopted. As an example, geometrical scaling of the electric conductivity σ along the transverse direction x is obtained by the following formula

$$\sigma(x) = \sigma_0 g^{x/\Delta x} \tag{5.119}$$

where σ_0 is the electric conductivity at the interface between the computational domain and the PML layer, g is the scaling factor and Δx is the space discretisation. Modifying Equation (5.118) in order to take into account the variation of σ along the transverse direction, x, it is possible to calculate the reflection error

$$R(\theta) = e^{-2\eta\sigma_0\Delta x (g^N - 1)\cos(\theta)/\ln(g)} \tag{5.120}$$

where N is the number of cells of the PML layer. From Equation (5.120) it is possible to derive a relationship for the value of the electric conductivity at the interface between the PML and the computational domain σ_0

$$\sigma_0 = -\frac{\ln(R(0))\ln(g)}{2\eta\varepsilon_r\Delta x\left(g^N - 1\right)} \tag{5.121}$$

In practical applications, the values of the maximum reflection error at normal incidence, $R(0)$, and the scaling factor, g, need to be fixed in order to calculate the value σ_0. Once σ_0 is calculated then the scaling of the electric conductivity σ is performed inside the PML layer using Equation (5.119).

5.10 PML Boundary Conditions for CE-ADI-FDTD

In conventional FDTD, PMLs have been extensively used because of their excellent absorption properties that give a robust way to terminate the computational domain. PMLs have been also incorporated into the CE-ADI-FDTD method, which has been used to simulate integrated photonic devices. However, problems with the numerical stability of the CE-ADI-FDTD algorithm have been reported due to the accumulation of reflection coming from the PML in the computational domain [28]. Here, a different PML approach will be considered which has been previously proposed in [39] in the context of the ADI-FDTD algorithm. Considering Equations (5.9)–(5.14) and applying a 2D case for TE$_z$ polarisation, the following 2D Maxwell's equations are obtained

$$\mu_0\mu_r\frac{\partial H_x}{\partial t} + \sigma^* H_x = -\frac{\partial E_z}{\partial y} \tag{5.122}$$

$$\mu_0\mu_r\frac{\partial H_y}{\partial t} + \sigma^* H_y = \frac{\partial E_z}{\partial x} \tag{5.123}$$

$$\varepsilon_0\varepsilon_r\frac{\partial E_z}{\partial t} + \sigma E_z = \frac{\partial H_y}{\partial x} - \frac{\partial H_x}{\partial y} \tag{5.124}$$

Applying the CE formulation to Equations (5.122)–(5.124) yields

$$\mu_0\mu_r\frac{\partial H_{xa}}{\partial t} + \left(\sigma^* + j\omega_c\mu_0\mu_r\right) H_{xa} = -\frac{\partial E_{za}}{\partial y} \tag{5.125}$$

$$\mu_0\mu_r\frac{\partial H_{ya}}{\partial t} + \left(\sigma^* + j\omega_c\mu_0\mu_r\right) H_{ya} = \frac{\partial E_{za}}{\partial x} \tag{5.126}$$

$$\varepsilon_0\varepsilon_r\frac{\partial E_{za}}{\partial t} + \left(\sigma + j\omega_c\varepsilon_0\varepsilon_r\right) E_{za} = \frac{\partial H_{ya}}{\partial x} - \frac{\partial H_{xa}}{\partial y} \tag{5.127}$$

Next step is to apply the PML formulation to Equations (5.125)–(5.127) in which the envelope of the electric field E_{za} is split in two sub-components as follows

$$E_{za} = E_{zax} + E_{zay} \tag{5.128}$$

The substitution of Equation (5.128) into Equations (5.125)–(5.127) yields

$$\mu_0\mu_r\frac{\partial H_{xa}}{\partial t} + \left(\sigma_y^* + j\omega_c\mu_0\mu_r\right)H_{xa} = -\frac{\partial E_{za}}{\partial y} \tag{5.129}$$

$$\mu_0\mu_r\frac{\partial H_{ya}}{\partial t} + \left(\sigma_x^* + j\omega_c\mu_0\mu_r\right)H_{ya} = \frac{\partial E_{za}}{\partial x} \tag{5.130}$$

$$\varepsilon_0\varepsilon_r\frac{\partial E_{zax}}{\partial t} + \left(\sigma_x + j\omega_c\varepsilon_0\varepsilon_r\right)E_{zax} = \frac{\partial H_{ya}}{\partial x} \tag{5.131}$$

$$\varepsilon_0\varepsilon_r\frac{\partial E_{zay}}{\partial t} + \left(\sigma_y + j\omega_c\varepsilon_0\varepsilon_r\right)E_{zay} = -\frac{\partial H_{xa}}{\partial y} \tag{5.132}$$

Applying the ADI-FDTD discretisation scheme to Equations (5.129)–(5.132), the following set of equations can be obtained for the first half time step

$$H_{xa}|_{i,j+1/2}^{n+1/2} = \alpha_{xh}|_{i,j+1/2}H_{xa}|_{i,j+1/2}^{n} - \beta_{xh}|_{i,j+1/2}\left(E_{za}|_{i,j+1}^{n} - E_{za}|_{i,j}^{n}\right) \tag{5.133a}$$

$$H_{ya}|_{i+1/2,j}^{n+1/2} = \alpha_{yh}|_{i+1/2,j}H_{ya}|_{i+1/2,j}^{n} + \beta_{yh}|_{i+1/2,j}\left(E_{za}|_{i+1,j}^{n+1/2} - E_{za}|_{i,j}^{n+1/2}\right) \tag{5.133b}$$

$$E_{zxa}|_{i,j}^{n+1/2} = \alpha_{xe}|_{i,j}E_{zxa}|_{i,j}^{n} + \beta_{xe}|_{i,j}\left(H_{ya}|_{i+1/2,j}^{n+1/2} - H_{ya}|_{i-1/2,j}^{n+1/2}\right) \tag{5.133c}$$

$$E_{zya}|_{i,j}^{n+1/2} = \alpha_{ye}|_{i,j}E_{zya}|_{i,j}^{n} - \beta_{ye}|_{i,j}\left(H_{xa}|_{i,j+1/2}^{n} - H_{xa}|_{i,j-1/2}^{n}\right) \tag{5.133d}$$

The coefficients of the PML equations are calculated using either a forward or a backward differencing approximation instead of the linear approximation used in the conventional implementation of the PML scheme. In this way it is possible to collocate the field component on the LHS of the equation at the same time step of the field component on the RHS of same equation. This procedure yields the following coefficients for the first half time step:

$$\alpha_{xe}|_{i,j} = \frac{4 - j\omega_c\Delta t}{4 + \left(j\omega_c + 2\dfrac{\sigma_x}{\varepsilon_r\varepsilon_0}\right)\Delta t} \tag{5.134a}$$

$$\alpha_{xh}|_{i,j+1/2} = \frac{4 - \left(j\omega_c + 2\dfrac{\sigma_y^*}{\mu_r\mu_0}\right)\Delta t}{4 + j\omega_c\Delta t} \tag{5.134b}$$

$$\alpha_{ye}|_{i,j} = \frac{4 - \left(j\omega_c + 2\dfrac{\sigma_y}{\varepsilon_r\varepsilon_0}\right)\Delta t}{4 + j\omega_c\Delta t} \tag{5.134c}$$

$$\alpha_{yh}|_{i+1/2,j} = \frac{4 - j\omega_c\Delta t}{4 + \left(j\omega_c + 2\dfrac{\sigma_x^*}{\mu_r\mu_0}\right)\Delta t} \tag{5.134d}$$

$$\beta_{xe}|_{i,j} = \cfrac{2\Delta t}{\left(4 + \left(j\omega_c + 2\dfrac{\sigma_x}{\varepsilon_r\varepsilon_0}\right)\Delta t\right)\varepsilon_r\varepsilon_0 h_{xi}} \tag{5.134e}$$

$$\beta_{xh}|_{i,j+1/2} = \cfrac{2\Delta t}{(4 + j\omega_c\Delta t)\,\mu_r\mu_0\Delta y_j} \tag{5.134f}$$

$$\beta_{ye}|_{i,j} = \cfrac{2\Delta t}{(4 + j\omega_c\Delta t)\,\varepsilon_r\varepsilon_0 h_{yj}} \tag{5.134g}$$

$$\beta_{yh}|_{i+1/2,j} = \cfrac{2\Delta t}{\left(4 + \left(j\omega_c + 2\dfrac{\sigma_x^*}{\mu_r\mu_0}\right)\Delta t\right)\mu_r\mu_0\Delta x_i} \tag{5.134h}$$

Similar expressions can be derived for the coefficients of the PML equations for the second half time step

$$\alpha_{xe}|_{i,j} = \cfrac{4 - \left(j\omega_c + 2\dfrac{\sigma_x}{\varepsilon_r\varepsilon_0}\right)\Delta t}{4 + j\omega_c\Delta t} \tag{5.135a}$$

$$\alpha_{xh}|_{i,j+1/2} = \cfrac{4 - j\omega\Delta t}{4 + \left(j\omega_c + 2\dfrac{\sigma_y^*}{\mu_r\mu_0}\right)_c\Delta t} \tag{5.135b}$$

$$\alpha_{ye}|_{i,j} = \cfrac{4 - j\omega_c\Delta t}{4 + \left(j\omega_c + 2\dfrac{\sigma_y}{\varepsilon_r\varepsilon_0}\right)\Delta t} \tag{5.135c}$$

$$\alpha_{yh}|_{i+1/2,j} = \cfrac{4 - \left(j\omega_c + 2\dfrac{\sigma_x^*}{\mu_r\mu_0}\right)\Delta t}{4 + j\omega_c\Delta t} \tag{5.135d}$$

$$\beta_{xe}|_{i,j} = \cfrac{2\Delta t}{(4 + j\omega_c\Delta t)\,\varepsilon_r\varepsilon_0 h_{xi}} \tag{5.135e}$$

$$\beta_{xh}|_{i,j+1/2} = \cfrac{2\Delta t}{\left(4 + \left(j\omega_c + 2\dfrac{\sigma_y^*}{\mu_r\mu_0}\right)\Delta t\right)\mu_r\mu_0\Delta y_j} \tag{5.135f}$$

$$\beta_{ye}|_{i,j} = \cfrac{2\Delta t}{\left(4 + \left(j\omega_c + 2\dfrac{\sigma_y}{\varepsilon_r\varepsilon_0}\right)\Delta t\right)\varepsilon_r\varepsilon_0 h_{yj}} \tag{5.135g}$$

$$\beta_{yh}|_{i+1/2,j} = \cfrac{2\Delta t}{(4 + j\omega_c\Delta t)\,\mu_r\mu_0\Delta x_i} \tag{5.135h}$$

The arrangement proposed here for CE-ADI-FDTD equations with PML boundary conditions leads to a stable algorithm even with large Courant numbers, as will be clearly shown in the examples presented in the next section.

5.11 PhC Resonant Cavities

Before proceeding to the analysis of photonic crystal (PhC) cavities, the modified PML boundary conditions will be tested in order to verify the effectiveness of their absorption properties, thus avoiding numerical instability. The structure considered for this test is a 5×5 square PhC cavity consisting of dielectric rods with refractive index $n_{\text{rods}} = 3.4$ in air, as shown in Figure 5.9.

First, simulations have been carried out with a uniform mesh and with different values of time step in order to test the effect of the time-step value on the stability of the developed CE-ADI-FDTD code. The discretisation step was fixed at 17.73 nm and 10 cells of the PML layer have been used to truncate the computational domain on all sides of the PhC cavity, as shown in Figure 5.9. Although the CE-ADI-FDTD code has been developed to rely on nonuniform mesh, to test the stability properties of the proposed method a uniform mesh has been utilised only in this test. Using the Courant criterion formula and the discretisation steps for the x- and y-directions utilised for this simulation, the maximum time step that it is possible to use with a conventional FDTD is calculated to be $\Delta t_{CL} \cong 0.042$ fs. The source used to excite the

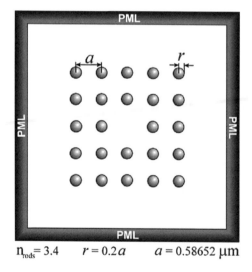

$n_{\text{rods}} = 3.4$ $r = 0.2a$ $a = 0.58652\ \mu m$

Figure 5.9 Schematic diagram of the simulated 5×5 dielectric rods photonic crystal cavity with PML boundary conditions.

PBG cavity is a Gaussian-shaped source in space and in time and it is defined as

$$E_{zs}(x, y, t) = e^{-\left(\frac{t-t_0}{T_0}\right)^2} e^{-\left(\frac{x-x_0}{X_0}\right)^2} e^{-\left(\frac{y-y_0}{Y_0}\right)^2} \tag{5.136}$$

where t_0 and T_0 are the delay and the width of the Gaussian pulse in time, respectively, x_0 and X_0 are the displacement and the width of the Gaussian pulse along x direction, respectively, and y_0 and Y_0 are the displacement and the width of the Gaussian pulse along y direction, respectively. For all simulations T_0 and t_0 were set to 30 and 90 fs, respectively, x_0 and y_0 were set to the coordinates of the centre of the PBG cavity, X_0 and Y_0 were fixed to $a/2$, where a is the lattice constant of the PhC as shown in Figure 5.9, and the angular frequency, ω_c, of the source was fixed to 1.256×10^{15} rad/s ($\lambda = 1.5$ µm). The soft-source technique has been used in order to insert the source in the computational domain [15]. The time-domain variation of the electric field at the centre of the cavity was recorded. In Figure 5.10 the time-domain responses of the envelope of the electric field obtained with the developed CE-ADI-FDTD and the approach used in [28] are plotted, both obtained with a time-step size fixed to 20 times the Courant limit. As can be clearly seen from this figure, the approach used in [28] gives rise to instability at the very early stages of the simulation, while

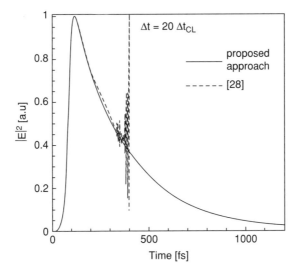

Figure 5.10 Time variation of the envelopes of the electric field recorded at the centre of the 5 × 5 square lattice cavity obtained with the new CE-ADI-FDTD approach and with the approach in [28]. (Reproduced with permission from Pinto, D. and Obayya, S.S.A (2007) Improved complex envelope alternative direction implicit finite difference time domain method for photonic bandgap cavities. *IEEE J. Lightwave Technol.*, **25** (1), 440–447. © 2007 IEEE.)

the proposed CE-ADI-FDTD leads to the formation of the fundamental resonant mode of the cavity and looks very stable. Further simulations carried out with the same computational domain and utilising the approach used in [28] have shown that the maximum Courant number which gives stable results is equal to 5, while with the proposed CE-ADI-FDTD, larger Courant numbers can be applied, obtaining at the same time faster results and reduction in required computational resources.

An explanation of this behaviour can be attributed to the superior performance in terms of absorption properties of the PML layers developed with the new approach. The increased absorption of the PML layers, even for very large Courant numbers, avoids the accumulation of reflection in the computational domain, which can iteratively add to build up instability in the simulation, and leads to an unconditionally stable algorithm. In order to verify the effect of the time-step size on the numerical dispersion of the results, the fast Fourier transform (FFT) of the time-domain responses of the electric field inside the PhC cavity for all three simulations have been computed. Figure 5.11 shows the results of this procedure. As can be seen from this figure, all the results obtained are in very good agreement with the data obtained with a conventional FDTD simulation from which the normalised frequency, a/λ, of the fundamental resonant mode of the 5×5 PhC cavity has been calculated to be 0.378.

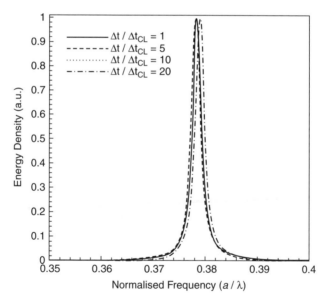

Figure 5.11 Spectral distributions of the resonant mode energy for the 5×5 square lattice cavity obtained with different simulations using different time steps. (Reproduced with permission from Pinto, D. and Obayya, S.S.A (2007) Improved complex envelope alternative direction implicit finite difference time domain method for photonic bandgap cavities. *IEEE J. Lightwave Technol.*, **25** (1), 440–447. © 2007 IEEE.)

For simulations carried out with time steps fixed to 5 and 10 times the Courant limit, respectively, the shift of the calculated normalised frequency of the resonant mode has been found to be negligibly small. Moreover, for a Courant number of 20, the proposed CE-ADI-FDTD gives a normalised frequency of the resonant mode that is shifted by only 0.2% from that obtained using conventional FDTD.

5.12   5 × 5 Rectangular Lattice PhC Cavity

The fundamental TE resonant mode of the 5 × 5 cavity, shown in Figure 5.9, will be considered. For all simulations, nonuniform mesh has been utilised in such a way that the computational domain grid contains more points in the centre of the cavity where the electromagnetic field of the resonant mode of the cavity is trapped. The structure has been discretised by a nonuniform mesh of 153 cells along the x- and y-directions in such a way that the minimum step size considered to discretise the structure is fixed at 17.73 nm. Furthermore, 10 cells of PML have been used to terminate the computational domain. The structure is excited with an electric-field profile given by Equation (5.136). In Figure 5.10, the time variation of the envelope of the electric field inside the cavity is shown and Figure 5.11 shows the spectral distribution of the resonant mode energy. The normalised frequency, a/λ, can be easily determined from Figure 5.11 to be 0.3789. The electric-field distribution of the resonant mode in the cavity is shown in Figure 5.12, obtained after running the code for 1024 fs.

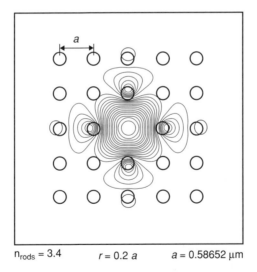

$n_{rods} = 3.4$ $r = 0.2\,a$ $a = 0.58652\ \mu m$

Figure 5.12   Electric-field profile of the resonant mode inside the 5 × 5 square lattice cavity. (Reproduced with permission from Pinto, D. and Obayya, S.S.A (2007) Improved complex envelope alternative direction implicit finite difference time domain method for photonic bandgap cavities. *IEEE J. Lightwave Technol.*, **25** (1), 440–447. © 2007 IEEE.)

The quality factor, Q, of the resonant mode can be calculated from the time-variation of the field, E_t, as the ratio of the energy stored to the energy lost after one cycle using [40]

$$Q = 2\pi \frac{|E_t|^2}{|E_t|^2 - |E_{t+T}|^2} \qquad (5.137)$$

where T represents the time cycle of the resonant mode. For this structure, the resonant wavelength is $\lambda_{res} = 1.548$ µm and the time cycle is $T = 5.16$ fs, so the quality factor has been found to be $Q = 184$. The mode area, A_{mod}, is also calculated using [41]

$$A_{mod} = \frac{\int (\varepsilon E^* \cdot E)\, ds}{[\varepsilon E^* \cdot E]_{max}} \qquad (5.138)$$

where ε is the permittivity of the dielectric medium in which the cavity is formed, $[\varepsilon E^* \cdot E]_{max}$ represents the maximum energy stored inside the cavity, and the integration is performed over the entire computational window. For this structure, the mode area was found to be $(0.31\lambda)^2$. These results of the resonant wavelength, the quality factor and the mode area, obtained with the newly developed CE-ADI-FDTD code are in excellent agreement with their counterparts reported in [40, 42] and obtained using different finite-element time-domain methods. Furthermore, the CE-ADI-FDTD can easily run on normal desktop computers. For example, it took 15 minutes to run the 5×5 cavity for a mesh of 153 cells in both the x- and y-directions and for 1200 time steps on a PC (Pentium IV, 3 GHz with 1 GB of RAM). In order to clarify this point, it should be mentioned that for a 7×7 photonic crystal cavity the proposed CE-ADI-FDTD was only about twice as fast as the conventional FDTD algorithm, even when the Courant number employed for the former was about 20. One would expect an execution time much faster than that, but it should be noted that the CE-ADI-FDTD algorithm relies on the computation of complex numbers, while the FDTD algorithm is based on real numbers. Thus, the employment of a time step much larger than the maximum fixed by the CFL limit can greatly reduce the total number of time steps necessary to run the simulation; on the other hand, the use of complex numbers increases the execution time of each single time step. For this reason, the total gain in execution time is slightly less than expected.

Figure 5.13 shows the calculation of the quality factor, Q, and the normalised frequency of the resonant mode for different sizes of the square cavity. It has to be mentioned that for all simulations a Courant number equal to 10 has been used, which has been shown to be a perfect compromise between accuracy and execution time.

It can be see that there is good agreement between the results reported here using CE-ADI-FDTD and their counterparts reported in [42, 43] using a finite-element frequency-domain approach. As noted from this figure, for cavities bigger than 5×5, the normalised frequency is almost the same (≈ 0.378), while the Q factor

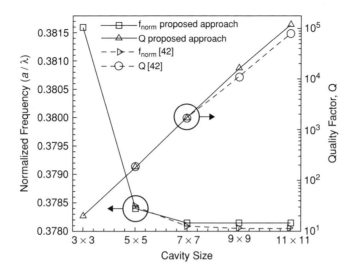

Figure 5.13 Variation of the normalised resonant frequency and quality factor of the resonant mode with the cavity size. (Reproduced with permission from Pinto, D. and Obayya, S.S.A (2007) Improved complex envelope alternative direction implicit finite difference time domain method for photonic bandgap cavities. *IEEE J. Lightwave Technol.*, **25** (1), 440–447. © 2007 IEEE.)

is exponentially increased. Also, the mode area variation with cavity size is shown in Figure 5.14. As may be observed from this figure, the mode area tends to be almost unchanged when the cavity size is bigger than 3×3 rods. This phenomenon can be explained by considering that the electric-field profile of the resonant mode is mostly contained in the defect of the photonic crystal for cavity sizes bigger than 5×5. For this reason, adding external layers of rods does not affect the field distribution inside the cavity.

5.13 Triangular Lattice PhC Cavity

Next, the developed CE-ADI-FDTD code will be tested for different PhC cavities. In the following examples, a hexagonal cavity arrangement, shown in Figure 5.15, is considered. The lattice constant of the PhC is $a = 0.7254$ μm, the radius of the dielectric rods is $r = 0.378\,a$ and the refractive index of the rods is $n_{\mathrm{rods}} = 3.0$ [40].

As first example, a four-ring PhC cavity without a central rod will be investigated. The source used to excite all the structures considered from now on was a Gaussian-shaped pulse expressed by Equation (5.136). The coordinates of the peak of the Gaussian shape x_0 and y_0 were set to the cavity centre coordinates, while the parameters representing the width of the Gaussian shape along the x- and y-directions, X_0 and Y_0, were both fixed to $a/2$, with a the lattice constant of the PhC. The parameters of

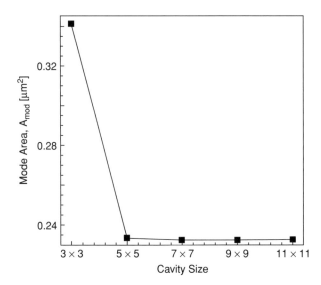

Figure 5.14 Variation of the mode area of the resonant mode with cavity size. (Reproduced with permission from Pinto, D. and Obayya, S.S.A (2007) Improved complex envelope alternative direction implicit finite difference time domain method for photonic bandgap cavities. *IEEE J. Lightwave Technol.*, **25** (1), 440–447. © 2007 IEEE.)

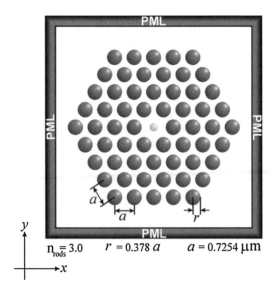

Figure 5.15 Schematic diagram of a four-ring hexagonal photonic crystal cavity [40].

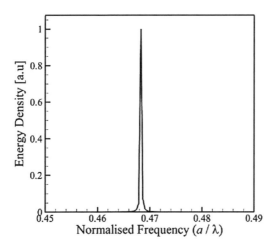

Figure 5.16 Spectral distribution of the resonant mode energy for the four-ring hexagonal photonic crystal cavity. (Reproduced with permission from Pinto, D. and Obayya, S.S.A (2007) Improved complex envelope alternative direction implicit finite difference time domain method for photonic bandgap cavities. *IEEE J. Lightwave Technol.*, **25** (1), 440–447. © 2007 IEEE.)

the Gaussian pulse at time t_0 and T_0 were fixed to 90 fs and 30 fs, respectively, while the angular frequency ω_c was set to 1.256×10^{15} rad/s ($\lambda = 1.5$ μm). A detector was placed in the cavity centre in order to store the time-domain variation of the electric field and, upon using the FFT, the spectral energy density of the resonant mode was computed. The result of this procedure is shown in Figure 5.16.

The narrow spectral width of that resonance curve shows the high selectivity and quality factor of this hexagonal cavity compared to the square case, a property that is also confirmed by the tight confinement of the electric-field profile of the resonant mode in the PhC cavity, shown in Figure 5.17.

This high selectivity property of this type of cavity is confirmed by Figure 5.18, which shows the quality factor and the normalised frequency of the resonant mode for different numbers of rings surrounding the cavity.

From this figure, it can be noted that the normalised frequency of the resonant mode remains nearly unchanged as the number of rings is greater than two. Further, a huge increase in the quality factor of the hexagonal ring cavity can also be observed thanks to the arrangement of a large number of dielectric rods. Figure 5.19 shows the mode area computed for different cavity sizes using Equation (5.138).

As may be noted from this figure, the mode area decreases as the size of the cavity is increased. However the rate of decrease of the mode area becomes smaller as the cavity size increases.

Next, a hexagonal four-ring cavity with a dielectric rod located in the centre of the cavity is considered. The radius of the central rod is $r_{rod} = 0.1a$, where a is the lattice

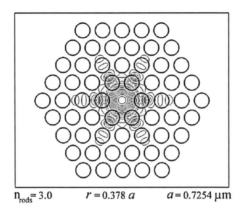

$n_{rods} = 3.0$ $r = 0.378\,a$ $a = 0.7254\,\mu m$

Figure 5.17   Electric-field profile of the resonant mode inside the four-rings triangular lattice PhC cavity.

constant of the PhC, and the refractive index of the rod is varied from 1.2 to 3.4. For each cavity type, the normalised frequency of the resonant mode, and the quality factor were computed. Figure 5.20 shows the variation of the normalised frequency and the quality factor with the refractive index of the central rod, calculated for these cavities.

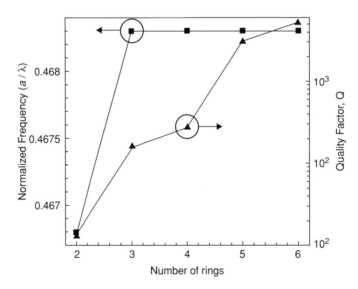

Figure 5.18   Variation of the normalised resonant frequency and quality factor of the resonant mode with the number of rings. (Reproduced with permission from Pinto, D. and Obayya, S.S.A (2007) Improved complex envelope alternative direction implicit finite difference time domain method for photonic bandgap cavities. *IEEE J. Lightwave Technol.*, **25** (1), 440–447. © 2007 IEEE.)

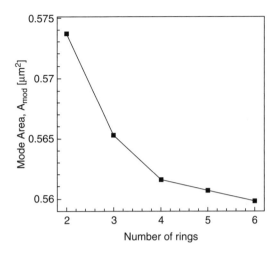

Figure 5.19   Variation of the mode area of the resonant mode with the number of rings. (Reproduced with permission from Pinto, D. and Obayya, S.S.A (2007) Improved complex envelope alternative direction implicit finite difference time domain method for photonic bandgap cavities. *IEEE J. Lightwave Technol.*, **25** (1), 440–447. © 2007 IEEE.)

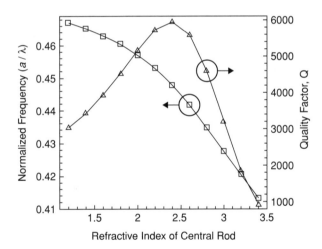

Figure 5.20   Variation of the normalised resonant frequency and quality factor of the resonant mode with the refractive index of the central defect rod. (Reproduced with permission from Pinto, D. and Obayya, S.S.A (2007) Improved complex envelope alternative direction implicit finite difference time domain method for photonic bandgap cavities. *IEEE J. Lightwave Technol.*, **25** (1), 440–447. © 2007 IEEE.)

From this figure, a shift in the normalised frequency of the resonant mode towards lower frequency as the refractive index of the central rod increases can be observed. These results can be intuitively justified as follows. The localised mode obtained by using a lattice defect consisting of removing dielectric material ('air defect') presents a normalised resonance frequency near the 'air band' (upper edge of the bandgap). On the other hand, the localised mode obtained by using lattice defect consisting of adding dielectric material ('dielectric defect') presents a normalised resonance frequency near the 'dielectric band' (lower edge of the bandgap) [44]. This property can be used to tune the resonant mode of the cavity inside the range of the bandgap by properly choosing the refractive index of the central rod, as can be seen from Figure 5.20.

5.14 Wavelength Division Multiplexing

Wavelength division multiplexing (WDM) is a technology used for multiplexing signals in optical fibre. The technology is based on separating the light in the optical fibre into distinctive channels according to the colour of the light, in other words, distinctive wavelength channels. The idea is that every channel transmits the same amount of data as a single fibre that has not been multiplexed.

WDM and frequency division multiplex (FDM) operate on similar principles, where WDM corresponds to wavelengths of light in optical fibre and FDM corresponds to electrical analogue transmission. As opposed to electrical FDM, WDM of optical fibre is highly reliable as it is completely passive.

A WDM system consists of a transmitter and receiver, a multiplexer and a demultiplexer, respectively. The transmitter takes several signals and sends them across a single channel, while the receiver separates these signals into distinctive channels. Ideally, such a system would have a switching device that simultaneously transmits and receives signals; such a device is known as an add-drop multiplexer.

The main advantage of the WDM technique in telecommunication is that it allows the capacity of the network to be increased without the need to change the backbone of the fibre network. This is made possible through implementing WDM and deploying optical amplifiers throughout the optical network. This capacity increase is achieved by upgrading the transmitters and receivers of the network, thereby allowing for many generations of technological advancement in the optical infrastructure without laying more fibre.

The WDM optical spectrum is divided into several distinct wavelengths that do not overlap, and each wavelength corresponds to a single communication channel, thus providing several WDM channels in the same fibre and greatly utilising the fiber's huge bandwidth. With such large bandwidth potential, research on WDM devices has increased with the aim of employing WDM-based optical backbones for the Internet.

The conventional WDM systems were dual-channel 1.31/1.55 µm systems including both the minimum dispersion window and minimum attenuation window. The WDM consists mainly of two types:

- *Coarse WDM* (CWDM), where the wavelengths are spaced well apart. This results in lower costs of optical transmitters and receivers, however this is at the cost of the number of wavelengths, which is relatively small.
- *Dense WDM* (DWDM), where the wavelengths are tightly spaced, providing a large number of wavelengths, but greatly increasing the cost of transmitters and receivers.

DWDM and CWDM are based on the same principle of using multiple wavelengths in a single fibre, differing mainly in the spacing between wavelengths, and the number of channels.

In communication systems, WDM devices show the ability to improve coherence without losing the quality of transmission, are tightly compact (micrometre scale) and practical to fabricate on integrated optical circuits. This is where photonic crystals have much potential, as PhC-based WDM for different wavelength selective-filtering techniques have been recently realised. Such devices include filters adjacent to waveguides, using coupling techniques [45, 46] or cavities [47–52] for the purpose of achieving PhC-based wavelength multiplexing and demultiplexing. Figure 5.21

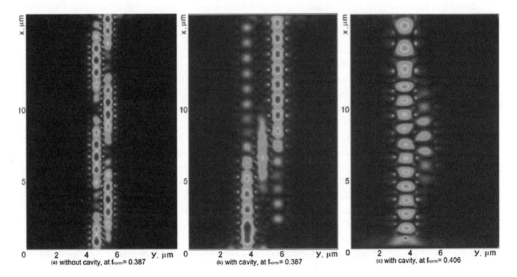

Figure 5.21 (a) Selected frequency ($f_{\text{norm}} = 0.387$) propagates in both waveguides WG1 and WG2. (b) Selected frequency ($f_{\text{norm}} = 0.387$) propagates in only in waveguide 2. (c) Other frequencies, ($f_{\text{norm}} = 0.406$) only propagate in waveguide 1. (Reproduced with permission from IET from 'Improved Design of Photonoic Crystal Based Multiplexer/Demultiplexer Devices', Special Issue of *IET Optelectronics* © IET 2010.)

illustrates an electric field propagating along a PhC waveguide, where both coupling length and cavity techniques have been used carefully to select the widely used communication wavelength $\lambda = 1.55$ μm.

References

[1] Yee, K.S. (1966) Numerical solution of initial boundary value problems involving Maxwell's equations in isotropic media. *IEEE Trans. Antennas Propagat.*, **14** (3), 302–307.

[2] Taflove, A. and Brodwin, M.E. (1975) Numerical solution of steady-state electromagnetic scattering problems using the time-dependent Maxwell's equations. *IEEE Trans. Microwave Theory Tech.*, **23** (8), 623–630.

[3] Holland, R. (1977) Threde: a free-field EMP coupling and scattering code. *IEEE Trans. Nuclear Sci.*, **24** (6), 2416–2421.

[4] Mur, G. (1981) Absorbing boundary conditions for the finite-difference approximation of the time-domain electromagnetic field equations. *IEEE Trans. Electromagn. Compat.*, **23** (4), 377–382.

[5] Choi, D.H. and Hoefer, W.J. (1986) The finite-difference time-domain method and its application to eigenvalue problems. *IEEE Trans. Microwave Theory Tech.*, **34** (12), 1464–1470.

[6] Kashiwa, T. and Fukai, I. (1990) A treatment by FDTD method of dispersive characteristics associated with electronic polarization. *Microwave Opt. Tech. Lett.*, **3** (6), 203–205.

[7] Joseph, R.M., Hagness, S.C. and Taflove, A. (1991) Direct time integration of Maxwell's equations in linear dispersive media with absorption for scattering and propagation of femtosecond electromagnetic pulses. *Opt. Lett.*, **16** (18), 1412–1414.

[8] Goorjian, P.M. and Taflove, A. (1992) Direct time integration of Maxwell's equations in nonlinear dispersive media for propagation and scattering of femtosecond electromagnetic solitons. *Opt. Lett.*, **17** (3), 180–182.

[9] Ziolkowski, R.W. and Judkins, J.B. (1993) Full-wave vector Maxwell's equations modeling of self-focusing of ultra-short optical pulses in a nonlinear Kerr medium exhibiting a finite response time. *J. Optical Soc. America B*, **10** (2), 186–198.

[10] Toland, B., Houshmand, B. and Itoh, T. (1993) Modeling of nonlinear active regions with the FDTD method. *IEEE Microwave Guided Wave Lett.*, **3** (9), 333–335.

[11] Berenger, J.P. (1994) A perfectly matched layer for the absorption of electromagnetic waves. *J. Comput. Phys.*, **114** (2), 185–200.

[12] Sacks, Z.S., Kingsland, D.M., Lee, R. and Lee, J.F. (1995) A perfectly matched anisotropic absorber for use as an absorbing boundary condition. *IEEE Trans. Antennas Propagat.*, **43** (12), 1460–1463.

[13] Gedney, S.D. (1996) An anisotropic perfectly matched layer absorbing media for the truncation of FDTD lattices. *IEEE Trans. Antennas Propagat.*, **44** (12), 1630–1639.

[14] Chang, S.H. and Taflove, A. (2004) Finite-difference time-domain model of lasing action in a four-level two-electron atomic system. *Opt. Express*, **12** (16), 3827–3833.

[15] Taflove, A. and Hagness, S.C. (2005) *Computational Electrodynamics: the Finite-Difference Time-Domain Method*, Artech House.

[16] Schneider, J.B. and Wagner, C.L. (1999) FDTD dispersion revisited: faster-than-light propagation. *IEEE Microw. Guided Wave Lett.*, **9** (2), 54–56.

[17] Zygiridis, T.T. and Tsiboukis, T.D. (2005) Phase error reduction in general FDTD methods via optimum configuration of material parameters. *J. Mater. Process Technol.*, **161** (1–2), 186–192.

[18] Juntunen, J.S. and Tsiboukis, T.D. (2000) Reduction of numerical dispersion in FDTD method through artificial anisotropy. *IEEE Trans. Microw. Theory Tech.*, **48** (4), 582–588.

[19] Suzuki, K., Kashiwa, T. and Hosoya, Y. (2002) Reducing the numerical dispersion in the analysis by modifying anisotropically the speed of light. *Electron. Comm. Jpn.*, **85** (1), 50–58.

[20] Xiao, F., Tang, X. and Wang, L. (2007) An explicit fourth-order accurate FDTD method based on the time staggered Adams-Bashforth time integrator. *Microw. Opt. Technol. Lett.*, **49** (4), 910–912.

[21] Hwang, K.P. and Ihm, J.Y. (2006) A stable fourth-order FDTD method for modeling electrically long dielectric waveguides. *J. Lightwave Technol.*, **24** (2), 1048–1056.

[22] Namiki, T. (2000) 3-D ADI-FDTD method – unconditionally stable time-domain algorithm for solving full vector Maxwell's equations. *IEEE Trans. Microwave Theory Tech.*, **48** (10), 1743–1748.

[23] Namiki, T. and Ito, K. (2000) Investigation of numerical errors of the two-dimensional ADI-FDTD method. *IEEE Trans. Microwave Theory Tech.*, **48** (11), 1950–1956.

[24] Zheng, F. and Chen, Z. (2001) Numerical dispersion analysis of the unconditionally stable 3-D ADI-FDTD method. *IEEE Trans. Microwave Theory Tech.*, **49** (5), 1006–1009.

[25] Zheng, F., Chen, Z. and Zhang, J. (2000) Toward the development of a three-dimensional unconditionally stable finite-difference time-domain method. *IEEE Trans. Microwave Theory Tech.*, **48** (9), 1550–1558.

[26] Zhao, A.N. (2002) Two special notes on the implementation of the unconditionally stable ADI-FDTD method. *Microwave Opt. Technol. Lett.*, **33** (4), 273–277.

[27] García, S.G., Lee, T.W. and Hagness, S.C. (2002) On the accuracy of the ADI-FDTD method. *IEEE Antennas Wireless Propag. Lett.*, **1**, 31–34.

[28] Rao, H., Scarmozzino, R. and Osgood, R.M. (2002) An improved ADI-FDTD method and its application to photonic simulations. *IEEE Photon. Technol. Lett.*, **14** (4), 477–479.

[29] Bayliss, A., Gunzburger, M. and Turkel, E. (1982) Boundary conditions for the numerical solution of elliptic equations in exterior regions. *SIAM J. Applied Math.*, **42** (2), 430–451.

[30] Trefethen, L.N. and Halpern, L. (1986) Well-posedness of one-way wave equations and absorbing boundary conditions. *Math. Comp.*, **47** (176), 421–435.

[31] Higdon, R.L. (1987) Numerical absorbing boundary conditions for the wave equation. *Math. Comp.*, **49** (179), 65–90.

[32] Liao, Z.P., Wong, H.L., Yang, B.P. and Yuan, Y.F. (1984) A transmitting boundary for transient wave analyses. *Scientia Sinica (series A)*, **XXVII** (10), 1063–1076.

[33] Ramahi, O.M. (1998) The concurrent complementary operators method for FDTD mesh truncation. *IEEE Trans. Antennas Propagat.*, **46** (10), 1475–1482.

[34] Zhao, L. and Cangellaris, A.C. (1996) A general approach for the development of unsplit-field time-domain implementations of perfectly matched layers for FDTD grid truncation. *IEEE Microwave Guided Wave Lett.*, **6** (5), 209–211.

[35] Sullivan, D.M. (1997) An unsplit step 3-D PML for use with the FDTD method. *IEEE Microwave Guided Wave Lett.*, **7** (7), 184–186.

[36] Katz, D.S., Thiele, E.T. and Taflove, A. (1994) Validation and extension to three dimensional of the berenger PML absorbing boundary condition for FDTD meshes. *IEEE Microwave Guided Wave Lett.*, **4** (4), 268–270.

[37] Pérez, I.V., García, S.G., Martín, R.G. and Olmedo, B.G. (1998) Generalization of Berenger's absorbing boundary conditions for 3-D magnetic and dielectric anisotropic media. *Microwave Opt. Technol. Lett.*, **18** (2), 126–130.

[38] Ramadan, O. (2008) Accuracy improved multi-stage ADI-PML algorithm for dispersive open region FDTD problems. *Electron. Lett.*, **44** (20), 1168–1170.

[39] Wang, S. and Teixeira, F.L. (2003) An efficient PML implementation for the ADI-FDTD method. *IEEE Microwave Wireless Comp. Lett.*, **13** (2), 72–74.

[40] Rodriguez-Esquerre, V.F., Koshiba, M. and Figueroa, H. (2004) Finite-element time-domain analysis of 2-D photonic crystal resonant cavities. *IEEE Photon. Technol. Lett.*, **16**, 816–818.

[41] Watts, M.R., Johnson, S.G., Haus, H.A. and Joannopoulos, J.D. (2002) Electromagnetic cavity with arbitrary Q and small modal volume without a complete photonic bandgap. *Opt. Lett.*, **27** (20), 1785–1787.

[42] Obayya, S.S.A. (2005) Finite element time domain solution of resonant modes in photonic bandgap cavities. *Opt. Quant. Electr.*, **37**, 865–873.

[43] Hwang, J.K., Hyun, S.B., Ryu, H.Y. and Lee, Y.H. (1998) Resonant modes of two-dimensional photonic bandgap cavities determined by the finite element method and by use of the anisotropic perfectly matched layer boundary condition. *J. Opt. Soc. Amer. B*, **15** (8), 2316–2324.

[44] Joannopoulos, J.D., Meade, R.D. and Winn, J.N. (1995) *Photonic Crystals: Molding the Flow of Light*, Princeton University Press, Princeton, NJ.

[45] Zimmermann, J., Kamp, M., Forchel, A. and Marz, R. (2004) Photonic crystal waveguide directional couplers as wavelength selective optical filters. *Opt. Commun.*, **230**, 387–392.

[46] Chien, F.S., Hsu, Y., Hsieh, W. and Cheng, S. (2004) Dual wavelength demultiplexing by coupling and decoupling of photonic crystal waveguides. *Opt. Express*, **12**, 1119–1125.

[47] Centeno, E., Guizal, B. and Felbacq, D. (1999) Multiplexing & demultiplexing with photonic crystal. *J. Opt. A: Pure Appl. Opt.*, **1**, L10–L13.

[48] Jin, C., Fan, S., Han, S. and Zhang, D. (2003) Reflectionless multichannel wavelength demultiplexer in a transmission resonator configuration. *IEEE J. Quantum Electron.*, **39**, 160–165.

[49] Kim, S., Park, I., Lim, H. and Kee, C.-S. (2004) Highly efficient photonic crystal-based multichannel drop filters of three-port system with reflection feedback. *Opt. Express.*, **12**, 5518–5525.

[50] Sharkawy, A., Shi, S. and Prather, D.W. (2001) Multichannel wavelength division multiplexing with photonic crystals. *Appl. Opt.*, **40**, 2247–2252.

[51] Koshiba, M. (2001) Wavelength division multiplexing and demultiplexing with photonic crystal waveguide couplers. *J. Lightwave Technol.*, **19**, 1970–1975.

[52] Pustai, D., Sharkawy, A., Shouyuan, S. and Prather, D.W. (2002) Tunable photonic crystal microcavities. *Appl. Opt.*, **41**, 5574–5579.

6

Finite-Volume Time-Domain Method

6.1 Introduction

In this chapter a detailed description of the finite-volume time-domain (FVTD) numerical modelling technique will be presented. The FVTD method has attracted a great deal of attention in numerical modelling, but it has been mainly used for computational fluid dynamics applications and very limited research efforts have been directed towards the use of this technique in computational electromagnetic problems. The beauty of FVTD is that it combines the versatile and flexible meshing capabilities of the finite-element time domain (FETD) method, in addition to being explicit (no solution of a large system of equations is required), where only field updates are performed at each time step, as in the finite-difference time-domain (FDTD) method. For this reason, a numerically efficient FVTD formulation based on the nondiffusive scheme for the calculation of the flux interaction will be suggested for the analysis of optical wave propagation in photonic bandgap (PBG) devices. The uniaxial perfectly matched layer (UPML) absorbing boundary condition will be rigorously incorporated into the FVTD formulation to mimic 'reflectionless' boundaries of the computational domain. In the last part of the chapter, a brief description of nonlinear optical phenomena will be given with a detailed analysis of the inclusion of nonlinear modelling in the FVTD and the FDTD methods.

Computational Photonics   Salah Obayya
© 2011 John Wiley & Sons, Ltd

6.2 Numerical Analysis

The analysis starts from the coupled Maxwell's equations for an isotropic, lossless medium without electric or magnetic sources

$$\varepsilon_0\varepsilon_r\frac{\partial \bar{E}}{\partial t} - \nabla \times \bar{H} = 0 \tag{6.1}$$

$$\mu_0\mu_r\frac{\partial \vec{H}}{\partial t} + \nabla \times \vec{E} = 0 \tag{6.2}$$

where ε_0 is the electric permittivity of free space, ε_r is the relative permittivity of the medium considered, μ_0 is the magnetic permeability of the free space, μ_r is the relative permeability of the medium considered, $\bar{E} = \begin{bmatrix} E_x & E_y & E_z \end{bmatrix}^T$ is the electric field, $\bar{H} = \begin{bmatrix} H_x & H_y & H_z \end{bmatrix}^T$ is the magnetic field and T is the matrix transpose operation. Equations (6.1) and (6.2) represent a system of six equations expressed as

$$\varepsilon_0\varepsilon_r\frac{\partial E_x}{\partial t} - \frac{\partial H_z}{\partial y} + \frac{\partial H_y}{\partial z} = 0 \tag{6.3}$$

$$\varepsilon_0\varepsilon_r\frac{\partial E_y}{\partial t} - \frac{\partial H_x}{\partial z} + \frac{\partial H_z}{\partial x} = 0 \tag{6.4}$$

$$\varepsilon_0\varepsilon_r\frac{\partial E_z}{\partial t} - \frac{\partial H_y}{\partial x} + \frac{\partial H_x}{\partial y} = 0 \tag{6.5}$$

$$\mu_0\mu_r\frac{\partial H_x}{\partial t} + \frac{\partial E_z}{\partial y} - \frac{\partial E_y}{\partial z} = 0 \tag{6.6}$$

$$\mu_0\mu_r\frac{\partial H_y}{\partial t} + \frac{\partial E_x}{\partial z} - \frac{\partial E_z}{\partial x} = 0 \tag{6.7}$$

$$\mu_0\mu_r\frac{\partial H_z}{\partial t} + \frac{\partial E_y}{\partial x} - \frac{\partial E_x}{\partial y} = 0 \tag{6.8}$$

Equations (6.3)–(6.8) can be rewritten in conservative form as [1]

$$\alpha\frac{\partial \bar{U}}{\partial t} + \frac{\partial F_1(\bar{U})}{\partial x} + \frac{\partial F_2(\bar{U})}{\partial y} + \frac{\partial F_3(\bar{U})}{\partial z} = 0 \tag{6.9}$$

where, $\alpha = \begin{bmatrix} \varepsilon I & 0 \\ 0 & \mu I \end{bmatrix}$ with I the 3×3 identity matrix, and $\bar{U}$, $F_1\left(\bar{U}\right)$, $F_2\left(\bar{U}\right)$, $F_3\left(\bar{U}\right)$ are written as

$$
\bar{U} = \begin{bmatrix} E_x \\ E_y \\ E_z \\ H_x \\ H_y \\ H_z \end{bmatrix}, \quad
F_1\left(\bar{U}\right) = \begin{bmatrix} 0 \\ H_z \\ -H_y \\ 0 \\ -E_z \\ E_y \end{bmatrix}, \quad
F_2\left(\bar{U}\right) = \begin{bmatrix} -H_z \\ 0 \\ H_x \\ E_z \\ 0 \\ -E_x \end{bmatrix}, \quad
F_3\left(\bar{U}\right) = \begin{bmatrix} H_y \\ -H_x \\ 0 \\ -E_y \\ E_x \\ 0 \end{bmatrix}
$$

Equation (6.9) can be rewritten in condensed form as

$$
\alpha \frac{\partial \bar{U}}{\partial t} + div F\left(\bar{U}\right) = 0 \tag{6.10}
$$

where $F\left(\bar{U}\right) = \left[\, F_1\left(\bar{U}\right)\; F_2\left(\bar{U}\right)\; F_3\left(\bar{U}\right)\,\right]$. This notation is fundamental for the derivation of the FVTD scheme. Considering a generic volume, V, Equation (6.10) is then integrated over this volume obtaining

$$
\int_V \alpha \frac{\partial \bar{U}}{\partial t} dV + \int_V div F\left(\bar{U}\right) dV = 0 \tag{6.11}
$$

Through the use of the divergence theorem, Equation (6.11) can be rewritten as

$$
\int_V \alpha \frac{\partial \bar{U}}{\partial t} dV + \int_S F\left(\bar{U}^*\right) \cdot \vec{a}_n dS = 0 \tag{6.12}
$$

where S represents the surface enclosing the volume V, $\vec{a}_n$ is the outward pointing normal unit vector of the surface S, and $\bar{U}^* = \left[E_x^* \;\; E_y^* \;\; E_z^* \;\; H_x^* \;\; H_y^* \;\; H_z^* \right]^{\mathrm{T}}$ denotes the electromagnetic field components at the surface S. In order to make possible the integration of Equation (6.12) into practical applications, the entire volume V is partitioned into small volumes V_i. In the literature there are two main formulations used for the partitioning of a domain when a triangulation is given: the cell-centred formulation and the cell-vertex formulation [1, 2]. In the former, the electromagnetic field components are collocated at the barycentres of the cells, while in the latter the electromagnetic field components are placed at the nodes of the cells. Figure 6.1 shows an example of a 2D cell-centred formulation. From this figure, it can be clearly seen that the single volume V_i corresponds to the single cell (a triangle in this case) of the triangulation.

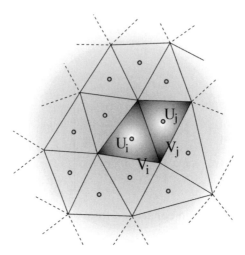

Figure 6.1 Finite-volume partition for a 2D computational domain using a cell-centred formulation.

In Figure 6.2, an example of a cell-vertex formulation for a 2D computational domain is shown. From this figure it can be clearly seen that the single volume V_i corresponds to polygons obtained by connecting the barycentres of the cells to which the node belongs. Each of these formulations used to partition the computational domain has advantages and disadvantages. As an example, the boundary conditions in the cell-centred formulation are taken into account in a more straightforward

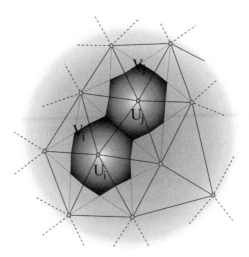

Figure 6.2 Finite-volume partition for a 2D computational domain using a cell-vertex formulation.

manner because the boundary corresponds to a face of the volume into which the computational domain is partitioned, while for the cell-vertex formulation particular faces have to be defined. On the other hand, in the cell-vertex formulation the extension of the FVTD scheme to a higher order is done in a much simpler way than for the cell-centred formulation.

Once the computational domain is partitioned in volumes V_i, which possess a boundary surface S_i consisting of a number of planar faces, Equation (6.12) is applied to each of these volumes in a discretised form expressed as

$$\alpha \, |V_i| \, \frac{\partial \bar{U}_i}{\partial t} = \sum_{k=1}^{m_i} |S_k| \, F\left(\bar{U}_k^*\right) \cdot \bar{n}_{ik} \qquad (6.13)$$

where $|V_i|$ is the volume of the ith volume, $|S_k|$ is the area of the kth planar surface that surrounds the ith element, m_i is the number of planar surfaces surrounding the volume V_i, $\bar{U}_i$ is the value of $\bar{U}$ at the centre of the ith element, $\bar{U}_k^*$ is the value of $\bar{U}$ at the centre of the kth planar surface, $\bar{n}_{ik}$ represents the unit normal of the kth planar surface of the ith volume, and $F\left(\bar{U}_k^*\right) \cdot \bar{n}_{ik}$ represents the flux at the centre of the kth planar surface. Figure 6.3 showns a schematic of the flux interactions for an elementary volume in a 2D cell-centred formulation.

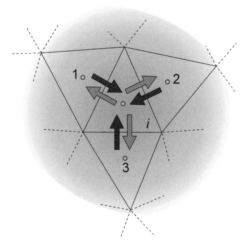

Figure 6.3 Flux interaction for a volume, V_i, in a 2D computational domain discretised with a cell-centred formulation. (Reproduced with permission from Pinto, D. and Obayya, S.S.A. (2008) Nonlinear finite volume time domain analysis of photonic crystal based resonant cavities. *IET Optoelectron.*, **2** (6), 254–261. © 2008 IET.)

6.3 The UPWIND Scheme for Flux Calculation

Different solutions have been proposed in the literature in order to evaluate the flux interaction for each volume into which the computational domain is discretised. One of the most employed schemes is the UPWIND scheme [1]. In this scheme, a relationship between the electromagnetic-field components at the centre of the element and the electromagnetic-field components at the centre of the corresponding boundary planar surface is formulated. In order to do so, the set of unit normal vectors of a single volume V_i is considered in order to parameterise R^3, so as to express each point of R^3 as a linear combination of this set. In formulas this is expressed as

$$Rm_i \ni \left(\xi_1, \cdots, \xi_{m_i}\right) \mapsto x = \sum_{k=1}^{m_i} \xi_k \bar{n}_{ik} \in R^3$$

In this way Equation (6.9) can be expressed as

$$\frac{\partial \bar{U}}{\partial t} = -\alpha^{-1} \sum_{k=1}^{m_i} \left(\frac{\partial F_1}{\partial \bar{U}} \frac{\partial \xi_k}{\partial x} + \frac{\partial F_2}{\partial \bar{U}} \frac{\partial \xi_k}{\partial y} + \frac{\partial F_3}{\partial \bar{U}} \frac{\partial \xi_k}{\partial z}\right) \frac{\partial \bar{U}}{\partial \xi_k} \qquad (6.14)$$

Equation (6.14) can be rewritten in a more compact form as

$$\frac{\partial \bar{U}}{\partial t} = -\alpha^{-1} \sum_{k=1}^{m_i} A\left(\bar{n}_{ik}\right) \frac{\partial \bar{U}}{\partial n_k} \qquad (6.15)$$

where the 6×6 matrix $A\left(\bar{n}_{ik}\right)$ is defined as

$$A\left(\bar{n}_{ik}\right) = \frac{\partial F_1}{\partial \bar{U}} \frac{\partial \xi_k}{\partial x} + \frac{\partial F_2}{\partial \bar{U}} \frac{\partial \xi_k}{\partial y} + \frac{\partial F_3}{\partial \bar{U}} \frac{\partial \xi_k}{\partial z} \qquad (6.16)$$

In this way the matrix $A\left(\bar{n}_i\right)$ is given by

$$A\left(\bar{n}_i\right) = \begin{bmatrix} 0 & 0 & 0 & 0 & n_z & -n_y \\ 0 & 0 & 0 & -n_z & 0 & n_x \\ 0 & 0 & 0 & n_y & -n_x & 0 \\ 0 & -n_z & n_y & 0 & 0 & 0 \\ n_z & 0 & -n_x & 0 & 0 & 0 \\ -n_y & n_x & 0 & 0 & 0 & 0 \end{bmatrix} \qquad (6.17)$$

It can be clearly seen that the product $A\left(\bar{n}_i\right) \bar{U}$ defines, in this case, the flux interaction of the face S_i associated with the unit normal $\bar{n}_i = \left(n_x\ n_y\ n_z\right)$. Introducing the notation $\tilde{A}\left(\bar{n}_i\right) = \alpha^{-1} A\left(\bar{n}_i\right)$, for each matrix $\tilde{A}\left(\bar{n}_i\right)$ it is possible to calculate the six real eigenvalues given by

$$\left(0 \quad 0 \quad \frac{1}{\sqrt{\varepsilon\mu}} \quad \frac{1}{\sqrt{\varepsilon\mu}} \quad -\frac{1}{\sqrt{\varepsilon\mu}} \quad -\frac{1}{\sqrt{\varepsilon\mu}}\right) \tag{6.18}$$

The matrix $\tilde{A}(\bar{n}_i)$ can be written in the form $\tilde{A} = P \Lambda P^{-1}$ where

$$\Lambda = \begin{bmatrix} 0 & 0 & 0 & 0 & 0 & 0 \\ 0 & 0 & 0 & 0 & 0 & 0 \\ 0 & 0 & \dfrac{1}{\sqrt{\varepsilon\mu}} & 0 & 0 & 0 \\ 0 & 0 & 0 & \dfrac{1}{\sqrt{\varepsilon\mu}} & 0 & 0 \\ 0 & 0 & 0 & 0 & -\dfrac{1}{\sqrt{\varepsilon\mu}} & 0 \\ 0 & 0 & 0 & 0 & 0 & -\dfrac{1}{\sqrt{\varepsilon\mu}} \end{bmatrix} \tag{6.19}$$

and

$$P = \begin{bmatrix} n_x & 0 & \dfrac{n_x n_z}{c\varepsilon} & -\dfrac{n_x n_y}{c\varepsilon} & -\dfrac{n_x n_z}{c\varepsilon} & \dfrac{n_x n_y}{c\varepsilon} \\ n_y & 0 & \dfrac{n_y n_z}{c\varepsilon} & \dfrac{n_x^2 + n_z^2}{c\varepsilon} & -\dfrac{n_y n_z}{c\varepsilon} & -\dfrac{n_x^2 + n_z^2}{c\varepsilon} \\ n_z & 0 & -\dfrac{n_x^2 + n_y^2}{c\varepsilon} & -\dfrac{n_y n_z}{c\varepsilon} & \dfrac{n_x^2 + n_y^2}{c\varepsilon} & \dfrac{n_y n_z}{c\varepsilon} \\ 0 & n_x & -n_y & -n_z & -n_y & -n_z \\ 0 & n_y & n_x & 0 & n_x & 0 \\ 0 & n_z & 0 & n_x & 0 & n_x \end{bmatrix} \tag{6.20}$$

where $c = 1/\sqrt{\varepsilon\mu}$ is the speed of light in the medium.

From Equation (6.15), considering the variation of the vector $\bar{U}$ along the direction of $\bar{n}_i$ gives

$$\frac{\partial \bar{U}}{\partial t} = -\tilde{A}(\bar{n}_i)\frac{\partial \bar{U}}{\partial n} \tag{6.21}$$

With the help of the decomposition used for $\tilde{A}(\bar{n}_i)$ and introducing the vector $\bar{V} = P^{-1}\bar{U}$, Equation (6.21) can be written as a system of six independent scalar equations

$$\frac{\partial V_j}{\partial t} + \lambda_i \frac{\partial V_j}{\partial n} = 0 \tag{6.22}$$

where $j = 1, \ldots, 6$. The solutions for each of these equations are any differentiable functions of the type $f\left(\lambda_j t - \xi n\right)$. This means that the value of V_j at $\xi = \xi^*$, the boundary face of the cell, along the direction $\bar{n}_i$ at time t is equal at the value of V_j at $\xi = \xi_0$, the centre of the cell, at time $t - s/\lambda_j$, where $s = |\xi^* - \xi_0|$. On the other hand, if the cell size is very small compared to the shortest wavelength involved, the propagation time needed to cover the distance s can be neglected with very good approximation. This estimation implies

$$V_j(t) = V_j^*(t) \tag{6.23}$$

This relationship holds for any direction of propagation and hence for all unit normals. For stability reasons [1, 3], the only waves considered in the following analysis are those propagating in the positive direction of $\bar{n}_i$ which implies that only positive eigenvalues need to be considered. In fact the matrix $\tilde{A}(\bar{n}_i)$ can be rewritten as

$$\tilde{A}(\bar{n}_i) = P\Lambda P^{-1} = P\left(\Lambda^+ + \Lambda^-\right)P^{-1} = P\Lambda^+ P^{-1} + P\Lambda^- P^{-1}$$
$$= \tilde{A}(\bar{n}_i)^+ + \tilde{A}(\bar{n}_i)^- \tag{6.24}$$

where Λ^+ is the diagonal matrix of positive eigenvalues of $\tilde{A}(\bar{n}_i)$, and Λ^- is the diagonal matrix of the negative eigenvalues. With this notation, the matrix $\tilde{A}(\bar{n}_i)^+$ is given by

$$\tilde{A}(\bar{n}_i)^+ = \frac{1}{2}$$
$$\begin{bmatrix}
\left(n_y^2 + n_z^2\right)c & -n_x n_y c & -n_x n_z c & 0 & n_z/\varepsilon & -n_y/\varepsilon \\
-n_x n_y c & \left(n_x^2 + n_z^2\right)c & -n_y n_z c & -n_z/\varepsilon & 0 & n_x/\varepsilon \\
-n_x n_z c & -n_y n_z c & \left(n_x^2 + n_y^2\right)c & n_y/\varepsilon & -n_x/\varepsilon & 0 \\
0 & -n_z/\mu & n_y/\mu & \left(n_y^2 + n_z^2\right)c & -n_x n_y c & -n_x n_z c \\
n_z/\mu & 0 & -n_x/\mu & -n_x n_y c & \left(n_x^2 + n_z^2\right)c & -n_y n_z c \\
-n_y/\mu & n_x/\mu & 0 & -n_x n_z c & -n_y n_z c & \left(n_x^2 + n_y^2\right)c
\end{bmatrix} \tag{6.25}$$

It is possible to note that by substituting c with $-c$, an explicit expression for $\tilde{A}(\bar{n}_i)^-$ can be obtained and, furthermore, the following relationship is obtained

$$\tilde{A}(\bar{n}_i)^- = -\tilde{A}(-\bar{n}_i)^+ \tag{6.26}$$

Finally, considering Equation (6.23), noticing that $\bar{V} = P^{-1}\bar{U}$ and multiplying both sides of Equation (6.23) by $P\Lambda^+$, the relationship between the values of the electromagnetic field components at the centre of the cell, $\bar{U}$, and the corresponding values at the surface S_i, $\bar{U}^*$ is obtained

$$\tilde{A}(\bar{n}_i)^+ \bar{U} = \tilde{A}(\bar{n}_i)^+ \bar{U}^* \tag{6.27}$$

Equation (6.27) can be expressed in terms of electromagnetic-field components, obtaining

$$Y\bar{n}_i \times \bar{E} - \bar{n}_i \times \left(\bar{n}_i \times \bar{H}\right) = Y\bar{n}_i \times \bar{E}^* - \bar{n}_i \times \left(\bar{n}_i \times \bar{H}^*\right) \qquad (6.28)$$

where $Y = \sqrt{\varepsilon/\mu}$. Equation (6.28) has to be applied at both of volume elements whose common interface is the surface S_i. In this way, considering volume 1 on the left side and volume 2 on the right side, the following equations are obtained

$$Y^1\bar{n}_i \times \bar{E}^* - \bar{n}_i \times \left(\bar{n}_i \times \bar{H}^*\right) = Y^1\bar{n}_i \times \bar{E}^1 - \bar{n}_i \times \left(\bar{n}_i \times \bar{H}^1\right) \qquad (6.29)$$

$$Y^2\bar{n}_i \times \bar{E}^{**} + \bar{n}_i \times \left(\bar{n}_i \times \bar{H}^{**}\right) = Y^2\bar{n}_i \times \bar{E}^2 + \bar{n}_i \times \left(\bar{n}_i \times \bar{H}^2\right) \qquad (6.30)$$

where Y^1 and Y^2 are the characteristic admittances of the left and right volumes, respectively, $\bar{E}^1$ and $\bar{H}^1$ are the electric and magnetic fields at the centre of the left volume, respectively, $\bar{E}^2$ and $\bar{H}^2$ are the electric and magnetic fields at the centre of the right volume, respectively, $\bar{E}^*$ and $\bar{H}^*$ are the electric and magnetic fields at the boundary surface of the left volume, respectively, and $\bar{E}^{**}$ and $\bar{H}^{**}$ are the electric and magnetic fields at the boundary surface of the right volume, respectively. Equations (6.29)–(6.30), in conjunction with appropriate continuity conditions can be used for the calculation of the fluxes necessary for the updating process of the FVTD algorithm.

6.3.1 Dielectric Contrast

For a dielectric contrast, Equations (6.29)–(6.30) are used with the continuity relationships of the tangential components of the electric and magnetic fields expressed as

$$\bar{n}_i \times \bar{E}^* = \bar{n}_i \times \bar{E}^{**} \qquad (6.31)$$

$$\bar{n}_i \times \bar{H}^* = \bar{n}_i \times \bar{H}^{**} \qquad (6.32)$$

Considering Equations (6.31)–(6.32) together with Equations (6.29)–(6.30) the following equations are obtained

$$\bar{n}_i \times \bar{E}^* = \bar{n}_i \times \bar{E}^{**} = \frac{1}{Y^1 + Y^2}\left(\bar{n}_i \times \left(Y^1\bar{E}^1 + Y^2\bar{E}^2\right) - \bar{n}_i \times \left(\bar{n}_i \times \left(\bar{H}^1 - \bar{H}^2\right)\right)\right)$$
$$(6.33)$$

$$\bar{n}_i \times \bar{H}^* = \bar{n}_i \times \bar{H}^{**} = \frac{1}{Z^1 + Z^2}\left(\bar{n}_i \times \left(Z^2\bar{H}^2 + Z^1\bar{H}^1\right) - \bar{n}_i \times \left(\bar{n}_i \times \left(\bar{E}^2 - \bar{E}^1\right)\right)\right)$$
$$(6.34)$$

where Y and Z are the characteristic admittance and the characteristic impedance of the volume, respectively, and the superscripts 1 and 2 refer to the left and the right volumes, respectively. Taking into account that

$$F\left(\bar{U}\right)\cdot\bar{n}_i = \begin{bmatrix} -\bar{n}_i \times \bar{H} \\ \bar{n}_i \times \bar{E} \end{bmatrix} \tag{6.35}$$

Equations (6.33)–(6.34) can be rewritten using the flux splitting formalism as

$$F\left(\bar{U}^*\right)\cdot\bar{n}_i = F\left(\bar{U}^{**}\right)\cdot\bar{n}_i = \alpha^1 T^1 \tilde{A}^1 \left(\bar{n}_i\right)^+ \bar{U}^1 + \alpha^2 T^2 \tilde{A}^2 \left(\bar{n}_i\right)^- \bar{U}^2 \tag{6.36}$$

where T is a 6×6 transmission matrix expressed as

$$T^{1,2} = \begin{bmatrix} \dfrac{2Z^{1,2}}{Z^1 + Z^2} & 0 \\ 0 & \dfrac{2Y^{1,2}}{Y^1 + Y^2} \end{bmatrix} \tag{6.37}$$

It can be noted that Equation (6.36) represents a relationship between the fluxes and the electromagnetic-field components at the centre of the volumes. Taking this into account Equation (6.13) can be finally expressed as

$$\alpha\frac{\partial \bar{U}_i}{\partial t} = -\frac{1}{|V_i|}\sum_{k=1}^{m_i} |S_k|\left(\alpha^1 T^1 \tilde{A}\left(\bar{n}_{ik}\right)^+ \bar{U}_k^1 + \alpha^2 T^2 \tilde{A}\left(\bar{n}_{ik}\right)^- \bar{U}_k^2\right) \tag{6.38}$$

6.3.2   Perfect Electric Conductor

In this case Equation (6.28) is used in conjunction with the condition

$$\bar{n}_i \times \bar{E}^* = 0 \tag{6.39}$$

which gives

$$\bar{n}_i \times \bar{H}^* = Y\bar{n}_i \times \left(\bar{n}_i \times \bar{E}\right) + \bar{n}_i \times \bar{H} \tag{6.40}$$

From Equation (6.40) is then possible to obtain the relation between the fluxes and the electromagnetic-field components at the centre of the volume

$$F\left(\bar{U}\right)\cdot\bar{n}_i = \alpha 1 T_{\text{PEC}} 1 \tilde{A}\left(\bar{n}_i\right) + \bar{U}1$$

where T_{PEC}^1 is the 6×6 transmission matrix obtained from T^I considering that the impedance of a perfect electric conductor tends to zero, or in formula

$$T_{\text{PEC}}^1 = \lim_{Z^1 \to 0} T^1 \tag{6.41}$$

The 6×6 transmission matrix T_{PEC}^1 is explicitly expressed as

$$T_{\text{PEC}}^1 = \begin{bmatrix} 2 & 0 & 0 & 0 & 0 & 0 \\ 0 & 2 & 0 & 0 & 0 & 0 \\ 0 & 0 & 2 & 0 & 0 & 0 \\ 0 & 0 & 0 & 0 & 0 & 0 \\ 0 & 0 & 0 & 0 & 0 & 0 \\ 0 & 0 & 0 & 0 & 0 & 0 \end{bmatrix} \tag{6.42}$$

6.3.3 Perfect Magnetic Conductor

In this case Equation (6.28) is used in conjunction with the condition

$$\bar{n}_i \times \bar{H}^* = 0 \tag{6.43}$$

which gives

$$\bar{n}_i \times \bar{E}^* = Z\bar{n}_i \times \left(\bar{n}_i \times \bar{H} \right) + \bar{n}_i \times \bar{E} \tag{6.44}$$

From Equation (6.44) is then possible to obtain the relation between the fluxes and the electromagnetic field components at the centre of the volume

$$F \left(\bar{U} \right) \cdot \bar{n}_i = \alpha 1 T_{\text{PMC}} 1 \tilde{A} \left(\bar{n}_i \right) + \bar{U} 1$$

where T_{PMC}^1 is the 6×6 transmission matrix obtained from T^I considering that the admittance of a perfect magnetic conductor tends to zero, or in formula

$$T_{\text{PMC}}^1 = \lim_{Y^1 \to 0} T^1 \tag{6.45}$$

The 6×6 transmission matrix T_{PMC}^1 is explicitly expressed as

$$T_{PMC}^1 = \begin{bmatrix} 0 & 0 & 0 & 0 & 0 & 0 \\ 0 & 0 & 0 & 0 & 0 & 0 \\ 0 & 0 & 0 & 0 & 0 & 0 \\ 0 & 0 & 0 & 2 & 0 & 0 \\ 0 & 0 & 0 & 0 & 2 & 0 \\ 0 & 0 & 0 & 0 & 0 & 2 \end{bmatrix} \tag{6.46}$$

6.3.4   Numerical Stability

As with every explicit numerical scheme, the FVTD method is not an unconditionally stable numerical method. Particular attention needs to be paid to the choice of the time step, Δt, in order to ensure the stability of the numerical scheme. In [1] a thorough stability analysis of the FVTD method is reported for structured and unstructured schemes. This analysis is based on the Von Neumann method [4] with which the evolution of the numerical error of the scheme is studied over time. The aim of this study is to find the appropriate values of the parameters of the scheme in order to obtain a decreasing behaviour of the error or, at least, to maintain it at a constant value. Essentially, the stability of the numerical scheme is guaranteed if the following relationship is verified

$$\frac{\left\| \varepsilon^{n+1} \right\|}{\left\| \varepsilon^{n} \right\|} \leq 1 \tag{6.47}$$

where ε indicates the error. For a 3D structured grid with discretisation steps $\Delta x = \Delta y = \Delta z = \Delta$, the application of the Von Neumann method leads to the following relationship for the maximum time step possible for the numerical scheme to be stable

$$dt \leq \frac{\Delta}{2c} \tag{6.48}$$

where c is the speed of light in the medium.

For a 3D unstructured grid, the Von Neumann method is applied on an explicit scheme of spatial order 1 [1], leading to the following relationship between the time step dt and the cell size of the discretised computational domain

$$dt \leq \frac{1}{c} \min_{i} \left(\frac{|V_i|}{\sum_{k=1}^{m_i} |S_k|} \right) \tag{6.49}$$

where V_i is the volume of the elementary cell, S_k is the area of the kth planar boundary face of the volume V_i, m_i is the number of planar faces surrounding the volume V_i and c is the speed of light in the medium. The verification of Equation (6.48) for FVTD schemes developed on structured grids, or Equation (6.49) for FVTD schemes developed for unstructured grid is sufficient to guarantee the stability of the numerical scheme for all time steps.

6.4   Nondiffusive Scheme for the Flux Calculation

The upwind scheme described in the previous section has been shown to suffer from numerical diffusion [5], which is necessary in the context of computational

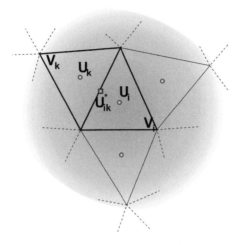

Figure 6.4 Schematic diagram of a 2D computational domain with the position of the electromagnetic-field components on the elementary volumes.

fluid dynamics in order to avoid the exponential growth of nonphysical oscillations artificially introduced by the numerical scheme. On the other hand, in the context of computational electromagnetics, this numerical diffusion creates a nonphysical dissipation of the electromagnetic energy inside the computational domain, which can lead to inaccurate results for simulations requiring a running time equivalent to several cycles of the main frequency signal. In [5], a different scheme for the calculation of the flux interaction between cells in the FVTD scheme has been proposed for the simulation of electromagnetic problems. This scheme doesn't suffer from numerical diffusion, making it suitable for the simulation of electromagnetic structures and problems involving highly resonant structures. The core of this scheme consists in a different approach in the calculation of the electromagnetic-field components at the centre of the boundary faces surrounding the elementary cell necessary for the flux calculation, as shown in Figure 6.4 for a 2D computational domain.

The electromagnetic-field component $\bar{U}_{ik}^*$ at the centre of the segment interface between the elements i and j is calculated using a linear approximation

$$\bar{U}_{ik}^* = \frac{\bar{U}_i + \bar{U}_k}{2} \tag{6.50}$$

In this way, applying this scheme to Equation (6.13), the following equations are obtained

$$\varepsilon_i \, |V_i| \, \frac{\bar{E}_i^{n+1} - \bar{E}_i^n}{\Delta t} + \sum_{k=1}^{m_i} \bar{F}_{ik}^{n+1/2} = 0 \tag{6.51}$$

$$\mu_i \, |V_i| \, \frac{\bar{H}_i^{n+3/2} - \bar{H}_i^{n+1/2}}{\Delta t} - \sum_{k=1}^{m_i} \bar{G}_{ik}^{n+1} = 0 \tag{6.52}$$

with

$$\bar{F}_{ik}^{n+1/2} = \bar{N}_{ik}\frac{\bar{H}_i^{n+1/2} + \bar{H}_k^{n+1/2}}{2} \tag{6.53}$$

$$\bar{G}_{ik}^{n+1} = \bar{N}_{ik}\frac{\bar{E}_i^{n+1} + \bar{E}_k^{n+1}}{2} \tag{6.54}$$

where

$$\bar{N}_{ik} = |S_k|\,\bar{N}_{\bar{n}_{ik}} \tag{6.55}$$

with

$$\bar{N}_{\bar{n}_{ik}} = n_{ikx}N_x + n_{iky}N_y + n_{ikz}N_z \tag{6.56}$$

where $\vec{n}_{ik} = \begin{pmatrix} n_{ikx} & n_{iky} & n_{ikz} \end{pmatrix}^{\mathrm{T}}$ is the outward normal unit vector of the kth planar face and the matrices N_x, N_y and N_z are explicitly expressed as

$$\bar{N}_x = \begin{pmatrix} 0 & 0 & 0 \\ 0 & 0 & 1 \\ 0 & -1 & 0 \end{pmatrix}, \quad \bar{N}_y = \begin{pmatrix} 0 & 0 & -1 \\ 0 & 0 & 0 \\ 1 & 0 & 0 \end{pmatrix}, \quad \bar{N}_z = \begin{pmatrix} 0 & 1 & 0 \\ -1 & 0 & 0 \\ 0 & 0 & 0 \end{pmatrix}$$

Equations (6.51)–(6.52) represent three equations each of which represent the updating equation for each of the six electromagnetic-field components. These equations can be explicitly solved and furthermore it can be seen that Equations (6.51)–(6.52) represent a scheme which employs a leapfrog pattern for updating in time of the electromagnetic-field components. In fact, from Equation (6.51) it can be clearly seen that in order to obtain the electric-field components at the time step $n + 1$, only the magnetic-field components at the time step $n + 1/2$ and the electric-field components at the time step n are used. Likewise from Equation (6.52), it is possible to note that to update the magnetic-field components at the time step $n + 3/2$, only the electric-field component at the time step $n + 1$ and the magnetic-field components at the time step $n + 1/2$ are employed.

6.5   2D Formulation of the FVTD Method

In this paragraph a 2D formulation of the FVTD using the nondiffusive scheme is derived. The scheme is based on a cell-centred formulation and the elementary cells are constituted by triangles. In this way, elementary volumes, V_i, are replaced by elementary surfaces, S_i, while the boundary planar surfaces, S_k, surrounding the

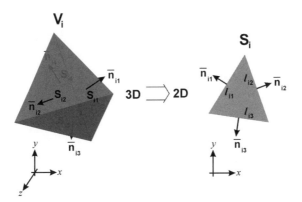

Figure 6.5 Elementary triangular volumes employed for the FVTD method. On the left, an elementary volume for a 3D computational domain is shown. On the right, an elementary volume for a 2D computational domain is derived.

volumes, V_i, are replaced by segments, l_k, which are the sides of the elementary triangles, as shown in Figure 6.5.

The 2D space considered here for this derivation can be obtained from a 3D space in which the z-direction is homogeneous in such a way that all derivatives along the z-axis are identically equal to zero. This condition is reflected in Equations (6.51)–(6.52) by forcing all elements of the matrix N_z to 0

$$\frac{\partial}{\partial z} \equiv 0 \quad \Rightarrow \quad N_z = \begin{pmatrix} 0 & 0 & 0 \\ 0 & 0 & 0 \\ 0 & 0 & 0 \end{pmatrix}$$

The substitution of Equations (6.53)–(6.54) into Equations (6.51)–(6.52), considering the condition previously expressed about the 2D space, leads to the following equations

$$\varepsilon_i \, |S_i| \, \frac{E_{xi}^{n+1} - E_{xi}^n}{\Delta t} + \sum_{k=1}^{3} |l_k| \left(-n_{iky} \frac{H_{zi}^{n+1/2} + H_{zk}^{n+1/2}}{2} \right) = 0 \qquad (6.57)$$

$$\varepsilon_i \, |S_i| \, \frac{E_{yi}^{n+1} - E_{yi}^n}{\Delta t} + \sum_{k=1}^{3} |l_k| \left(n_{ikx} \frac{H_{zi}^{n+1/2} + H_{zk}^{n+1/2}}{2} \right) = 0 \qquad (6.58)$$

$$\varepsilon_i \, |S_i| \, \frac{E_{zi}^{n+1} - E_{zi}^n}{\Delta t} + \sum_{k=1}^{3} |l_k| \left(n_{iky} \frac{H_{xi}^{n+1/2} + H_{xk}^{n+1/2}}{2} - n_{ikx} \frac{H_{yi}^{n+1/2} + H_{yk}^{n+1/2}}{2} \right) = 0$$
$$(6.59)$$

$$\mu_i \, |S_i| \, \frac{H_{xi}^{n+3/2} - H_{xi}^{n+1/2}}{\Delta t} - \sum_{k=1}^{3} |l_k| \left(-n_{iky} \frac{E_{zi}^{n+1} + E_{zk}^{n+1}}{2} \right) = 0 \qquad (6.60)$$

$$\mu_i \, |S_i| \, \frac{H_{yi}^{n+3/2} - H_{yi}^{n+1/2}}{\Delta t} - \sum_{k=1}^{3} |l_k| \left(n_{ikx} \frac{E_{zi}^{n+1} + E_{zk}^{n+1}}{2} \right) = 0 \qquad (6.61)$$

$$\mu_i \, |S_i| \, \frac{H_{zi}^{n+3/2} - H_{zi}^{n+1/2}}{\Delta t} - \sum_{k=1}^{3} |l_k| \left(n_{iky} \frac{E_{xi}^{n+1} + E_{xk}^{n+1}}{2} - n_{ikx} \frac{E_{yi}^{n+1} + E_{yk}^{n+1}}{2} \right) = 0$$

$$(6.62)$$

As can be clearly seen, Equations (6.57)–(6.62) can be collected in two groups of dependent equations with common electromagnetic-field components. For instance, in Equations (6.59)–(6.61) only the electromagnetic-field components E_z, H_x, H_y are involved, while in Equations (6.57), (6.58), (6.62) only E_x, E_y, H_z electromagnetic-field components are considered. The existence of these two independent sets of equations reflects the fact that in a 2D space two distinct polarisations are involved, which concurrently exist without any mutual interaction, if considered in structures composed of isotropic materials. In the literature, these two distinct polarisations are referred as transverse electric (TE) and transverse magnetic (TM) polarisations. At this point, clarification of the classification of TE and TM polarization is required. In the literature there are two different ways to associate the TE or TM polarisations for a certain set of equations. Considering z as homogeneous direction, in most textbooks on microwave theory TE polarisation is usually associated with electromagnetic fields whose main components are E_x, E_y, H_z, while TM polarisation is associated with electromagnetic fields whose main components are H_x, H_y, E_z. In textbooks on PhCs, the classification of TE and TM polarisation is linked to the axis of periodicity of the crystal itself. For instance, considering a 2D PhC whose axes of periodicity are x and y, TE polarisation refers to electromagnetic fields whose electric-field component is perpendicular to the plane formed by the two axes of periodicity, while TM polarisation refers to electromagnetic fields whose magnetic-field component is perpendicular to the periodicity plane. Following this definition of the two different polarisations, for a PhC a TE-polarised wave is an electromagnetic field whose main components are H_x, H_y, E_z. Consequently, a TM-polarised wave contains E_x, E_y, H_z as the main components. It is clear now that these two different ways to classify TE and TM polarisations are in contrast with each other. Therefore, in order to avoid confusion, throughout the whole book the convention used for PhCs will be followed.

Equations (6.59)–(6.61) are considered for TE polarisation and can be rewritten in explicit form as

$$E_{zi}^{n+1} = E_{zi}^{n} - \frac{\Delta t}{\varepsilon_i \, |S_i|} \sum_{k=1}^{3} |l_k| \left(n_{iky} \frac{H_{xi}^{n+1/2} + H_{xk}^{n+1/2}}{2} - n_{ikx} \frac{H_{yi}^{n+1/2} + H_{yk}^{n+1/2}}{2} \right)$$

$$(6.63)$$

$$H_{xi}^{n+3/2} = H_{xi}^{n+1/2} - \frac{\Delta t}{\mu_i \, |S_i|} \sum_{k=1}^{3} |l_k| \left(n_{iky} \frac{E_{zi}^{n+1} + E_{zk}^{n+1}}{2} \right) \tag{6.64}$$

$$H_{yi}^{n+3/2} = H_{yi}^{n+1/2} + \frac{\Delta t}{\mu_i \, |S_i|} \sum_{k=1}^{3} |l_k| \left(n_{ikx} \frac{E_{zi}^{n+1} + E_{zk}^{n+1}}{2} \right) \tag{6.65}$$

The update process starts with the computation of the electric field E_z using the value of the same field at the previous time step and the old values of the magnetic-field components H_x and H_y. Once the electric field has been updated, the calculation of the new values of the magnetic-field components is performed using the magnetic-field components values at the previous time step and the newly updated electric field. This process is repeated either for a fixed number of time steps or until steady state is reached.

Equations (6.57), (6.58)–(6.62) are considered for TM polarisation and can be rewritten in explicit form as

$$E_{xi}^{n+1} = E_{xi}^{n+1} + \frac{\Delta t}{\varepsilon_i \, |S_i|} \sum_{k=1}^{3} |l_k| \left(n_{iky} \frac{H_{zi}^{n+1/2} + H_{zk}^{n+1/2}}{2} \right) \tag{6.66}$$

$$E_{yi}^{n+1} = E_{yi}^{n} - \frac{\Delta t}{\varepsilon_i \, |S_i|} \sum_{k=1}^{3} |l_k| \left(n_{ikx} \frac{H_{zi}^{n+1/2} + H_{zk}^{n+1/2}}{2} \right) \tag{6.67}$$

$$H_{zi}^{n+3/2} = H_{zi}^{n+1/2} + \frac{\Delta t}{\mu_i \, |S_i|} \sum_{k=1}^{3} |l_k| \left(n_{iky} \frac{E_{xi}^{n+1} + E_{xk}^{n+1}}{2} - n_{ikx} \frac{E_{yi}^{n+1} + E_{yk}^{n+1}}{2} \right)$$
$$\tag{6.68}$$

The update of Equations (6.66)–(6.68) is performed in a similar way to the update process for the TE polarisation case.

6.5.1   Numerical Stability

The analysis of the numerical stability of the method is based on the determination of the variation of the total discrete electromagnetic energy of the scheme. In [5] it has been proven that the scheme conserves exactly the total discrete electromagnetic energy when a metallic cavity is simulated. This implies that the scheme is effectively stable and nondiffusive. Furthermore, a stability analysis has been carried out when first-order boundary conditions are applied to the computational domain [5]. For this case the stability and nondiffusivity of the scheme have also been proven. However, in order to maintain the stability of the scheme, some restrictions on the maximum time step it is possible to employ for a given problem have to be applied. These restrictions

take into account different scenarios which can be represented by the computational
domain, and are expressed as

$$
\begin{cases}
\forall \text{ internal interface} & \Delta t^2 < 16\dfrac{V_i V_k}{S_i S_k}\min\left(\varepsilon_k\mu_i,\ \varepsilon_i\mu_k\right) \\[2ex]
\forall \text{ metallic interface} & \Delta t < 4\dfrac{V_i}{c_i S_i} \\[2ex]
\forall \text{ absorbing interface} & \Delta t < 4\dfrac{V_i}{c_i S_i}
\end{cases}
\tag{6.69}
$$

6.6 Boundary Conditions

The analysis of scattering problems of electromagnetic waves propagating in
optical waveguides is usually studied in infinitely extended regions. On the other
hand, the FVTD method works only with a finite number of volumes which represent
the computational domain. This introduces a problem for the simulation of scattering problems because in a computer memory only finite regions can be simulated.
This problem has been brilliantly solved with the advent of the boundary conditions
which can artificially simulate open-space regions within a finite discretised space.
In the context of the FVTD method, the most used boundary conditions were the
'Silver–Müller' boundary conditions also known as 'radiation boundary conditions'
until the advent of the PML boundary conditions, which present better performance
in terms of absorption of the outer propagating waves.

6.6.1 Silver–Müller Boundary Conditions

In the FVTD method, the most frequently used boundary conditions applied on the
outer boundary are the Silver–Müller conditions [5, 6]. These boundary conditions
belong to the family of boundary conditions referred to in the literature as local non-
reflecting or radiation boundary conditions. These local conditions are less accurate
than exact or global outer boundary conditions based on boundary integral equations
[7]. However, they are very easy to implement and require less computer resources.
The computational volume is truncated by introducing a surface, S_0, enclosing the
computational domain. Consider a cell, V, with a face, S, which belongs to the surface,
S_0, and the electromagnetic fields, $\bar{E}^*$ and $\bar{H}^*$, defined on S, with unit normal, $\bar{n}$, as
shown in Figure 6.6.

The Silver–Müller conditions, applied to the outer boundary, force the flux on the
outer boundary to apply to outgoing waves only. This condition is represented by the
following:

$$
\sqrt{\frac{\varepsilon_0}{\mu_0}}\,\bar{n} \times \bar{E}^* + \bar{n} \times \left(\bar{n} \times \bar{H}^*\right) = 0
\tag{6.70}
$$

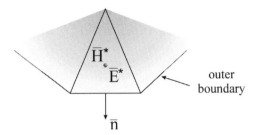

Figure 6.6 Schematic diagram of an outer boundary for the application of the Silver–Müller boundary condition.

This condition ensures that waves which are normal to the artificial boundary are not reflected, whereas other waves are only partially reflected. For this reason, in order to avoid high reflections, the outer boundary has to be placed sufficiently far away from the core of the computational domain so as to assume the incident waves on the boundary are local plane waves.

6.6.2   PML Boundary Conditions

The UPML has been previously implemented in the context of conventional FDTD, and was shown to be very efficient and robust [8]. In this section, the incorporation of the UPML boundary conditions has been considered in the context of FVTD scheme. The FVTD formulation can be applied to the Maxwell's equations for the flux density, D, and the magnetic flux, B

$$\frac{\partial \bar{D}}{\partial t} - \nabla \times \bar{H} = 0 \tag{6.71}$$

$$\frac{\partial \bar{B}}{\partial t} + \nabla \times \bar{E} = 0 \tag{6.72}$$

Applying the nondiffusive FVTD scheme to Equations (6.71)–(6.72) the following discretised equations are obtained

$$|V_i| \frac{D_i^{n+1} - D_i^n}{\Delta t} + \sum_{k=1}^{m_i} \bar{N}_{ik} \frac{H_i^{n+1/2} - H_k^{n+1/2}}{2} = 0 \tag{6.73}$$

$$|V_i| \frac{B_i^{n+3/2} - B_i^{n+1/2}}{\Delta t} - \sum_{k=1}^{m_i} \bar{N}_{ik} \frac{E_i^{n+1} - E_k^{n+1}}{2} = 0 \tag{6.74}$$

where $|V_i|$ represents the volume of the ith elementary cell and $\bar{N}_{ik}$ is defined by Equation (6.55).

The update of the electric field, E, and the magnetic field, H, is obtained through the proper constitutive relationships as defined in [8] in the context of the UPML formulation

$$\breve{D}_x = \varepsilon_0 \varepsilon_r \frac{s_z}{s_x} \breve{E}_x \tag{6.75}$$

$$\breve{D}_y = \varepsilon_0 \varepsilon_r \frac{s_x}{s_y} \breve{E}_y \tag{6.76}$$

$$\breve{D}_z = \varepsilon_0 \varepsilon_r \frac{s_y}{s_z} \breve{E}_z \tag{6.77}$$

$$\breve{B}_x = \mu_0 \mu_r \frac{s_z}{s_x} \breve{H}_x \tag{6.78}$$

$$\breve{B}_y = \mu_0 \mu_r \frac{s_x}{s_y} \breve{H}_y \tag{6.79}$$

$$\breve{B}_z = \mu_0 \mu_r \frac{s_y}{s_z} \breve{H}_z \tag{6.80}$$

where s_x, s_y, s_z are defined as

$$s_x = 1 + \frac{\sigma_x}{j\omega\varepsilon_0} \tag{6.81}$$

$$s_y = 1 + \frac{\sigma_y}{j\omega\varepsilon_0} \tag{6.82}$$

$$s_z = 1 + \frac{\sigma_z}{j\omega\varepsilon_0} \tag{6.83}$$

It needs to be noted that Equations (6.75)–(6.80) are defined in the frequency domain. In order to make them useful for a time-domain method, they have to be transformed into their time-domain equivalents. Following the procedure suggested in [8], the transformation of Equations (6.75)–(6.80) in their equivalent time-domain equations is easily obtained with a few mathematical manipulations. For instance, considering Equation (6.75), with the explicit expression for the quantities s_z and s_x, and multiplying both sides by $j\omega$, the following equation is obtained

$$\left(j\omega + \frac{\sigma_x}{\varepsilon_0} \right) \breve{D}_x = \varepsilon_0 \varepsilon_r \left(j\omega + \frac{\sigma_x}{\varepsilon_0} \right) \breve{E}_x \tag{6.84}$$

The transformation of Equation (6.84) into time domain leads to

$$\frac{\partial D_x}{\partial t} + \frac{\sigma_x}{\varepsilon_0} D_x = \varepsilon_0 \varepsilon_r \left(\frac{\partial E_x}{\partial t} + \frac{\sigma_z}{\varepsilon_0} E_x \right) \tag{6.85}$$

Following the same procedure for Equations (6.76)–(6.80), similar equations are obtained. Their discretisation in time is then performed using a leapfrog arrangement, which leads to the following equations:

$$E_i^{n+1} = \frac{2\varepsilon_0 - \sigma_k \Delta t}{2\varepsilon_0 + \sigma_k \Delta t} E_i^n + \frac{1}{(2\varepsilon_0 + \sigma_k \Delta t)\,\varepsilon}$$
$$\left[(2\varepsilon_0 + \sigma_j \Delta t)\, D_i^{n+1} - (2\varepsilon_0 - \sigma_j \Delta t)\, D_i^n \right] \qquad (6.86)$$

$$H_i^{n+3/2} = \frac{2\varepsilon_0 - \sigma_k \Delta t}{2\varepsilon_0 + \sigma_k \Delta t} H_i^{n+1/2} + \frac{1}{(2\varepsilon_0 + \sigma_k \Delta t)\,\mu}$$
$$\left[(2\varepsilon_0 + \sigma_j \Delta t)\, B_i^{n+3/2} - (2\varepsilon_0 - \sigma_j \Delta t)\, B_i^{n+1/2} \right] \qquad (6.87)$$

with $k = z, x, y, j = x, y, z$ and where σ is the conductivity of the UPML layers. For the conductivity in the UPML layers, a geometrical grading has been chosen [8], and the maximum value of the conductivity $\sigma_{\max}$ of the UPML absorber has been calculated using the relationship [9]

$$\sigma_{\max} = \left(\frac{\varepsilon_0 c_0}{w_{\text{upml}}} \right) \left(\frac{p+1}{2} \right) \left(\ln \left(\frac{1}{\Gamma_{\text{th}}} \right) \right) \qquad (6.88)$$

where ε_0 and c_0 are the free space permittivity and the light velocity in free space, respectively, w_{upml} is the width of the UPML layer, p is the order of geometrical grading and Γ_{th} is the theoretical reflection coefficient of the UPML absorber.

6.7 Nonlinear Optics

A brief overview on nonlinear optics is given here. In Chapter 10 a complete analysis of nonlinear optics will be given. Nonlinear optics is a branch of optic science which deals with nonlinear phenomena in material systems due to the presence of light, which modifies the optical properties of the material itself. In order to observe non-linear phenomena, high-intensity electromagnetic fields are needed and this explains why nonlinear optics is a relatively young branch of optic science. In fact, the first demonstration of a working laser was given by Maiman in 1960 [10] and soon after the first work on nonlinear optics was carried out by Franken et al. in 1961, who observed and discovered the first second-harmonic generation (SHG) nonlinear process [11]. It is understood that the term nonlinear refers to the fact that the phenomena depend in a nonlinear manner upon the intensity of the applied electromagnetic field. As an example, the intensity of the field generated at the second harmonic frequency, due to the SHG nonlinear phenomenon, depends quadratically on the strength of the applied field at the fundamental frequency (FF).

The description of nonlinear phenomena is usually done with the introduction of the polarisation $\tilde{P}(t)$ of a material which is related to the applied optical field $\tilde{E}(t)$, where the symbol '$\sim$' represents fast-varying fields in time. Considering $\tilde{P}(t)$ and $\tilde{E}(t)$ to be scalar quantities, the polarisation $\tilde{P}(t)$ in the optical response is usually described as a power series in the field $\tilde{E}(t)$

$$\tilde{P}(t) = \chi^{(1)}(t)\tilde{E}(t) + \chi^{(2)}(t)\tilde{E}^2(t) + \chi^{(3)}(t)\tilde{E}^3(t) + \ldots \equiv \tilde{P}^{(1)}(t) + \tilde{P}^{(2)}(t)$$
$$+ \tilde{P}^{(3)}(t) + \ldots \tag{6.89}$$

where $\chi^{(1)}(t)$ is the linear susceptibility, $\chi^{(2)}(t)$ is the second-order nonlinear susceptibility and $\chi^{(3)}(t)$ is the third-order nonlinear susceptibility. $\tilde{P}^{(2)}(t)$ and $\tilde{P}^3(t)$ are also known as the second-order nonlinear polarisation and the third-order nonlinear polarisation, respectively. It should be mentioned that because of the vectorial nature of $\tilde{E}(t)$ and $\tilde{P}(t)$, the quantities $\chi^{(1)}(t)$, $\chi^{(2)}(t)$ and $\chi^{(3)}(t)$ are not scalars, but tensors. Infact, $\chi^{(1)}(t)$ is a second-rank tensor, $\chi^{(2)}(t)$ is a third-rank tensor, $\chi^{(3)}(t)$ is a fourth-order tensor, and so on. Furthermore, in Equation (6.89) it has been assumed that the polarization, $\tilde{P}(t)$, depends only upon the instantaneous values of the electric field $\tilde{E}(t)$, which implies lossless and dispersionless properties of the medium involved in the nonlinear optical process [12].

The importance of the polarization, $\tilde{P}(t)$, in nonlinear optics can be fully understood considering that the wave equation in nonlinear optics media is usually expressed as [12]

$$\nabla^2 \tilde{E}(t) - \frac{n^2}{c^2}\frac{\partial^2 \tilde{E}}{\partial t^2} = \frac{4\pi}{c^2}\frac{\partial^2 \tilde{P}^{NL}(t)}{\partial t^2} \tag{6.90}$$

where n is the linear refractive index of the medium and c is the speed of light in free space. Equation (6.90) is an inhomogeneous differential equation in which it can be clearly seen that the term $\tilde{P}^{NL}(t)$ acts as source of new components of the electromagnetic field.

6.8 Nonlinear Optical Interactions

The interaction between incident light and optical nonlinear media can give rise to a variety of different nonlinear phenomena. Each of these phenomena is related to the applied electromagnetic field through Equation (6.89) by means of the different susceptibilities. For this reason, each susceptibility is responsible for a particular class of nonlinear phenomena. For instance, the second-order nonlinear susceptibility $\chi^{(2)}(t)$ accounts for nonlinear phenomena such as second harmonic generation (SHG), sum-frequency generation (SFG), difference-frequency generation (DFG) and optical parametric oscillation.

6.8.1 Second Harmonic Generation (SHG)

SHG describes a nonlinear optical process in which two photons interact with a nonlinear medium generating a new photon with double the energy, and thus light travelling with double the frequency. This nonlinear process is characteristic of optical materials which possess nonzero second-order nonlinear susceptibility $\chi^{(2)}(t)$.

A laser beam is used as a source to excite a nonlinear material which presents a second-order nonlinearity described by $\chi^{(2)}(t)$. The source can be described as

$$\tilde{E}(t) = Ee^{-j\omega t} + c.c. \tag{6.91}$$

where E is the amplitude of the electric field and $e^{-j\omega t}$ describes the fast-varying time domain part of the electric field with angular frequency ω. From Equation (6.89), it is possible to derive the following expression

$$\tilde{P}^{(2)}(t) = 2\chi^{(2)}EE^* + \left(\chi^{(2)}E^2e^{-2j\omega t} + c.c\right) \tag{6.92}$$

From Equation (6.92), it is possible to note that $\tilde{P}^{(2)}(t)$ is composed of two different terms, the first at zero frequency and the second at angular frequency 2ω. The first term leads to the generation of an electric static field within the nonlinear material and this process is also known as optical rectification (OR). The second term leads to the generation of an electromagnetic radiation at angular frequency 2ω and this process is also known as SHG. It has to be mentioned that in studying optical propagation, $\tilde{P}^{(2)}(t)$ in Equation (6.92) has to be substituted into Equation (6.90), which represents the wave equation. For this reason, it is clear that $\tilde{P}^{(2)}(t)$ acts as a source in Equation (6.90) which implies that an electric field at angular frequency 2ω is travelling inside the nonlinear material. Figure 6.7 shows a schematic diagram of the SHG process, explained in terms of a black box and energy levels.

6.8.2 Third Harmonic Generation

Third harmonic generation describes a nonlinear optical process in which three photons interact with a nonlinear medium generating a new photon with triple the

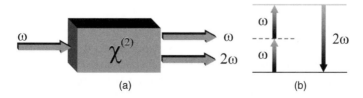

(a) (b)

Figure 6.7 (a) Block diagram and (b) energy-level description of the SHG nonlinear process.

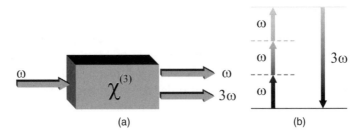

Figure 6.8 (a) Block diagram and (b) energy-level description of the third harmonic generation nonlinear process.

energy, and thus light with triple the frequency. This nonlinear process is characteristic of optical materials which possess nonzero second-order nonlinear susceptibility $\chi^{(3)}(t)$. Figure 6.8 shows a schematic diagram of the third harmonic generation process explained in terms of a black box and energy levels.

In order to show the nonlinear process of third harmonic generation associated with third-order nonlinear susceptibility, $\chi^{(3)}(t)$, a monochromatic field is applied to the nonlinear material system

$$\tilde{E}(t) = E \cos(\omega t) \tag{6.93}$$

Assuming the material possesses only nonzero $\chi^{(3)}(t)$, the polarization, $\tilde{P}(t)$ within the nonlinear material consists of only the third-order nonlinear polarisation $\tilde{P}^{(3)}(t)$

$$\tilde{P}^{(3)}(t) = \chi^{(3)} \tilde{E}^3(t) \tag{6.94}$$

Using the identity

$$\cos^3(\omega t) = \frac{1}{4}\cos(3\omega t) + \frac{3}{4}\cos(\omega t) \tag{6.95}$$

and substituting in Equation (6.94) yields

$$\tilde{P}^{(3)}(t) = \frac{1}{4}\chi^{(3)} E^3 \cos(3\omega t) + \frac{3}{4}\chi^{(3)} E^3 \cos(\omega t) \tag{6.96}$$

The first term of Equation (6.96) leads to the generation of an electromagnetic radiation at angular frequency 3ω and this process is also known as third harmonic generation. Similarly to the SHG process, it should be mentioned that $\tilde{P}^{(3)}(t)$ in Equation (6.96) has to be substituted into Equation (6.90), which represents the wave equation. For this reason, it is clear that $\tilde{P}^{(3)}(t)$ acts as a source in Equation (6.90), which implies that an electric field at angular frequency 3ω is travelling inside the nonlinear material.

6.8.3   Intensity-Dependent Refractive Index

The second term in Equation (6.96) describes a nonlinear contribution at the frequency of the source field. This nonlinear effect leads to a phenomenon also known as the intensity-dependent refractive index, which is usually described by

$$n = n_0 + n_2 I \tag{6.97}$$

where n_0 is the linear refractive index of the material, n_2 is the nonlinear refractive index and I is the intensity of the incident wave. The nonlinear refractive index n_2 is linked to the third-order nonlinear susceptibility, $\chi^{(3)}(t)$, through the following relationship

$$n_2 = \frac{12\pi}{n_0^2 c} \chi^{(3)}(t) \tag{6.98}$$

where c is the speed of light in a vacuum. The intensity of the incident field is related to the amplitude of the electric field, E, through the relationship

$$I = \frac{n_0 c}{8\pi} E^2 \tag{6.99}$$

It is clear from Equation (6.97) that the intensity-dependent refractive index is a local phenomenon in the sense that only the part of the material which is affected by the incident electric field experiences a change in the value of the refractive index. This leads to a local inhomogeneous refractive index which affects the propagating electromagnetic field. In fact, from Equation (6.97), if $n_2 > 0$, it can be seen that the refractive index of the material increases as the intensity of the propagating electromagnetic field increases. This particular phenomenon is also known as self-focusing and it consists of focusing lens behaviour of the nonlinear material with respect to electromagnetic radiations, as shown in Figure 6.9.

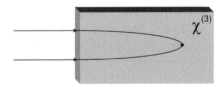

Figure 6.9   Schematic diagram of the intensity-dependent refractive index phenomenon. The two arrows represent two light rays propagating inside a nonlinear material. If $\chi^{(3)} > 0$, the two rays are focusing.

6.9 Extension of the FDTD Method to Nonlinear Problems

In order to consider a 2D problem, the z-axis has been chosen as the homogeneous direction, while the x-axis has been considered as the direction of propagation. With these assumptions, for TE modes, the principal electromagnetic field components are E_z, H_x and H_y. Under the scalar approximation, the following 2D TE equations can be derived, as reported in [8, 13]:

$$\frac{\partial H_x}{\partial t} = -\frac{1}{\mu_0}\frac{\partial E_z}{\partial y} \tag{6.100}$$

$$\frac{\partial H_y}{\partial t} = \frac{1}{\mu_0}\frac{\partial E_z}{\partial x} \tag{6.101}$$

$$\frac{\partial D_z}{\partial t} = \frac{\partial H_y}{\partial x} - \frac{\partial H_x}{\partial y} \tag{6.102}$$

$$D_z = \varepsilon E_z \tag{6.103}$$

where μ_0 is the permeability of free space, and ε is the permittivity of the dielectric medium. The constitutive relationship (6.103) is considered here to model the nonlinearity, this being an instantaneous Kerr-type nonlinearity and this is considered in the permittivity of the dielectric medium

$$\varepsilon = \varepsilon_0 \varepsilon_r = \varepsilon_0 n^2 = \varepsilon_0 \left(n_0 + \alpha \left|E\right|^2\right)^2 \tag{6.104}$$

where ε_0 is the permittivity of free space, n_0 is the linear part of the nonlinear refractive index of the nonlinear dielectric medium, α is the nonlinear coefficient of the nonlinear refractive index of the dielectric medium, and in this relationship it has units of m^2/V^2, and where E is the electric field. Substituting Equation (6.104) into Equation (6.103), the following relationship can be derived

$$E_z = \frac{D_z}{\varepsilon_0 \left(n_0 + \alpha \left|E_z\right|^2\right)^2} \tag{6.105}$$

Next, the discretisation in space is performed based on the unit cell of the Yee space lattice, as shown in Figure 6.10, while the discretisation in time is obtained following the leapfrog arrangement [8]. In this way, a set of discretised equations in space and in time are obtained. Using the notation introduced in [8] and considering the position of the electromagnetic-field components in a 2D grid, as shown in Figure 6.10, the discretisation of Equations (6.100)–(6.103) yields

$$H_x\big|_{i+1/2,j}^{n+1} = H_x\big|_{i+1/2,j}^{n} - \frac{\Delta t}{\mu_0}\frac{E_z\big|_{i+1/2,j+1/2}^{n+1/2} - E_z\big|_{i+1/2,j-1/2}^{n+1/2}}{\Delta y} \tag{6.106}$$

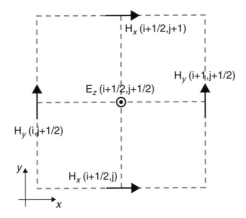

Figure 6.10   Electromagnetic field components placed in a 2D Yee's cell.

$$H_y\big|_{i,j+1/2}^{n+1} = H_y\big|_{i,j+1/2}^{n} + \frac{\Delta t}{\mu_0} \frac{E_z\big|_{i+1/2,j+1/2}^{n+1/2} - E_z\big|_{i-1/2,j+1/2}^{n+1/2}}{\Delta y} \tag{6.107}$$

$$D_z\big|_{i+1/2,j+1/2}^{n+1/2} = D_z\big|_{i+1/2,j+1/2}^{n-1/2} +$$

$$\Delta t \left(\frac{H_y\big|_{i+1,j+1/2}^{n} - H_y\big|_{i,j+1/2}^{n}}{\Delta x} - \frac{H_x\big|_{i+1/2,j+1}^{n} - H_x\big|_{i+1/2,j}^{n}}{\Delta y} \right) \tag{6.108}$$

$$E_z\big|_{i+1/2,j+1/2}^{n+1/2} = \frac{D_z\big|_{i+1/2,j+1/2}^{n+1/2}}{\varepsilon_0 \left(n_0\big|_{i+1/2,j+1/2} + \alpha\big|_{i+1/2,j+1/2} \left| E_z\big|_{i+1/2,j+1/2}^{n-1/2} \right|^2 \right)^2} \tag{6.109}$$

The updating process starts with the calculation of the new values of the magnetic-field components at all the grid points of the computational domain, using Equations (6.106)–(6.107). By means of the values obtained, the D_z field component is computed in the same way using Equation (6.108). Subsequently, new values of the electric-field component are calculated with reference to Equation (6.109). This approach avoids the necessity of updating the values of the dielectric material of the simulated device, arising due to the nonlinear effects caused by the applied electromagnetic field, because the nonlinear effects are directly applied to the electric-field component E_z through the computation of Equation (6.109). In this equation it should be noted that the value of E_z from the previous time step is used to calculate the value of E_z in the subsequent time step through a Newton-iteration procedure. This approach is justified because the former value of E_z is a very good approximation of the new value of E_z, arising because of the application of the Courant–Friedrichs–Lewy (CFL) condition for the FDTD time-stepping procedure. Furthermore, solving all the electromagnetic-field components of the TE mode instead of solving only the main electric-field component derived from the wave equation produces a stable method for the simulation of nonlinear devices, as will be clearly demonstrated in the Chapter 7. Additionally, this approach avoids the extra computational effort necessary to solve the nonlinear matrix equation that would be obtained if the value of E_z determined at the same time step were used on both sides of Equation (6.109), as proposed in the literature [15].

6.10 Extension of the FVTD Method to Nonlinear Problems

Starting from Maxwell's equations and the constitutive relationships between electric flux density and electric field, and magnetic flux density and magnetic field

$$\frac{\partial \vec{D}}{\partial t} - \nabla \times \vec{H} = 0 \tag{6.110}$$

$$\frac{\partial \vec{B}}{\partial t} + \nabla \times \vec{E} = 0 \tag{6.111}$$

$$\vec{D} = \varepsilon_0 \varepsilon_r \vec{E} \tag{6.112}$$

$$\vec{B} = \mu_0 \mu_r \vec{H} \tag{6.113}$$

it is possible to rewrite the Equations (6.110) and (6.111) in a conservative form as in [1]

$$\frac{\partial \bar{M}}{\partial t} + \frac{\partial F_1\left(\bar{U}\right)}{\partial x} + \frac{\partial F_2\left(\bar{U}\right)}{\partial y} + \frac{\partial F_3\left(\bar{U}\right)}{\partial z} = 0 \tag{6.114}$$

where $\bar{M} = [D_x \quad D_y \quad D_z \quad B_x \quad B_y \quad B_z]^T$, $\bar{U} = [E_x \quad E_y \quad E_z \quad H_x \quad H_y$ $H_z]^T$, $F_1(\bar{U}) = [0 \quad H_z \quad -H_y \quad 0 \quad -E_z \quad E_y]^T$, $F_2\left(\bar{U}\right) = -H_z \quad 0 \quad H_x \quad E_z$ $0 \quad -E_x]^T$, $F_3\left(\bar{U}\right) = [H_y \quad -H_x \quad 0 \quad -E_y \quad E_x \quad 0]^T$, and T represents the transpose matrix operator. Equation (6.114) can be rewritten in condensed form as

$$\frac{\partial \bar{M}}{\partial t} + \mathrm{div} F\left(\bar{U}\right) = 0 \tag{6.115}$$

where $F\left(\bar{U}\right) = \left[F_1\left(\bar{U}\right) \; F_2\left(\bar{U}\right) \; F_3\left(\bar{U}\right)\right]$. This notation is fundamental for the derivation of the FVTD scheme. Considering a generic volume, V, Equation (6.115) is then integrated and through the use of the divergence theorem the following equation is obtained

$$\int_V \frac{\partial \bar{M}}{\partial t} dV + \int_S F\left(\bar{U}^*\right) \cdot \vec{a}_n dS = 0 \tag{6.116}$$

where S represents the surface enclosing the volume, V, $\vec{a}_n$ is the outward pointing normal unit vector of the surface S, and $\bar{U}^* = \begin{bmatrix} E_x^* & E_y^* & E_z^* & H_x^* & H_y^* & H_z^* \end{bmatrix}^T$ denotes the electromagnetic-field components at the surface, S. In order to make possible the integration of Equation (6.116) in practical applications, the entire volume, V, is partitioned into small volumes, V_i, each of one with a boundary surface, S_i. Each of these surfaces is composed of a number, m_i, of planar faces with outward unit normal, $\vec{n}_{ik}$. The partition of the entire volume has been performed using a 'cell-centred' formulation [1]. In a 2D computational domain each volume is represented by a triangle, while planar faces are represented by its sides. Applying Equation (6.116) to the discretised computational domain leads to the following discretised equation

$$|V_i| \frac{\partial \bar{M}_i}{\partial t} = \sum_{k=1}^{m_i} |S_k| \, F\left(\bar{U}_k^*\right) \cdot \vec{n}_{ik} \tag{6.117}$$

where $|V_i|$ is the volume of the ith element, $|S_k|$ is the area of the kth planar surface that surrounds the ith element, $\bar{M}_i$ is the value of $\bar{M}$ at the centre of the ith element, $\bar{U}_k^*$ is the value of $\bar{U}$ at the centre of the kth planar surface, and $F\left(\bar{U}_k^*\right) \cdot \vec{n}_{ik}$ represents the flux at centre of the kth planar surface. In order to calculate the flux for each volume element, the scheme proposed in [5] has been employed so as assure the scheme does not suffer from numerical diffusion. Applying the aforementioned scheme to Equation (6.117) the following equations are obtained

$$|V_i| \frac{D_i^{n+1} - D_i^n}{\Delta t} + \sum_{k=1}^{m_i} \bar{N}_{\vec{n}_{ik}} \frac{H_i^{n+1/2} - H_k^{n+1/2}}{2} = 0 \tag{6.118}$$

$$|V_i| \frac{B_i^{n+3/2} - B_i^{n+1/2}}{\Delta t} - \sum_{k=1}^{m_i} \bar{N}_{\vec{n}_{ik}} \frac{E_i^{n+1} - E_k^{n+1}}{2} = 0 \tag{6.119}$$

where $\bar{N}_{\vec{n}_{ik}} = n_{ikx}\bar{N}_x + n_{iky}\bar{N}_y + n_{ikz}\bar{N}_z$, with

$$\bar{N}_x = \begin{pmatrix} 0 & 0 & 0 \\ 0 & 0 & 1 \\ 0 & -1 & 0 \end{pmatrix}, \quad \bar{N}_y = \begin{pmatrix} 0 & 0 & -1 \\ 0 & 0 & 0 \\ 1 & 0 & 0 \end{pmatrix}, \quad \bar{N}_z = \begin{pmatrix} 0 & 1 & 0 \\ -1 & 0 & 0 \\ 0 & 0 & 0 \end{pmatrix}$$

and with $\vec{n}_{ik} = \begin{pmatrix} n_{ikx} & n_{iky} & n_{ikz} \end{pmatrix}^T$ the outward normal unit vector of the kth planar face. The discretisation of Equations (6.112) and (6.113), equations for the calculation of the new values of the electric and magnetic field necessary for the update procedure of D and B, leads to the following two relationships

$$E_i^n = \frac{D_i^n}{\varepsilon_0 \varepsilon_r} \tag{6.120}$$

$$H_i^{n+1/2} = \frac{B_i^{n+1/2}}{\mu_0 \mu_r} \tag{6.121}$$

The constitutive relationship (6.112), and its discretised version (6.120), is used here to model the nonlinearity (instantaneous Kerr-type), through the permittivity of the dielectric medium [16]

$$\varepsilon_r = \varepsilon_l + \Delta\varepsilon_s \left(1 - e^{\left(-\dfrac{\gamma |E|^2}{\Delta\varepsilon_s} \right)} \right) \tag{6.122}$$

where ε_l is the linear relative permittivity of the nonlinear medium, $\Delta\varepsilon_s$ is the maximum variation of the nonlinear permittivity, γ is a nonlinear coefficient related to the nonlinear parameter n_2 through the relationship $\gamma = c\varepsilon_0 \varepsilon_l n_2$ in which c is the speed of the light and ε_0 is the free space permittivity, respectively, and E is the electric field. Substituting Equation (6.122) into Equation (6.120), the following relationship can be derived

$$E_i^n = \frac{D_i^n}{\varepsilon_0 \left(\varepsilon_l + \Delta\varepsilon_s \left(1 - e^{\left(-\dfrac{\gamma |E_i^n|^2}{\Delta\varepsilon_s} \right)} \right) \right)} \tag{6.123}$$

Equation (6.123) is then solved through an iterative procedure.

The updating process starts with the calculation of the new values of the magnetic flux density, B, components at all grid points of the computational domain using Equation (6.119), after which the new values of the magnetic field, H, components can be computed using Equation (6.121). By means of the values obtained, the electric flux density, D, components are updated using Equation (6.118). Subsequently, new values of the electric-field component are calculated with reference to Equation (6.123). This approach avoids the necessity of updating the values of the dielectric material of the simulated device, arising due to the nonlinear effects caused by the applied electromagnetic field, because the nonlinear effects are directly applied in the electric-field component, E, through the computation of Equation (6.123). It should be noted that even though Equation (6.123) is solved using a Newton-iteration procedure, the additional cost to the computational burden is negligible because in order to assure the numerical stability of the scheme, the time step is bounded by the relationship [1,5]

$$dt \leq \frac{1}{c} \min_i \left(\frac{|V_i|}{\sum_{k=1}^{m_i} S_k} \right) \tag{6.124}$$

where V_i is the volume of the ith element, S_k is the surface in which the volume, V_i, is enclosed, c is the speed of light in the dielectric medium contained in the volume, V_i, and m_i is the number of planar faces into which the surface, S_k, is divided. For all simulations, the time-step size was small enough to guarantee the solution of Equation (6.123) with only a few iterative steps.

References

[1] Rao, M. (1999) *Time Domain Electromagnetics*, Academic Press, London.
[2] Cioni, J.P. and Remaki, M. (1997) Comparaison de deux methods de volumes finis en electromagnetisme, Technical report RR-3166, INRIA.
[3] Steger, J.L. and Warming, R.F. (1981) Flux vector splitting of the inviscid gasdynamic equations with applications to finite-difference methods. *J. Comp. Phys.*, **40** (2), 263–293.
[4] Hirsch, C. (1988) *Numerical Computation of Internal and External Flows*, Vol. 1, John Wiley and Sons Ltd.
[5] Piperno, S., Remaki, M. and Fezoui, L. (2002) A nondiffusive finite volume scheme for 3-D Maxwell equations on unstructured meshes. *SIAM J. Numer. Anal.*, **39**, 2089–2108.
[6] Müller, C. (1969) *Foundations of the Mathematical Theory of Electromagnetic Waves*, Springer−Verlang, Berlin.
[7] McLean, W. (2000) *Strongly Elliptic Systems and Boundary Integral Equations*, Cambridge University Press, Cambridge.
[8] Taflove, A. and Hagness, S.C. (2005) *Computational Electrodynamics: the Finite-Difference Time-Domain Method*, Artech House.
[9] Sankaran, K., Fumeaux, C. and Vahldiek, R. (2006) Cell-centered finite-volume-based perfectly matched layer for time-domain Maxwell system. *IEEE Trans. Microwave Theory Tech.*, **54** (3), 1269–1276.
[10] Maiman, T.H. (1960) Stimulated optical radiation in ruby. *Nature*, **187** (4736), 493−494.
[11] Franken, P.A., Hill, A.E., Peters, C.W. and Weinreich, G. (1961) Generation of optical harmonics. *Phys. Rev. Lett.*, **7** (4), 118−119.
[12] Boyd, R.W. (2000) *Nonlinear Optics*, Academic Press, London.
[13] Joseph, R.M. and Taflove, A. (1997) FDTD Maxwell's equations models for nonlinear electrodynamics and optics. *IEEE Trans. Antennas Propagat.*, **45** (3), 364−374.
[14] Kelley, C.T. (2003) *Solving Nonlinear Equations with Newton's Method, No. 1 in Fundamentals of Algorithms*, SIAM.
[15] Van, V. and Chaudhuri, S.K. (1999) A hybrid implicit-explicit FDTD scheme for nonlinear optical waveguide modelling. *IEEE Trans. Microwave Theory Tech.*, **47** (5), 540–545.
[16] Obayya, S.S.A., Rahman, B.M.A., Grattan, K.T.V. and El-Mikati, H.A. (2002) Full vectorial finite-element solution of nonlinear bistable optical waveguides. *IEEE J. Quantum Electron.*, **38** (8), 1120−1125.

7

Numerical Analysis of Linear and Nonlinear PhC-Based Devices

7.1 Introduction

In this chapter the finite-volume time-domain (FVTD) method combined with the uniaxial perfectly matched layer (UPML) boundary conditions will be employed for simulations of photonic bandgap (PBG)-based devices. The formulation of the method relies on triangular elements that allow accurate representation of curved surfaces by means of an unstructured mesh, and avoiding the staircase problem always present for methods based on a Cartesian grid. Furthermore, the effectiveness of the UPML absorbing boundary rigorously incorporated into the FVTD formulation will be clearly demonstrated through examples. The aim of those results will be the assessment of the method and the demonstration of its powerful capabilities in dealing with photonic crystal devices. Moreover, in this chapter the FDTD and FVTD techniques have been used for the analysis of nonlinear devices. A brief theoretical background to nonlinear optical processes was introduced in the previous chapter, with a detailed analysis of how this nonlinearity has been inserted in the FVTD and FDTD techniques. Numerical examples will be presented in order to assess the accuracy of the two numerical techniques.

7.2 FVTD Method Assessment: PhC Cavity

The first structure analysed is a photonic crystal (PhC) cavity consisting of a 7×7 square lattice of dielectric rods with refractive index, $n_{rods} = 3.4$ in air ($n = 1.0$), with lattice constant, $a = 0.586\,52$ μm, and rod radius, $r = 0.2a$, as shown in Figure 7.1(a). From this figure it can be also clearly seen that the cavity defect consists of a missing rod in the crystal pattern. Furthermore, this PhC cavity can sustain only TE resonant

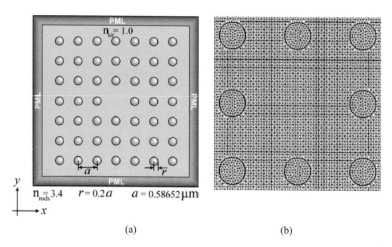

(a)   (b)

Figure 7.1   (a) Schematic diagram of a 7 × 7 PhC resonant cavity with (b) an example of the discretised computational domain. (Reproduced with permission from Pinto, D. and Obayya, S.S.A. (2008) 2D Analysis of multimode photonic crystal resonant cavities with the finite volume time domain method. *Opt. Quantum Electron.*, **40** (11–12), 875–890. Copyright 2008 Springer Science and Business Media.)

modes [1] where TE modes have the electric-field component parallel to the rods, while the magnetic-field components are in the plane of Figure 7.1(a). This structure has been considered in order to assess the accuracy of the FVTD method with mesh size. Different simulations have been carried out with the same structure discretised with different mesh sizes and for all these simulations the resonant wavelength, λ_{res}, and the quality factor, Q, of the resonant cavity have been computed. In Figure 7.1(b), an example of the discretised computational domain is shown.

It has to be mentioned that, in order to exclude any influence derived from possible reflections from the newly formulated UPML boundary conditions, whose absorption properties will be analysed in detail in the next section, the computational domain has been set large enough to ensure that any possible reflections are negligible. The source used to excite the structure was a Gaussian pulse modulated in time by a sinusoidal function with the shape of a 2D Gaussian bell, expressed as

$$E_z(x, y, t) = e^{-\left(\frac{x-x_0}{X_0}\right)^2} e^{-\left(\frac{y-y_0}{Y_0}\right)^2} e^{-\left(\frac{t-t_0}{T_0}\right)^2} \sin\left(2\pi f_s t\right) \qquad (7.1)$$

where x_0 and X_0 are the position of the peak and the width of the Gaussian bell in the x-direction, respectively, y_0 and Y_0 are the position of the peak and the width of the Gaussian bell in the y-direction, respectively, t_0 and T_0 are the time delay and the time width of the Gaussian pulse, respectively, and f_s

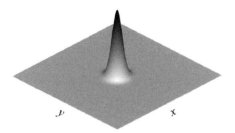

Figure 7.2 Gaussian bell used to excite the 7×7 PhC cavity.

is the central frequency. The time-domain parameters of the Gaussian pulse were fixed to $T_0 = 30$ fs and $t_0 = 3T_0$ in order to excite the structure with a signal with an appropriate bandwidth, while the central frequency was set to $f_s = 199.86$ THz ($\lambda_s = 1.5$ µm). For all simulations, x_0 and y_0 were set to the coordinates of the centre of the PhC cavity, while X_0 and Y_0 were fixed to $a/2$ where a is the lattice constant. Figure 7.2 shows an example of the Gaussian bell used to excite the PhC cavity.

A detector was inserted at the centre of the PhC cavity in order to store the time-domain evolution of the electric-field component from which the resonant wavelength, λ_{res}, and quality factor, Q, were calculated using [2]

$$Q = 2\pi \frac{|\Phi_t|^2}{|\Phi_t|^2 - |\Phi_{t+T}|^2} \tag{7.2}$$

where Φ_t represents the main electromagnetic field component of the resonance at time t, Φ_{t+T} represents the same component at time $t + T$ and T is the time cycle of the resonant mode. The results of all simulations are summarised in Figure 7.3.

From this figure it can be clearly seen that the results reach a stable behaviour when the mesh size is denser than 20 points per lattice constant.

In particular, for a mesh size of 35 points per lattice constant, the resonant wavelength has been calculated to be $\lambda_{\text{res}} = 1.549$ µm, while the quality factor has been computed to be $Q \cong 1450$. Figures 7.4 and 7.5 also show the envelope of the time-domain variation of the electric field inside the PhC cavity and its energy-density spectrum in the frequency domain, which has been obtained though FFT of the time-domain variation. The relatively high Q obtained for this 7×7 PhC cavity can be clearly seen from the slow decay of the time-domain variation of the electric field, shown in Figure 7.4 and from the strong confinement of the electric field inside the cavity itself, shown in Figure 7.6.

Furthermore, these results are in strong agreement with their counterparts already published in the literature in which other computational methods have been used [3]. Based on the results obtained from this analysis, all the simulations have been carried

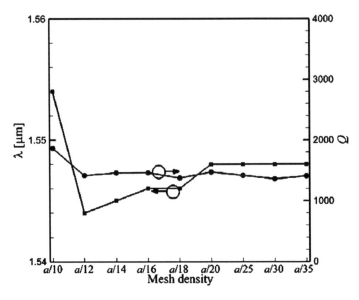

Figure 7.3 Variation of the resonant wavelength, λ_{res}, and quality factor, Q, with the mesh density of the discretised computational domain. (Reproduced with permission from Pinto, D. and Obayya, S.S.A. (2008) 2D Analysis of multimode photonic crystal resonant cavities with the finite volume time domain method. *Opt. Quantum Electron.*, **40** (11–12), 875–890. Copyright 2008 Springer Science and Business Media.)

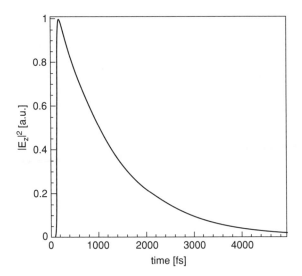

Figure 7.4 Time-domain variation of the electric field inside the 7×7 PhC resonant cavity. (Reproduced with permission from Pinto, D. and Obayya, S.S.A. (2008) 2D Analysis of multimode photonic crystal resonant cavities with the finite volume time domain method. *Opt. Quantum Electron.*, **40** (11–12), 875–890. Copyright 2008 Springer Science and Business Media.)

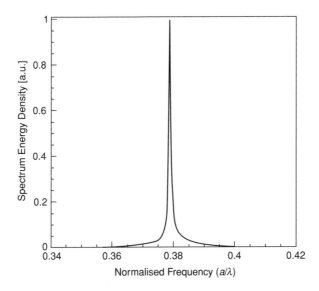

Figure 7.5 Energy-density spectrum of the time electric field inside the 7×7 PhC resonant cavity. (Reproduced with permission from Pinto, D. and Obayya, S.S.A. (2008) 2D Analysis of multimode photonic crystal resonant cavities with the finite volume time domain method. *Opt. Quantum Electron.*, **40** (11–12), 875–890. Copyright 2008 Springer Science and Business Media.)

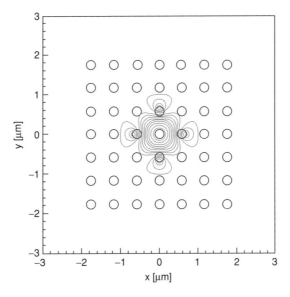

Figure 7.6 Electric-field distribution of the resonant mode of the 7×7 PhC resonant cavity. (Reproduced with permission from Pinto, D. and Obayya, S.S.A. (2008) 2D Analysis of multimode photonic crystal resonant cavities with the finite volume time domain method. *Opt. Quantum Electron.*, **40** (11–12), 875–890. Copyright 2008 Springer Science and Business Media.)

out with computational domains discretised using a mesh density of 20 points per lattice constant.

7.2.1   UPML Effectiveness

In further work, a thorough investigation on the effects of the UPML parameters on the resonant wavelength, λ_{res}, and quality factor, Q, has been carried out. As a first analysis, the distance between the PhC resonant cavity and the UPML layer has been varied from a minimum of 0.1 µm to a maximum of 2.0 µm. For the conductivity in the UPML layers a geometrical grading has been chosen and the maximum value of the conductivity σ_{max} of the UPML absorber has been calculated using the relationship

$$\sigma_{max} = \left(\frac{\varepsilon_0 c_0}{w_{upml}} \right) \left(\frac{p+1}{2} \right) \left(\ln \left(\frac{1}{\Gamma_{th}} \right) \right) \tag{7.3}$$

where ε_0 and c_0 are the free space permittivity and the light velocity in free space, respectively, w_{upml} is the width of the UPML layer, p is the order of geometrical grading and Γ_{th} is the theoretical reflection coefficient of the UPML absorber. The parameters of the UPML have been fixed to constant values: $w_{upml} = 0.75$ µm, $p = 3$ and $\Gamma_{th} = 10^{-5}$. The parameters of the source, expressed by Equation (7.1), have been fixed to the values used for the previous example and have been used for all the simulations carried out in this investigation. For all simulations, the resonant wavelength, λ_{res}, and quality factor, Q, have been calculated and the results are depicted in Figure 7.7.

As can be clearly seen from this figure, for distances higher than 0.5 µm, the influence of the UPML absorber on the resonant wavelength, λ_{res}, and the quality factor, Q, is negligible, as the cavity parameters assume almost a constant value.

Next, the effect of the UPML thickness on the cavity parameters has been investigated. In Figure 7.8, the conductivity of the UPML layer for different UPML thicknesses is depicted and, as can be clearly seen from this figure, the thicker the UPML layer the smoother the grading of the conductivity, σ. A different simulation had been run for each thickness of the UPML layer and for all of them the cavity parameters have been computed. All the other parameters of the UPML have been fixed to constant values throughout all simulations: the distance between the UPML and the structure, $d = 0.5$ µm, $p = 3$ and $\Gamma_{th} = 10^{-5}$. The results of this procedure are shown in Figure 7.9.

As can be observed from this figure, the convergence of the cavity parameters to the values obtained in the previous example is obtained for UPML layers thicker than 0.75 µm. An explanation of this behaviour can be given considering the shape of the grading of the conductivity, σ, in the UPML layer. For a thickness of 0.5 µm the conductivity, σ, assumes a faster transition from zero to the maximum value calculated

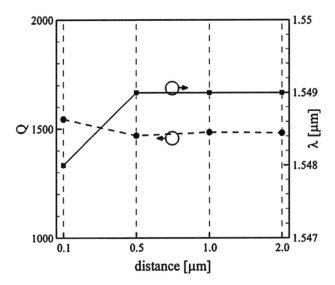

Figure 7.7 Variation of the resonant wavelength, λ_{res}, and quality factor, Q, with the distance between the UPML absorber and the PhC cavity. (Reproduced with permission from Pinto, D. and Obayya, S.S.A. (2008) 2D Analysis of multimode photonic crystal resonant cavities with the finite volume time domain method. *Opt. Quantum Electron.*, **40** (11–12), 875–890. Copyright 2008 Springer Science and Business Media.)

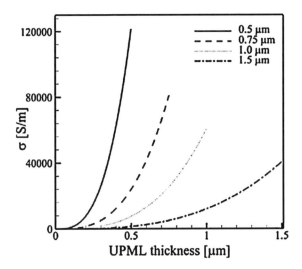

Figure 7.8 Variation of the conductivity of the UPML absorber for different thicknesses of the UPML absorber. (Reproduced with permission from Pinto, D. and Obayya, S.S.A. (2008) 2D Analysis of multimode photonic crystal resonant cavities with the finite volume time domain method. *Opt. Quantum Electron.*, **40** (11–12), 875–890. Copyright 2008 Springer Science and Business Media.)

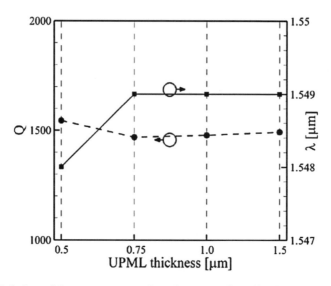

Figure 7.9   Variation of the resonant wavelength, λ_{res}, and quality factor, Q, with the UPML thickness. (Reproduced with permission from Pinto, D. and Obayya, S.S.A. (2008) 2D Analysis of multimode photonic crystal resonant cavities with the finite volume time domain method. *Opt. Quantum Electron.*, **40** (11–12), 875–890. Copyright 2008 Springer Science and Business Media.)

by Equation (7.3) which, as the last result, gives a higher discretisation error for the UPML layer in the computational domain for a fixed discretisation step.

The effects of the theoretical reflection coefficient Γ_{th} on the cavity parameters have been analysed below.

The theoretical reflection coefficient has been varied from a maximum value of 10^{-3} to a minimum value of 10^{-10}, while all the other parameters of the UPML have been set to constant values: $d = 0.5$ μm, $p = 3$ and $w_{upml} = 0.75$ μm. Figure 7.10 shows the conductivity of the UPML layer for different values of the theoretical reflection coefficient.

For all simulations the resonant wavelength, λ_{res}, and quality factor, Q, have been calculated and the result of this procedure is shown in Figure 7.11.

From this figure it can be clearly seen that a value of the theoretical reflection coefficient, Γ_{th}, lower than 10^{-5} can assure a good level of absorption of the radiated electromagnetic waves and, as a direct consequence, a good level of accuracy of the results. An explanation of this behaviour can be given by direct inspection of Figure 7.10. From this figure it can be seen that for the theoretical reflection coefficient, $\Gamma_{th} = 10^{-3}$, the maximum value of the conductivity, σ, is lower with respect to the other curves; this implies a lower absorption rate of the electromagnetic waves propagating inside the UPML layer which are not completely absorbed before reaching the end of the UPML layer and are reflected back to the computational domain.

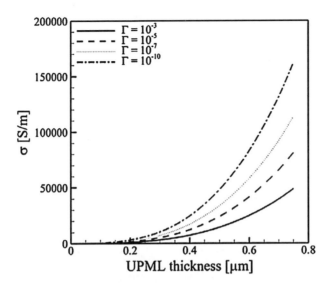

Figure 7.10 Variation of the conductivity of the UPML absorber for different theoretical reflection coefficients , Γ_{th}. (Reproduced with permission from Pinto, D. and Obayya, S.S.A. (2008) 2D Analysis of multimode photonic crystal resonant cavities with the finite volume time domain method. *Opt. Quantum Electron.*, **40** (11–12), 875–890. Copyright 2008 Springer Science and Business Media.)

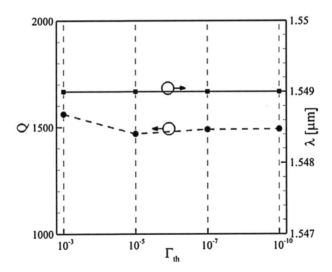

Figure 7.11 Variation of the resonant wavelength, λ_{res}, and quality factor, Q, with the theoretical reflection coefficient Γ_{th}. (Reproduced with permission from Pinto, D. and Obayya, S.S.A. (2008) 2D Analysis of multimode photonic crystal resonant cavities with the finite volume time domain method. *Opt. Quantum Electron.*, **40** (11–12), 875–890. Copyright 2008 Springer Science and Business Media.)

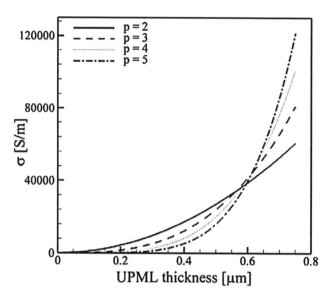

Figure 7.12 Variation of the conductivity of the UPML absorber for different order, p, of the polynomial grading. (Reproduced with permission from Pinto, D. and Obayya, S.S.A. (2008) 2D Analysis of multimode photonic crystal resonant cavities with the finite volume time domain method. *Opt. Quantum Electron.*, **40** (11–12), 875–890. Copyright 2008 Springer Science and Business Media.)

As a last example, the effects of the polynomial order, p, of the grading of the UPML layer on the cavity parameters have been investigated. Different values of the polynomial order, p, have been tested for the UPML layer while all the other parameters have been maintained at constant values: $d = 0.5$ μm, $= 10^{-5}$ and $w_{upml} = 0.75$ μm. Figure 7.12 shows the conductivity of the UPML layer for different values of the polynomial order.

For all simulations the quality factor, Q, and the resonant wavelength, λ_{res}, have been calculated and the results are shown in Figure 7.13.

From this figure it can be clearly seen that for a polynomial order, p, higher than 2 the cavity parameters converge to constant values. From the simulations carried out for the investigation of the effects of the UPML parameters on the PhC cavity characteristics, optimum settings for best performance of the UPML layers have been extracted and have been used for all simulations. The optimum values for the UPML layers have been set to $d = 0.5$ μm, $w_{upml} = 0.75$ μm, $\Gamma_{th} = 10^{-5}$ and $p = 3$.

7.3 FVTD Method Assessment: PhC Waveguide

As a last test for assessing the effectiveness of the absorption of the UPML boundary conditions (BC), a PBG waveguide has been investigated. The PBG waveguide consists of dielectric rods arranged by a square lattice with periodicity, $a = 0.58$ μm,

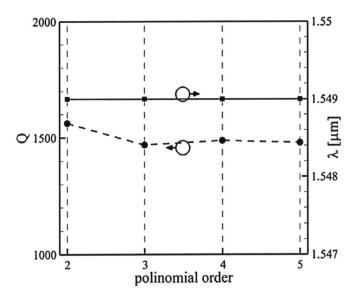

Figure 7.13 Variation of the resonant wavelength, λ_{res}, and quality factor, Q, with the order, p, of the polynomial grading. (Reproduced with permission from Pinto, D. and Obayya, S.S.A. (2008) 2D Analysis of multimode photonic crystal resonant cavities with the finite volume time domain method. *Opt. Quantum Electron.*, **40** (11–12), 875–890. Copyright 2008 Springer Science and Business Media.)

refractive index, $n_{rods} = 3.4$ in air ($n = 1.0$) and radius, $r = 0.18a$, which possesses a photonic bandgap for TE polarisation that extends from 0.302 to 0.443 in terms of normalised frequency (a/λ) [4]. A schematic of this structure is shown in Figure 7.14.

The computational domain has been divided in 150 888 volumes and the time step taken as d$t = 0.0154$ fs; the UPML thickness has been fixed to 1.5 µm and it has been divided into an approximate number of 30 volumes. The order of the profile function,

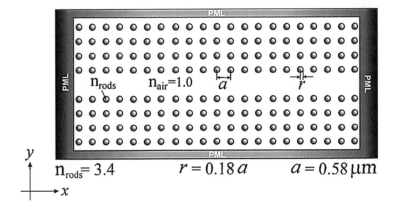

Figure 7.14 Schematic diagram of the PBG waveguide.

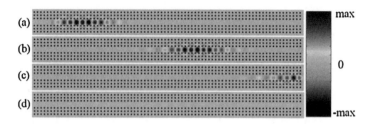

Figure 7.15 Electric-field profile of the TE mode inside the PBG waveguide (a) launched into waveguide at $t = 20$ fs, (b) propagating inside the PBG waveguide at $t = 80$ fs, (c) impinging on the UPML ABC at $t = 140$ fs and (d) at $t = 200$ fs absorbed by UPML ABC. (Reproduced with permission from Pinto, D. and Obayya, S.S.A. (2008) Accurate perfectly matched layer finite-volume time-domain method for photonic bandgap devices. *IEEE Photon. Technol. Lett.*, **20** (5), 339–365. © 2008 IEEE.)

p, has been set to 3, while the theoretical reflection coefficient, Γ_{th}, has been fixed to 10^{-5}.

A Gaussian-shaped pulse in space with normalised frequency $f_{norm} = 0.387$ ($\lambda = 1.5$ µm), modulated by a Gaussian pulse in time, has been launched into a PBG waveguide. The propagation of the TE mode inside the PBG waveguide at different times is shown in Figure 7.15. As may be clearly observed from these figures, the mode is nicely propagating inside the PBG waveguide, and as it approaches the UPML region, it is clearly absorbed. The reflection coefficient due to the UPML has been calculated for this structure and it is shown in Figure 7.16. As can be clearly seen from this figure, the reflection from the UPML is as low as −50 dB over the entire range of frequencies of the source, proving the excellent absorbing properties of the UPML boundary conditions.

7.4 FVTD Method Assessment: PBG T-Branch

Below, a PBG T-branch, whose schematic is shown in Figure 7.17, has beenis analysed with the PBG material having same the characteristics as the PBG waveguide considered in the previous example [4].

The source used was a Gaussian-shaped source in space and in time, as expressed in Equation (7.1). The centre normalised frequency was set to $f_{norm} = 0.387$ ($\lambda = 1.5$ µm) and the parameters of the Gaussian pulse in time were set in order to excite the structure with a wideband signal. This structure has been discretised using 109 608 volumes and the time step was about $dt = 0.0122$ fs. The entire simulation took almost 2 hours to run on a desktop PC (Pentium IV, 3 GHz with 1 GB of RAM). As can be clearly seen from Figure 7.18, the results obtained with the developed FVTD are in very good agreement with those published in [4].

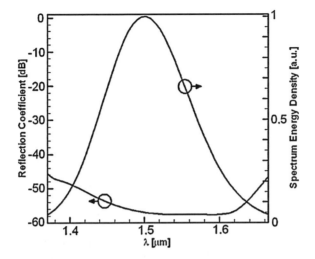

Figure 7.16 UPML reflection coefficient for a PBG waveguide and energy-density spectrum of the propagating pulse inside the PBG waveguide. (Reproduced with permission from Pinto, D. and Obayya, S.S.A. (2008) Accurate perfectly matched layer finite-volume time-domain method for photonic bandgap devices. *IEEE Photon. Technol. Lett.*, **20** (5), 339–365. © 2008 IEEE.)

7.5 PhC Multimode Resonant Cavity

In further work, a different PhC resonant cavity has been considered. The structure consists of a dielectric slab with refractive index, $n_2 = (11.4)^{1/2}$, in which air holes ($n = 1.0$) of radius, $r = 0.45a$ have been drilled with a triangular pattern with lattice constant, $a = 0.650\,225$ µm, as shown in Figure 7.19. From this figure it can be clearly

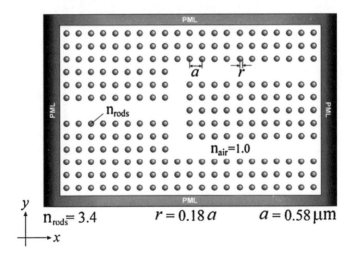

Figure 7.17   Schematic diagram of a PBG T-junction.

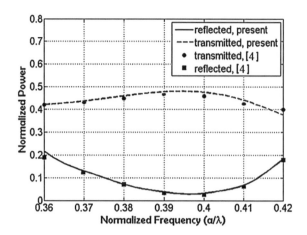

Figure 7.18 Spectra of the normalised reflected and transmitted power in the PBG T-brunch shown in the inset. (Reproduced with permission from Pinto, D. and Obayya, S.S.A. (2008) Accurate perfectly matched layer finite-volume time-domain method for photonic bandgap devices. *IEEE Photon. Technol. Lett.*, **20** (5), 339–365. © 2008 IEEE.)

observed that the cavity defect is represented by a missing air hole at the centre of the PhC. For all simulations, the structure has been discretised using an unstructured mesh with a density of points sufficient to place a minimum of 10 points along the diameters of the dielectric rods, while the UPML layers were set using the optimum values of the parameters obtained from the performance analysis of the previous section. The

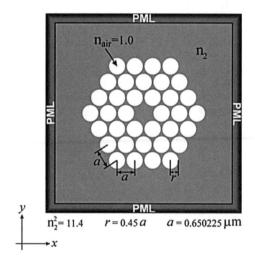

Figure 7.19 Schematic diagram of a three-ring hexagonal PhC resonant cavity. (Reproduced with permission from Pinto, D. and Obayya, S.S.A. (2008) 2D Analysis of multimode photonic crystal resonant cavities with the finite volume time domain method. *Opt. Quantum Electron.*, **40** (11–12), 875–890. Copyright 2008 Springer Science and Business Media.)

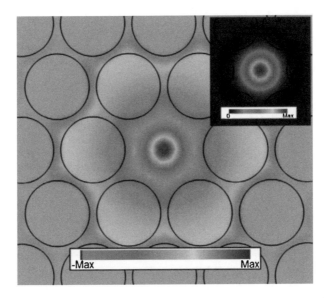

Figure 7.20   Magnetic-field distribution of the 'monopole' resonant mode inside the three-ring hexagonal PhC resonant cavity. In the inset, the profile of the source used to excite the resonant mode.

main characteristic of this PhC resonant cavity is that it can sustain different TM resonant modes [3]. In order to extract all the parameters of each resonant mode, it is necessary to isolate each single mode in order to avoid beating amongst the different modes, which, as an ultimate result, makes it impossible to obtain useful data to use for post-processing analysis. To accomplish this, a strategically chosen source profile has been used in order to excite each single resonant mode. The first resonant mode analysed was the 'monopole' resonant mode.

To excite this mode, the magnetic field H_z has been excited with a source, expressed by Equation (7.1), shaped like a Gaussian bell positioned at the centre of the cavity ($x_0 = y_0 = 0$) with the width along the x- and y-axes fixed to $X_0 = a/3$ and $Y_0 = a/3$, respectively, as shown in the inset of Figure 7.20. This source profile has been multiplied by a Gaussian pulse in time modulated by a sinusoidal carrier with frequency set to $f_s = 193.41$ THz ($\lambda_s = 1.55$ μm). The parameters of the Gaussian pulse, T_0 and t_0, were set to 30 fs and 90 fs, respectively, in order to excite the PhC resonant cavity with a wide range of frequencies. A detector has been inserted at the centre of the cavity to store the time-domain evolution of the magnetic field, H_z. Exciting the PhC cavity with this particular source assures the resonance of only one mode inside the cavity itself. This assumption is confirmed by Figure 7.20, in which the magnetic field profile, H_z, obtained from the simulation is shown.

As can be clearly seen from this figure, the resonant field profile presents a single peak at the centre of the PhC cavity from which the name 'monopole' derives. In

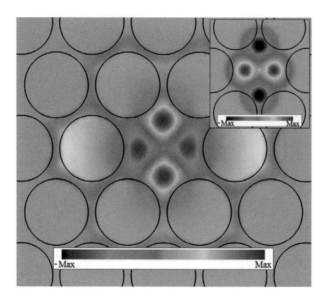

Figure 7.21 Magnetic field distribution of the 'quadrupole' resonant mode inside the three-ring hexagonal PhC resonant cavity. In the inset, profile of the source used to excite the resonant mode.

this way, from the time-domain data stored by the detector it has been possible to extract all the necessary parameters of the resonant mode. For the 'monopole' mode the resonant wavelength has been calculated to be $\lambda_{res} = 1.545$ µm, while the quality factor has been computed to be $Q = 779$.

A similar procedure has been followed for the extraction of the parameters of the 'quadrupole' resonant mode. For this mode, the source consists of four Gaussian bells, with alternate signs for the peak values, placed at specific points inside the PhC cavity, as shown in the inset of Figure 7.21. The parameters of the Gaussian bells were fixed to $x_0^1 = -a/3$, $y_0^1 = 0$, $X_0^1 = a/5$ and $Y_0^1 = a/5$ for the first Gaussian bell, $x_0^2 = 0$, $y_0^2 = a/2$, $X_0^2 = a/5$ and $Y_0^2 = a/5$ for the second Gaussian bell, $x_0^3 = a/3$, $y_0^3 = 0$, $X_0^3 = a/5$ and $Y_0^3 = a/5$ for the third Gaussian bell and $x_0^4 = 0$, $y_0^4 = -a/2$, $X_0^4 = a/5$ and $Y_0^4 = a/5$ for the fourth Gaussian bell.

The parameters of the Gaussian pulse in time and the carrier frequency were set to the same values of the previous example. In Figure 7.21 the magnetic field profile inside the computational domain is shown. From this figure, it can be clearly seen that the 'quadrupole' mode is the only resonant mode that has been excited. From the time-domain variation of the magnetic field it has been possible to extract the resonant wavelength and the quality factor of the selected mode, which have been calculated to be $\lambda_{res} = 1.639$ µm and $Q = 1660$, respectively.

The last mode to be selected from the PhC resonant cavity is the 'hexapole' resonant mode. The source used for this example is a superposition of six Gaussian bells, with

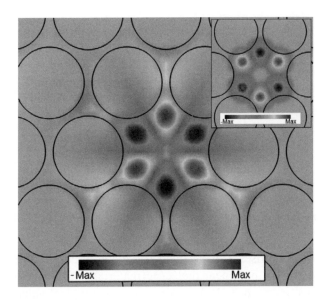

Figure 7.22 Magnetic-field distribution of the 'hexapole' resonant mode inside the three-ring hexagonal PhC resonant cavity. In the inset, profile of the source used to excite the resonant mode.

alternate signs for the peak values, placed at strategic points inside the PhC cavity, as shown in the inset of Figure 7.22. The parameters of the Gaussian bells were chosen to be $x_0^1 = -a/3$, $y_0^1 = a/4$, $X_0^1 = a/7$ and $Y_0^1 = a/7$ for the first Gaussian bell, $x_0^2 = 0$, $y_0^2 = a/2$, $X_0^2 = a/7$ and $Y_0^2 = a/7$ for the second Gaussian bell, $x_0^3 = a/3$, $y_0^3 = a/4$, $X_0^3 = a/7$ and $Y_0^3 = a/7$ for the third Gaussian bell, $x_0^4 = a/3$, $y_0^4 = -a/4$, $X_0^4 = a/7$ and $Y_0^4 = a/7$ for the fourth Gaussian bell, $x_0^5 = 0$, $y_0^5 = -a/2$, $X_0^5 = a/7$ and $Y_0^5 = a/7$ for the fifth Gaussian bell and $x_0^6 = -a/3$, $y_0^6 = -a/4$, $X_0^6 = a/7$ and $Y_0^6 = a/7$ for the sixth Gaussian bell. The time-domain parameters of the source were fixed to the same values used for the two previous examples. Figure 7.22 shows the magnetic field profile inside the PhC cavity.

From this figure, the profile of the 'hexapole' resonant mode can be observed, which confirms that the source used for this simulation has excited only the 'hexapole' mode. Furthermore, from the time-domain evolution of the magnetic field the resonant wavelength and the quality factor have been computed to be $\lambda_{res} = 1.422$ μm and $Q = 3223$, respectively.

7.6   FDTD Analysis of Nonlinear Devices

Before proceeding with the analysis of a nonlinear periodic structure, a nonlinear slab waveguide, shown in Figure 7.23, will first be considered [5].

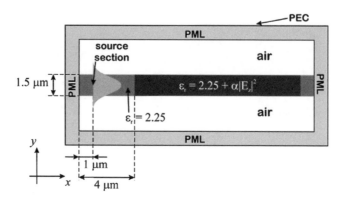

Figure 7.23   Schematic diagram of a nonlinear slab waveguide [5].

This waveguide comprises a linear dielectric material core with refractive index, $n_0 = 1.5$, followed by a nonlinear dielectric material with a nonlinear part of the nonlinear refractive index, $\alpha = 100$ m^2/V^2, which is a typical value for a nonlinear coefficient adopted in the literature [5], and width set to $w = 1.5$ µm. It should mentioned that for this example the nonlinearity has been modelled using

$$\varepsilon = \varepsilon_0 \varepsilon_r = \varepsilon_0 n^2 = \varepsilon_0 \left(n_0 + \alpha \, |E|^2\right)^2 \tag{7.4}$$

where ε_0 is the permittivity of the space, n_0 is the linear part of the nonlinear refractive index of the nonlinear dielectric medium, α is the nonlinear coefficient of the nonlinear refractive index of the dielectric medium, and in this relationship it has units of m^2/V^2, and where E is the electric field. The structure is surrounded by air ($n = 1.0$). The structure was excited, as shown below, using a CW source with the shape of the fundamental TE$_0$ mode profile

$$\Phi_s(y, t) = \Phi_0(y) \sin(2\pi f_0 t) \tag{7.5}$$

where Φ represents the main electromagnetic-field component, in this case the electric-field component E_z, $f_0 = 0.6 \times 10^{15}$ Hz is the modulation frequency and $\Phi_0(y)$ is the fundamental TE$_0$ mode profile. The power of the TE$_0$ mode launched in the structure was normalised to 1 W/m. Because of the relatively low value of the nonlinear coefficient of the nonlinear media and the low value of the power of the source employed for this example, a weak nonlinear effect is expected to influence the electromagnetic field of the fundamental mode propagating inside the nonlinear slab waveguide. Figure 7.24 shows the electric field component, E_z, inside the nonlinear slab waveguide obtained with the method developed here and with an FDTD code that solves the wave equation. As clearly observed, the FDTD approach developed here gives an electric-field distribution that is more accurate than the one obtained

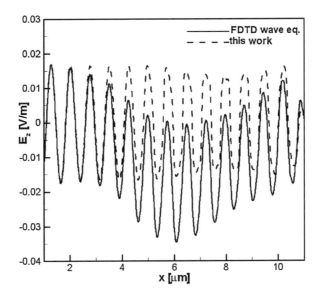

Figure 7.24 Electric-field component, E_z, in a nonlinear slab waveguide at $T = 50$ fs calculated with FDTD for wave equations and FDTD for all TE electromagnetic-field components. (Reproduced with permission from Pinto, D., Obayya, S.S.A., Rahman, B.M.A. and Grattan, K.T.V. (2006) FDTD analysis of nonlinear bragg grating based devices. *Opt. Quantum Electron.*, **38** (15), 1217–1235. Copyright 2006 Springer Science and Business Media.)

by solving the wave equation through an explicit FDTD procedure. The accuracy of the developed method can be observed in the absence of spurious oscillations of the electric field amplitude of the E_z component. As may be observed from the same figure, these spurious oscillations lead to amplitude fluctuations of the E_z component ,which is evident in the electric-field distribution obtained through explicit FDTD updating of the wave equation. On the other hand, the results of the FDTD approach developed here compare well with those obtained using a hybrid implicit–explicit FDTD scheme [5].

In order to emphasise the self-phase-modulation (SPM) effect, the structure shown in Figure 7.23 was excited with the fundamental TE$_0$ mode profile modulated by a Gaussian pulse in time, as shown below

$$\Phi_s(y, t) = \Phi_0(y) \sin(2\pi f_0 t) e^{-\left(\frac{2(t-t_0)}{T_0}\right)^2} \qquad (7.6)$$

where Φ represents the main electromagnetic-field component, in this case the electric-field component, E_z, $f_0 = 0.6 \times 10^{15}$ Hz is the modulation frequency, $T_0 = 5.0$ fs is the width in time (which also determines the bandwidth of the signal) and $t_0 = 3T_0$ is the time delay of the above-mentioned pulse. The time-domain variation of the propagating electric field was monitored at different points inside the waveguide.

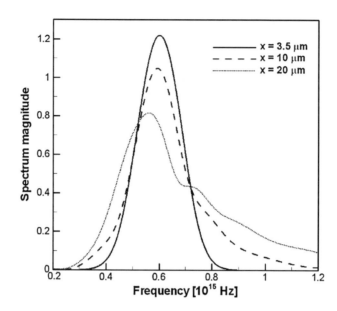

Figure 7.25 Spectral distribution of the electric field observed at three different points in the nonlinear slab waveguide. (Reproduced with permission from Pinto, D., Obayya, S.S.A., Rahman, B.M.A. and Grattan, K.T.V. (2006) FDTD analysis of nonlinear bragg grating based devices. *Opt. Quantum Electron.*, **38** (15), 1217–1235. Copyright 2006 Springer Science and Business Media.)

Following the use of a fast Fourier transform (FFT) of these data, spectra of the time-domain responses were obtained and are shown in Figure 7.25.

As can be seen from this figure, spectral broadening, arising due to the SPM effect, is evident. It can also be noted that this spectral broadening becomes larger and the spectral peak is shifted towards lower frequencies as the pulse propagates further inside the nonlinear waveguide. Figure 7.26 shows the electric-field profile at $t = 50$ fs at the centre of the waveguide, inside the nonlinear part, during its propagation. In this figure, the effect of the SPM can most noticeably be seen in the compression of the pulse at the back and the rarefaction at the front.

An analysis of nonlinear periodic media will next be performed. The nonlinear periodic structure considered first is a Bragg reflector (BR), consisting of alternating layers of linear and nonlinear media with a periodicity equal to Λ and number of periods equal to N, as shown in Figure 7.27. The waveguide width is $w = 1.6$ μm, the refractive index of the waveguide and the BR linear medium is $n_w = 2.0$, the linear part of the nonlinear refractive index of the BR is $n_0 = 2.03$ and the nonlinear parameter, α, is variable. Also for this example the nonlinearity has been modelled using Equation (7.4).

The structure is surrounded by air ($n = 1.0$). It is well known from previous work [6] that in the BR region, composed of linear dielectric media, waves travelling in

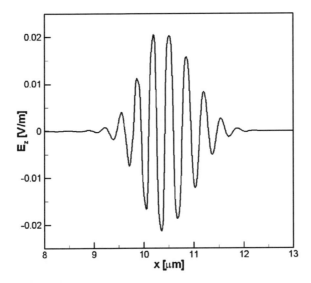

Figure 7.26 Electric-field profile at $t = 50$ fs inside the nonlinear part of the waveguide. (Reproduced with permission from Pinto, D., Obayya, S.S.A., Rahman, B.M.A. and Grattan, K.T.V. (2006) FDTD analysis of nonlinear bragg grating based devices. *Opt. Quantum Electron.*, **38** (15), 1217–1235. Copyright 2006 Springer Science and Business Media.)

opposite directions with propagation constants, $\pm\beta$, at an angular frequency, ω, are coupled. In this case the Bragg condition is expressed by

$$2\beta = K \equiv \frac{2\pi}{\Lambda} \tag{7.7}$$

where K is the grating wavenumber. The wavelength that satisfies Equation (7.7) is given by

$$\lambda = 2n_{\text{eff}}\Lambda \tag{7.8}$$

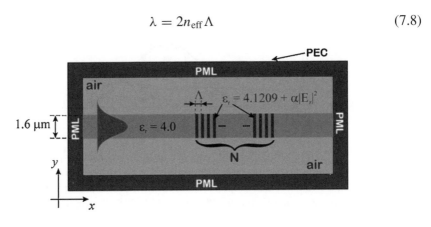

Figure 7.27 Schematic diagram of a nonlinear Bragg reflector (BR).

where n_{eff} is the modal index of the planar waveguide. The small reflections caused by each discontinuity in the layered media of the BR add constructively in the backward direction. At wavelength, λ, the coupling between the opposite travelling waves inside the BR is maximised, and, as a result, the transmission is minimised. In this example, λ is fixed at 0.99 μm, and for the waveguide considered here $n_{\text{eff}} = 1.9806$. With these data, using Equation (7.8) the periodicity of the BR is $\Lambda = 0.25$ μm. The linear case ($\alpha = 0$ m^2/V^2) was first considered for three different numbers of periods of the BR, $N = 10$, $N = 20$ and $N = 30$. The fundamental TE$_0$ mode profile, modulated by a Gaussian pulse in time whose expression is given by Equation (7.6), was used to excite the structure and the frequency f_0 was set to 300 THz ($\lambda = 0.99$ μm). The parameters of the Gaussian pulse were set to the values used for the previous example. The time-domain responses of the incident and reflected electric fields were monitored for each of the three cases given for the number of periods of the BR. The FFT of both the incident and the reflected time-domain responses was performed in the calculation of the reflection coefficient. Figure 7.28 shows the reflection coefficient spectra calculated for the three different cases mentioned above. In the same figure

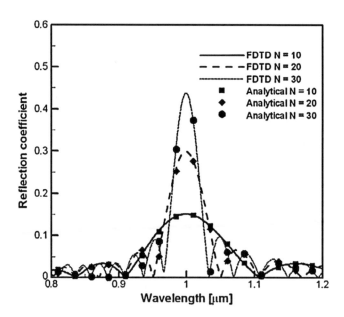

Figure 7.28 Variation of the reflection coefficient of a linear BR with wavelength for three different values of number of periods, N. (Reproduced with permission from Pinto, D., Obayya, S.S.A., Rahman, B.M.A. and Grattan, K.T.V. (2006) FDTD analysis of nonlinear bragg grating based devices. *Opt. Quantum Electron.*, **38** (15), 1217–1235. Copyright 2006 Springer Science and Business Media.)

analytical data points have been added in order to verify the accuracy of the FDTD approach developed here for the linear case. Very good agreement can be noted between the data points obtained from the FDTD simulations and the analytical data. As can be seen from this figure, the reflection coefficient and the selectivity of the BR both increase with the number of periods. This can be explained by noting that by increasing the number of periods, the coupling between the forward and the backward travelling waves inside the BR becomes stronger.

Next, nonlinear media were added to the previous BR structure with the nonlinear part of the nonlinear refractive index given by $\alpha = 500$ m^2/V^2. The fundamental TE$_0$ mode profile modulated by a Gaussian pulse with the same values of the parameters used for the linear structure was used as an excitation source, and its power was set to 5 W/m.

Simulations were carried out for three different values of the number of periods of the BR structure, $N = 10$, $N = 20$ and $N = 30$. The time-domain responses of the incident and the reflected electric field were monitored and the reflection coefficients for the three different structures were computed using the same procedure utilised for the linear BR. Figure 7.29 shows the reflection coefficient spectra calculated for

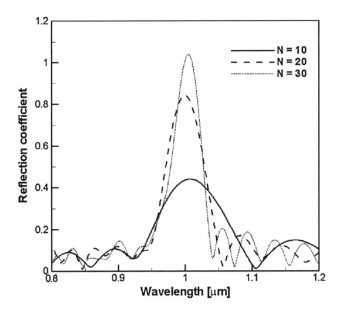

Figure 7.29 Variation of the reflection coefficient of a nonlinear BR with the wavelength for three different values of number of periods, N. (Reproduced with permission from Pinto, D., Obayya, S.S.A., Rahman, B.M.A. and Grattan,K.T.V. (2006) FDTD analysis of nonlinear bragg grating based devices. *Opt. Quantum Electron.*, **38** (15), 1217–1235. Copyright 2006 Springer Science and Business Media.)

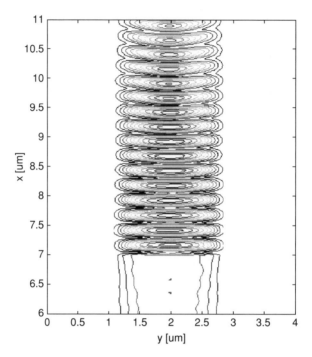

Figure 7.30 Sinusoidal steady-state amplitude of a nonlinear BR with $N = 20$ excited with
a continuous wave of frequency $f = 300$ THz and power $P = 5$ W/m. (Reproduced with
permission from Pinto, D., Obayya, S.S.A., Rahman, B.M.A. and Grattan, K.T.V. (2006)
FDTD analysis of nonlinear bragg grating based devices. *Opt. Quantum Electron.*, **38** (15),
1217–1235. Copyright 2006 Springer Science and Business Media.)

these structures. As can be noticed from this figure, the reflection coefficient and
the selectivity of the BR increase as the number of periods increases. Furthermore,
a shift in the peak wavelength of the reflection coefficient towards higher values
may also be noted as the number of periods, N, increases. It can be also observed
that the reflection coefficients obtained with the nonlinear BR are higher than their
counterparts obtained with a linear BR. Figures 7.30 and 7.31 show the contour plots
of the sinusoidal steady-state amplitude of the electric field at the interface between
the waveguide and the nonlinear BR.

The sinusoidal steady-state amplitude has been obtained through a numerical integration of the time-domain signal over one period, as shown in the work of De Pourcq
and Eng [7]. The characteristic of a sinusoidal source when its steady-state condition
is reached can be expressed as

$$f(t) = a + A \cos(\omega t + \phi) \tag{7.9}$$

where a represents the DC component, and A, ω, and ϕ represent the maximum amplitude, the angular frequency and the initial phase of the sinusoidal

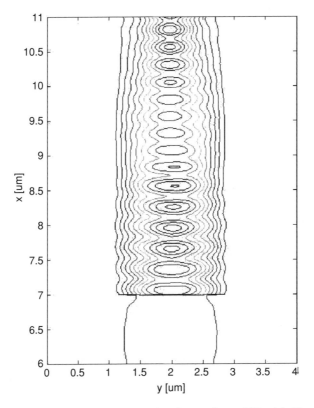

Figure 7.31 Sinusoidal steady-state amplitude of a nonlinear BR with $N = 20$ excited with a continuous wave of frequency $f = 254$ THz and power $P = 5$ W/m. (Reproduced with permission from Pinto, D., Obayya, S.S.A., Rahman, B.M.A. and Grattan, K.T.V. (2006) FDTD analysis of nonlinear bragg grating based devices. *Opt. Quantum Electron.*, **38** (15), 1217–1235. Copyright 2006 Springer Science and Business Media.)

steady-state signal, respectively. These values can be calculated from following relationships

$$\frac{1}{T} \int_{t_0}^{t_0+T} f(t)\, dt = a \tag{7.10}$$

$$\frac{1}{T} \int_{t_0}^{t_0+T} f(t) \cos(\omega t)\, dt = \frac{A}{2} \cos \phi \tag{7.11}$$

$$\frac{1}{T} \int_{t_0}^{t_0+T} f(t) \sin(\omega t)\, dt = -\frac{A}{2} \sin \phi \tag{7.12}$$

where T represents the period of the sinusoidal function, and t_0 represent an initial time for the integration. To obtain the steady state, the total field-scattered field approach for the excitation source was implemented [8]. The number of periods of BR considered in this example was $N = 20$, the power of the source was set to 5 W/m, and the frequency of the sinusoidal signal was fixed at $f = 300$ THz, a value at which the reflection coefficient of the BR has its maximum ($\cong 0.84$). The source section was placed at 7 μm, while the BR interface was placed at 9 μm. From Figure 7.30, the high reflectivity of the nonlinear BR can be clearly observed on the formation of a strong standing-wave pattern whose peaks are evident in the region where the incident and reflected waves are contrarily propagating.

Figure 7.31 shows a contour plot of the sinusoidal steady-state amplitude of the electric field obtained using sinusoidal signal frequency fixed at $f = 254$ THz, at which the reflection coefficient of the nonlinear BR has a low value ($\cong 0.12$). The standing wave is still present, but its peaks are less pronounced than was evident from the previous case. Next, the reflection coefficient for a nonlinear BR was calculated for different numbers of periods N, and for different levels of power of the incident pulse.

For all the simulations carried out, the TE_0 fundamental mode profile was used, as expressed by Equation (7.6). The parameters of the Gaussian pulse used to modulate the signal source in time were set to $T_0 = 5$ fs and $t_0 = 3T_0$, and the centre wavelength was fixed at $\lambda_0 = 0.99$ μm. Time-domain responses of the incident and reflected field were monitored for all time steps, and the FFT approach was applied to time-domain data. The reflection coefficient was calculated as the peak value of the division performed between the FFT of the reflected and incident fields. All results are summarised in Figure 7.32. From this figure, it may be seen that for a fixed value of the number of periods, N, the reflection coefficient value increases with the power of the incident source. Furthermore, the reflection coefficient seems to reach a saturation value once the power of the incident source becomes higher than 10 W/m. Moreover, for a fixed level of power it can be noted that the reflection coefficient increases as the number of periods, N, increases. Figure 7.33 shows the variation of the wavelength of the peak of the spectrum of the reflection coefficient with the power of the incident signal for three different numbers of periods, N, of the Bragg grating.

As can be seen from this figure, the shift towards lower frequency is greater as the nonlinear effect become stronger. An explanation for the values of the reflection coefficient being greater than unity can be founded in the SPM effect in nonlinear media. The spectral broadening, as can be seen from Figure 7.25, and the shift of the reflection coefficient peak towards lower frequencies, as can be noted in Figure 7.33, induces a band of frequencies for the reflected signal to gain in magnitude with respect to the same band of frequencies for the incident signal.

In further work, stacks of Bragg resonators have been investigated, and Figure 7.34 shows a schematic of the analysed structure. The resonator is composed of two BRs separated by a phase-shift area made up of nonlinear material.

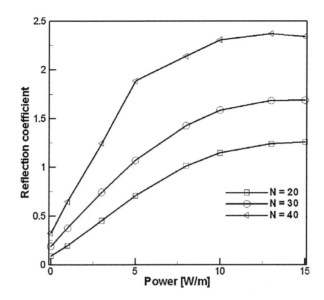

Figure 7.32 Reflection-coefficient variation with the power of the incident fundamental TE_0 mode for three different numbers of periods, N, of a nonlinear BR. (Reproduced with permission from Pinto, D., Obayya, S.S.A., Rahman, B.M.A. and Grattan, K.T.V. (2006) FDTD analysis of nonlinear bragg grating based devices. *Opt. Quantum Electron.*, **38** (15), 1217–1235. Copyright 2006 Springer Science and Business Media.)

The refractive index of the linear media is $n_w = 2.0$, the linear part of the nonlinear refractive index is $n_0 = 2.03$, while the nonlinear coefficient is $\alpha = 500 \text{ m}^2/\text{V}^2$; the number of periods of the two BRs is first set to $N_1 = N_2 = 20$, the periodicity of the BRs is $\Lambda = 0.25$ μm, and the width of the phase shift area is $l = \Lambda/2$. The structure is surrounded by air ($n = 1.0$). The TE_0 fundamental mode profile was used, while the parameters of the Gaussian pulse used to modulate the signal source in time and the centre wavelength were set to the values used for the analysis of the BR structure. Two different simulations were carried out using two different levels of power of the input signal, and in order to obtain the reflection coefficient for this structure the same procedure utilised for the BR was used.

Figure 7.35 shows the variation of the reflection coefficient with respect to the wavelength. The data obtained are in good agreement with their counterparts reported in the literature [5] using a hybrid implicit-explicit FDTD scheme, which requires the solution of a system of equations at each time step. In contrast, the FDTD approach presented employs an efficient and explicit way of updating the field quantities. Figure 7.36 shows the variation of the reflection coefficient with the wavelength for different values of the number of periods N_1 and N_2 of the two BRs excited with the fundamental

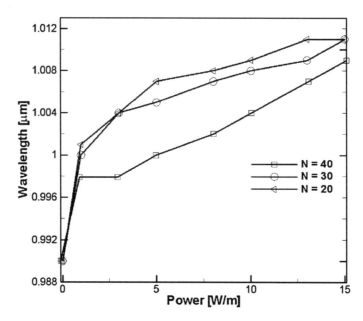

Figure 7.33 Wavelength variation of the reflection coefficient peak with the power of the incident fundamental TE_0 mode for three different numbers of periods, N, of a nonlinear BR. (Reproduced with permission from Pinto, D., Obayya, S.S.A., Rahman, B.M.A. and Grattan, K.T.V. (2006) FDTD analysis of nonlinear bragg grating based devices. *Opt. Quantum Electron.*, **38** (15), 1217–1235. Copyright 2006 Springer Science and Business Media.)

TE_0 mode with the power level set to 5 W/m. As can be seen from this figure, the reflectivity increases strongly with the addition of more periods to the BR structure. It can also be noted that a shift of the entire spectra towards higher wavelengths occurs as the number of periods increases, a result that can be related to a more intense SPM effect in the nonlinear material.

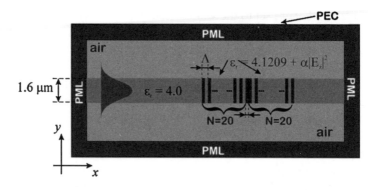

Figure 7.34 Schematic diagram of stacks of Bragg reflectors.

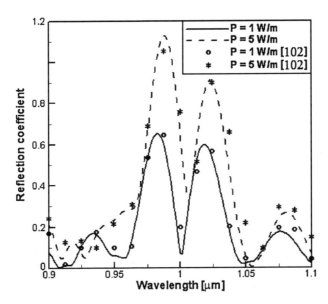

Figure 7.35 Variation of the reflection coefficient with the wavelength of stacks of BRs for two different values of power of the incident fundamental TE_0 mode. (Reproduced with permission from Pinto, D., Obayya, S.S.A., Rahman, B.M.A. and Grattan, K.T.V. (2006) FDTD analysis of nonlinear bragg grating based devices. *Opt. Quantum Electron.*, **38** (15), 1217–1235. Copyright 2006 Springer Science and Business Media.)

7.7 FVTD Analysis of Nonlinear Photonic Crystal Wires

The structure analysed will be a photonic crystal wire (PhW) (whose width has been fixed to $w = 0.38$ µm in order to ensure single mode operation) consisting of a dielectric waveguide with refractive index, $n_2 = 3.48$, surrounded by air ($n_1 = 1.0$). A row of eight air holes has been added to the waveguide to form the photonic crystal. The lattice constant is $a = 0.3162$ µm and the hole radius is $r = 0.23a$, while the distance between the two central holes is varied in order to form a cavity whose size is $c = a$, as shown in Figure 7.37.

Firstly, the structure was analysed in a linear regime. The source used to excite the structure was a Gaussian pulse modulated in time by a sinusoidal function with the shape of the fundamental mode profile of the waveguide, as expressed in Equation (7.6), where Φ represents the main electromagnetic-field component, which in this example is the magnetic-field component, H_z. The time-domain parameters of the Gaussian pulse were fixed to $T_0 = 30$ fs and $t_0 = 3T_0$ in order to excite the structure with a signal of an appropriate bandwidth, while the central frequency was set to $f = 193$ THz ($\lambda = 1.55$ µm). Different detectors were strategically inserted into the structure: a detector was inserted at the input section of the structure in order to record the time-domain variation of the incident field; another was inserted at the output

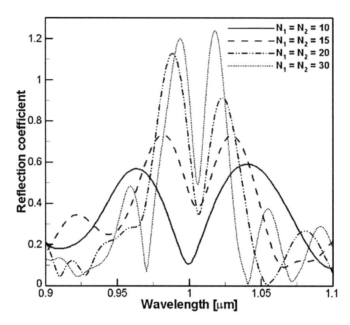

Figure 7.36 Variation of the reflection coefficient with the wavelength of stacks of BRs for four different values of the numbers of periods, N_1 and N_2. (Reproduced with permission from Pinto, D., Obayya, S.S.A., Rahman, B.M.A. and Grattan, K.T.V. (2006) FDTD analysis of nonlinear bragg grating based devices. *Opt. Quantum Electron.*, **38** (15), 1217–1235. Copyright 2006 Springer Science and Business Media.)

section in order to record the time domain variation of the transmitted field; the last detector was inserted at the centre of the cavity in order to store the time-domain variation of the resonant field. Figure 7.38 shows the time-domain variation of the magnetic field, H_z, inside the cavity of the structure.

From this figure the resonance of the electromagnetic field can be clearly seen, shown by the slow decrease of the magnetic field after reaching its maximum peak.

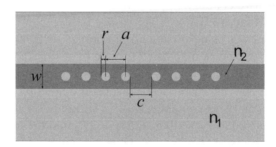

Figure 7.37 Schematic diagram of a linear PhW cavity. (Reproduced with permission from Pinto, D. and Obayya, S.S.A. (2008) Nonlinear finite volume time domain analysis of photonic crystal based resonant cavities. *IET Optoelectron.*, **2** (6), 254–261. © 2008 IET.)

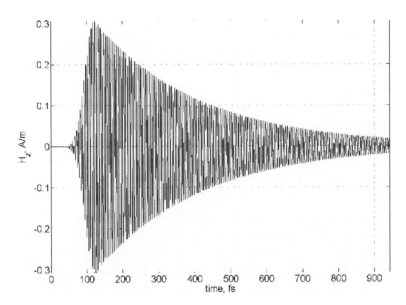

Figure 7.38 Time-domain variation of the magnetic field inside the linear PhW cavity. (Reproduced with permission from Pinto, D. and Obayya, S.S.A. (2008) Nonlinear finite volume time domain analysis of photonic crystal based resonant cavities. *IET Optoelectron.*, **2** (6), 254–261. © 2008 IET.)

Using FFT, it was possible to calculate the resonance frequency of the cavity. The result of this operation is shown in Figure 7.39. The results observed from Figure 7.38 are confirmed by this figure in which a clear sharp peak can be identified at wavelength $\lambda = 1.515$ µm. For the analysed structure the time cycle was $T = 5.05$ fs and, with the use of Equation (7.2), the quality factor was computed to be $Q = 190$.

The FFTs of the incident and transmitted fields have been used to compute the transmission coefficient (calculated as the ratio of the transmitted and incident fields) of the structure. The result of this operation is shown in Figure 7.40. From this figure it can be seen that the maximum transmission coefficient is about 0.5, which means that half of the energy stored in the cavity is lost in radiation. An explanation for this phenomenon can be found in the mode mismatch of the different sections of the structure [9].

Next, the previous structure was modified in order to insert nonlinear material into the centre of the cavity, as shown in Figure 7.41.

The reason why the nonlinearity has been considered only at the centre of the cavity is the fact that most of the electromagnetic field is strongly confined in the cavity itself, as can be clearly seen in Figures 7.42–7.44.

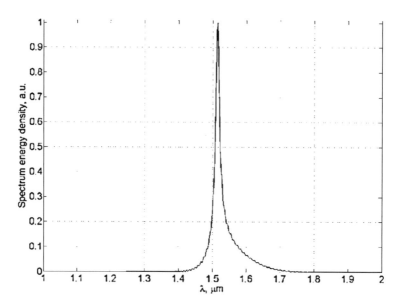

Figure 7.39 Energy-density spectrum of the resonant mode inside the linear PhW cavity. (Reproduced with permission from Pinto, D. and Obayya, S.S.A. (2008) Nonlinear finite volume time domain analysis of photonic crystal based resonant cavities. *IET Optoelectron.*, **2** (6), 254–261. © 2008 IET.)

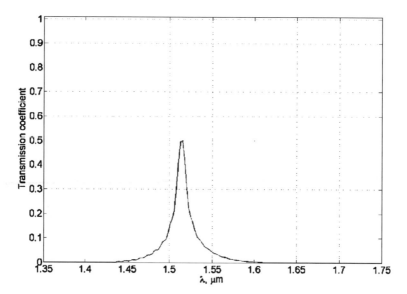

Figure 7.40 Spectrum of the transmission coefficient for the linear PhW cavity. (Reproduced with permission from Pinto, D. and Obayya, S.S.A. (2008) Nonlinear finite volume time domain analysis of photonic crystal based resonant cavities. *IET Optoelectron.*, **2** (6), 254–261. © 2008 IET.)

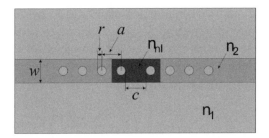

Figure 7.41 Schematic diagram of a nonlinear PhW cavity. (Reproduced with permission from Pinto, D. and Obayya, S.S.A. (2008) Nonlinear finite volume time domain analysis of photonic crystal based resonant cavities. *IET Optoelectron.*, **2** (6), 254–261. © 2008 IET.)

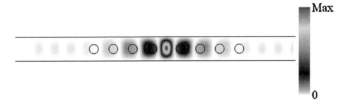

Figure 7.42 Field distribution of the magnetic-field component, H_z, inside the PhW cavity. (Reproduced with permission from Pinto, D. and Obayya, S.S.A. (2008) Nonlinear finite volume time domain analysis of photonic crystal based resonant cavities. *IET Optoelectron.*, **2** (6), 254–261. © 2008 IET.)

Figure 7.43 Field distribution of the electric-field component, E_x, inside the PhW cavity. (Reproduced with permission from Pinto, D. and Obayya, S.S.A. (2008) Nonlinear finite volume time domain analysis of photonic crystal based resonant cavities. *IET Optoelectron.*, **2** (6), 254–261. © 2008 IET.)

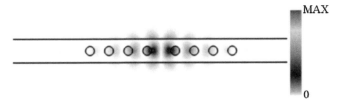

Figure 7.44 Field distribution of the electric-field component, E_y, inside the PhW cavity. (Reproduced with permission from Pinto, D. and Obayya, S.S.A. (2008) Nonlinear finite volume time domain analysis of photonic crystal based resonant cavities. *IET Optoelectron.*, **2** (6), 254–261. © 2008 IET.)

For this reason, the nonlinear interaction takes part in the cavity, while the other sections of the structure give negligible contributions to nonlinear effects. In this way, it is also possible to save computational resources for the update process of the electromagnetic-field components. The nonlinearity considered is a Kerr-like instantaneous nonlinearity modelled by

$$\varepsilon_r = \varepsilon_1 + \Delta\varepsilon_s \left(1 - e^{\left(-\frac{\gamma|E|^2}{\Delta\varepsilon_s}\right)}\right) \tag{7.13}$$

where ε_1 is the linear relative permittivity of the nonlinear medium, $\Delta\varepsilon_s$ is the maximum variation of the nonlinear permittivity, γ is a nonlinear coefficient related to the nonlinear parameter, n_2, through the relationship $\gamma = c\varepsilon_0 \varepsilon_1 n_{nl}$ in which c is the speed of the light and ε_0 is the free-space permittivity, respectively, and E is the electric field. The nonlinear parameters of the material have been set to $n_{nl} = 1.43 \times 10^{-17}$ m^2/W and $\Delta\varepsilon_{sat} = 0.31$ [10,11].

The expression of the source used to excite this structure is given by Equation (7.6), with the parameters of the Gaussian pulse and central frequency having the same values used for the linear case, while three different amplitudes of the magnetic field, H_z, were set. The incident and transmitted time-domain variation of the electromagnetic field for each case have been recorded in order to calculate the transmission coefficient of the structure. The result of this procedure is shown in Figure 7.45. From this figure, the shift of the transmission coefficient peak with the different source magnitudes can be clearly observed.

Finally, each peak corresponds to the resonant wavelength of the cavity mode of the structure excited with different amplitudes of input magnetic field. It can be also noted that the peak is shifted towards longer wavelengths (lower frequencies) as the amplitude of the magnetic field increases, while the peak of the transmission coefficients can be considered to be 0.5 with good approximation for all cases.

The resonant wavelength and the quality factor have been also computed for all cases. All these results are summarised in Figure 7.46.

This figure confirms the shift towards longer wavelengths of the cavity resonance as the amplitude of the exciting magnetic field increases, passing from $\lambda = 1.515$ μm for the linear case to $\lambda = 1.527$ μm for the maximum value of the magnetic field, $H_z = 5.4 \times 10^8$ A/m. The growing shift of the resonant wavelength towards longer values can be explained by a stronger nonlinear effect due to the increasing amplitude of the magnetic field applied to the structure. This deduction is also confirmed in [12], in which a different configuration of PhC cavity, realised with dielectric material exhibiting a Kerr-nonlinearity, was experimentally studied in the context of optical bistability. The quality factor, Q, on the other hand, decreases as the peak of the magnetic field increases, passing from a value of $Q = 190$ for the linear case to $Q = 180$ for the maximum magnetic field value, $H_z = 5.4 \times 10^8$ A/m. Furthermore, in order to assess the initial assumption of considering the nonlinearity only at the centre

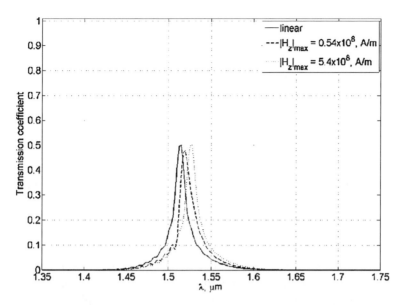

Figure 7.45 Spectra of the transmission coefficients of the nonlinear PhW cavity for three different amplitudes of the magnetic field. (Reproduced with permission from Pinto, D. and Obayya, S.S.A. (2008) Nonlinear finite volume time domain analysis of photonic crystal based resonant cavities. *IET Optoelectron.*, **2** (6), 254–261. © 2008 IET.)

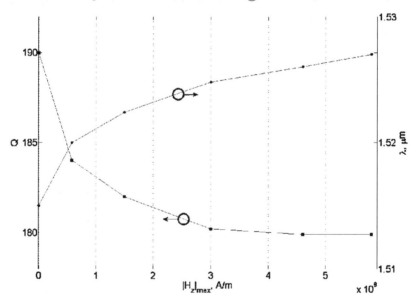

Figure 7.46 Resonant wavelength, λ, and quality factor, Q, for different amplitudes of H_z for the nonlinear PhW cavity. (Reproduced with permission from Pinto, D. and Obayya, S.S.A. (2008) Nonlinear finite volume time domain analysis of photonic crystal based resonant cavities. *IET Optoelectron.*, **2** (6), 254–261. © 2008 IET.)

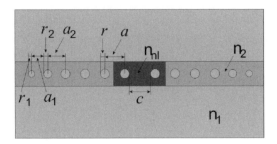

Figure 7.47   Schematic diagram of a nonlinear PhW cavity with external tapers. (Reproduced with permission from Pinto, D. and Obayya, S.S.A. (2008) Nonlinear finite volume time domain analysis of photonic crystal based resonant cavities. *IET Optoelectron.*, **2** (6), 254–261. © 2008 IET.)

of the cavity, a simulation was carried out extending the nonlinear material along the entire PhW structure. For this example, the source parameters have been fixed to the same values as the previous examples, while the maximum amplitude of the magnetic field has been set to $H_z = 5.4 \times 10^8$ A/m. The resonant wavelength, λ, and the quality factor, Q, have been computed to be 1.5271 μm and 179.9, respectively, results that differ from the ones previously obtained considering a smaller nonlinear region by less than 0.01%, which, ultimately, confirm the validity and the good accuracy of the data calculated when considering the nonlinearity to be only at the centre of the resonant cavity.

In further work, modifications were made to the previous structure in order to increase the transmission coefficient. Following the procedure suggested in [9], an external taper were added at the input and output sections of the structure, as shown in Figure 7.47.

The external taper consists of two air holes with gradually increasing radius and distance between hole centres. The radius of the first hole is fixed to $r_1 = 51.93$ nm and the distance from the centre of the second hole of the taper is $a_1 = 262$ nm, while the radius of the second hole is fixed to $r_2 = 63.94$ nm and the distance from the first holes of the structure is $a_2 = 280$ nm. Also for this example, the source used to excite the structure is given by Equation (7.6) with the parameters of the Gaussian pulse and central frequency of the same values as used for the previous example, while two different amplitudes of the magnetic field, H_z, have been set. Through the recorded incident and transmitted time-domain variations of the electromagnetic field, the resonant wavelength, quality factor and transmission coefficient have been calculated for both cases. In Figure 7.48 the spectra of the transmission coefficient for both cases is shown. From this figure, the resonant wavelength for the linear case can be identified at $\lambda = 1.536$ μm, while for the nonlinear case it is $\lambda = 1.543$ μm.

It can be also noted that for the linear case the transmission coefficient peak has been increased up to about 0.66 compared to the previous structure without taper,

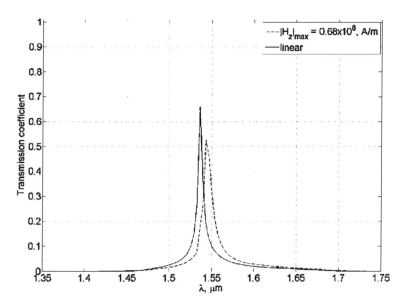

Figure 7.48 Spectra of the transmission coefficients of the nonlinear PhW cavity with external tapers for different amplitudes of H_z. (Reproduced with permission from Pinto, D. and Obayya, S.S.A. (2008) Nonlinear finite volume time domain analysis of photonic crystal based resonant cavities. *IET Optoelectron.*, **2** (6), 254–261. © 2008 IET.)

which shows a peak of about 0.5, as can be seen in Figure 7.45. This increase can be attributed to the added taper, which performs an adiabatic conversion of the mode from the fundamental mode of the waveguide to the (nonpropagating) Bloch mode of the mirror of the cavity [9]. Further increase in the transmission coefficient peak can be expected with increasing of number of sections of the external taper and with the realisation of taper sections inside the resonant cavity also. With the use of Equation (7.2), the quality factors, Q, have also been calculated. For the linear case the quality factor has been found to be $Q = 427$, while for the nonlinear case it has been calculated to be $Q = 419$. The increasing values of the quality factors can be also attributed to the external tapers added to the structure which also reduce, as the ultimate result, the overall losses of the cavity.

References

[1] Joannopoulos, J.D., Meade, R.D. and Winn, J.N. (1995) *Photonic Crystals: Molding the Flow of Light*, Princeton University Press, Princeton, NJ.
[2] Rodriguez-Esquerre, V.F., Koshiba, M. and Figueroa, H. (2004) Finite-element time-domain analysis of 2-D photonic crystal resonant cavities. *IEEE Photon. Technol. Lett.*, **16**, 816–818.
[3] Rodriguez-Esquerre, V.F., Koshiba, M. and Figueroa, H. (2005) Finite-element analysis of photonic crystal cavities: Time and frequency domain. *J. Lightwave Technol.*, **23**, 1514–1521.

[4] Koshiba, M. *et al.* (2000) Time-domain beam propagation method and its application to photonic crystal circuits. *J. Lightwave Technol.*, **18**, 102–110.

[5] Van, V. and Chaudhuri, S.K. (1999) A hybrid implicit-explicit FDTD scheme for nonlinear optical waveguide modelling. *IEEE Trans. Microwave Theory Tech.*, **47** (5), 540–545.

[6] Tamir, T. (1990) *Guided-Wave Optoelectronics*, Springler-Verlag.

[7] De Pourcq, M. and Eng, C. (1985) Field and power-density calculations in closed microwave systems by three-dimensional finite difference. *IEE Proc. H Microwave Antennas Propag.*, **132** (6), 360–368.

[8] Taflove, A. and Hagness, S.C. (2005) *Computational Electrodynamics: the Finite-Difference Time-Domain Method*, Artech House.

[9] Lalanne, P. and Hugonin, J.P. (2003) Bloch-wave engineering for high-Q, small-V microcavities. *IEEE J. Quantum Electron.*, **39** (11), 1430–1438.

[10] Obayya, S.S.A., Rahman, B.M.A., Grattan, K.T.V. and El-Mikati, H.A. (2002) Full vectorial finite-element solution of nonlinear bistable optical waveguides. *IEEE J. Quantum Electron.*, **38** (8), 1120–1125.

[11] Maksymov, I.S., Marsal, L.F. and Pallares, J. (2006) An FDTD analysis of nonlinear photonic crystal waveguides. *Opt. Quant. Electron.*, **38** (1–3), 149–160.

[12] Kawashima, H., Tanaka, Y., Ikeda, N. *et al.* (2008) Optical bistable response in AlGaAs-based photonic crystal microcavities and related nonlinearities. *IEEE J. Quantum Electron.*, **44** (9–10), 841–849.

8

Multiresolution Time Domain

8.1 Introduction

In the past, EM wave interactions with nonlinear optical materials have been modelled with a variety of different numerical methods such as the bi-directional beam propagation method (Bi-BPM) [1], the coupled-mode theory [2, 3] and the method of lines [4]. In the last decade, research efforts have addressed in particular the development of accurate and efficient finite-difference time-domain (FDTD) schemes that, unlike the just-cited techniques, can provide a full-wave description of all quadratic nonlinear phenomena in generic and even complex structures such as photonic crystals. In this sense, FDTD has proved to be a robust and flexible method that can be implemented in a relatively straightforward way. Its scheme can be easily adapted to the analysis of complex geometric features, dealing also with many different material properties. For all these reasons, various modifications and extensions of the conventional FDTD scheme have been proposed in order to optimise its performance and numerical efficiency. These include hybrid methods, higher-order FDTD methods and Berenger's perfectly matched layer (PML) method, which allow the computational window to be terminated with boundaries that are very close to the structure under investigation [5]. Significant progress has also been made in the development of FDTD algorithms for the analysis of electromagnetic propagation in nonlinear and dispersive materials [5–7].

However, despite its many capabilities, the FDTD technique suffers from a high numerical phase velocity error which puts a serious limit to the accurate modelling of phase-sensitive nonlinear frequency-conversion phenomena, such as second harmonic generation (SHG), which will be treated in Chapter 10. Being associated with high numerical dispersion, a fine grid resolution in space is needed to minimise phase error and thus ensure accurate results: a cell size of 15–20 times less than the minimum free space simulated wavelength ($\lambda_{min}/15$–20) is typically required [5]. The requirement of fine grid resolution also implies a stricter limit on the choice of time-step size, Δt,

Computational Photonics   Salah Obayya
© 2011 John Wiley & Sons, Ltd

as these two values are directly linked under the stability condition. Therefore, a large number of unknowns, together with a small Δt, results in an intensive use of computer resources. The FDTD limitations on the choice of the step size in space become particularly strict in nonlinear applications where a very large range of frequencies has to be investigated in one time simulation. Looking at the SHG problem, FDTD requires a discretisation in space of $\lambda_{sh}/15$–20, where λ_{sh} represents the second harmonic wavelength generated at half the wavelength of the pump. In this context, the need for an alternative time-domain technique that can alleviate some of the above problems has become stronger. The multiresolution time domain (MRTD) method shows excellent potential for fulfilling this requirement. The main purpose of the MRTD method is to reduce the computational burden required for a determined accuracy of the electromagnetic solution by use of a grid density close to the Nyquist sampling rate (2–3 grid points per wavelength). Applying the multiresolution analysis, with the use of scaling and wavelet functions, in the context of the method-of-moments-based discretisation of Maxwell's equations, is the foundation of the MRTD scheme. By relying on a higher order of approximation of the spatial derivatives, as it will be shown in this chapter, numerical dispersion in MRTD only arises from the approximation used for the derivatives in time, which is the same as in FDTD. Hence, MRTD is a computationally efficient solution to the numerical phase error associated with FDTD and allows a decrease in the points per wavelength into which the problem is discretised. Therefore, the basics of MRTD technique are presented here and used in Chapter 10 to provide an efficient and yet accurate modelling tool for phase-sensitive nonlinear optical problems. Besides the advantages in saving computational resources, MRTD provides also a unified and higher-order field-expansion scheme for both the FDTD and method-of-moments (MoM) methods. From this point of view FDTD can be considered as the simplest version of the MRTD scheme.

8.2   MRTD Basics

8.2.1   Multiresolution Analysis: Overview

The MRTD technique for solving the Maxwell's equations was first proposed by Krumpholz and Katehi in 1996 [8]. They proposed the use of wavelet expansion that, at that time, had become very popular in many fields as a way to increase both the efficiency and the accuracy of numerical methods, in the context of the method of moments for solving electromagnetic wave interaction problems. The technique that they originally proposed in [8] is very flexible and can be considered a generic discretisation scheme to be used with different types of wavelet. Typically, the wavelet scheme adopted in MRTD is orthonormal with scaling functions and a mother wavelet. The scaling functions generate the mother wavelet from which all the other wavelets are derived. The use of wavelet expansion makes it possible to talk about numerical accuracy in terms of 'levels' of wavelet resolution. Each level consists of a set of functions that is added to the expansions to increase the discretisation accuracy [9]. The scaling functions $\varphi_i(x)$ are defined by

$$\varphi_i(x) = \varphi\left(\frac{x}{\Delta x} - i\right) \tag{8.1}$$

A wavelet coefficient $\psi_{i,p}^r(x)$ is represented as

$$\psi_{i,p}^r(x) = 2^{r/2}\psi_0\left(2^{r/2}\left(\frac{x}{\Delta x} - i\right) - p\right) \tag{8.2}$$

where r is the wavelet resolution and p is an integer in the range $[0; 2^r{-}1]$. Each level of resolution, r, consists in 2^r wavelets that are misplaced in space by $\Delta x/2^r$.

The following relationships define scaling functions and wavelet coefficients [10]

$$\int \varphi_i(x)\varphi_j(x) = \delta_{i,j} \tag{8.3}$$

$$\int \varphi_i(x)\psi_{j,p}^r(x) = 0 \quad \forall i, j, r, p \tag{8.4}$$

$$\int \psi_{i,p}^r(x)\psi_{j,q}^s(x) = \delta_{i,j}\delta_{r,s}\delta_{p,q} \tag{8.5}$$

$$\delta_{i,j} = \begin{cases} 1 & i = j \\ 0 & i \neq j \end{cases} \tag{8.6}$$

From the theory of wavelet expansion and multiresolution decomposition, it is known that a set of subspaces $\{V_i\}_{i\in Z}$ is a multiresolution approximation of $L^2(R)$ if the following hold [11]

$$\ldots V_{-3} \subset V_{-2} \subset V_{-1} \subset V_0 \subset V_1 \subset V_2 \subset V_3 \cdots$$
$$\bigcup_{i\in Z} V_i = L^2(R)$$
$$\bigcap_{i\in Z} V_i = 0$$

where Z represents the set of all integers, while $L^2(R)$ is the set of all square integrable functions.

All of these properties allow us to achieve the wanted level of accuracy by increasing/decreasing the wavelet resolution.

8.2.2   MRTD Scheme

The basic concept of MRTD is to represent first the electric and magnetic fields as expansions in scaling/wavelet functions in space and time and then to apply the method of moments to the conventional Maxwell's equations.

For example, for a 1D scheme with arbitrary order of resolution up to r_{max}, the expression used to expand each field component in scaling and wavelet functions is the following

$$F_x(x) = \sum_{n,i} h_n(t) \left[{}_nF_i^{x,\phi}\varphi_i(x) + \sum_{r=0}^{r_{max}} \sum_{p=0}^{2^r-1} {}_nF_{i,r,p}^{x,\Psi}\Psi_{i,p}^r(x) \right] \quad (8.7)$$

where ${}_nF_i^{x,\phi}$ and ${}_nF_{i,r,p}^{x,\psi}$ are the expansion coefficients and represent the magnitudes of the scaling and wavelet functions. Discretisation is performed both in space and in time. In time, pulse functions $h_n(t)$ are used in order to ensure causality. These functions in fact don't overlap in representing a given time step and thus prevent a past event being determined by a future one. Finally, i indicates the position in space along the x-direction. When a 2D or 3D system is considered, the products of all scaling/wavelet functions in each dimension must be calculated.

For any wavelet basis adopted, the number of expansion coefficients for a given resolution r_{max}, is determined by the formula

$$\text{Number of coefficients} = 2^{D + \sum_{i=x,y,z} r_{max,i}} \quad (8.8)$$

where D represents the dimensionality of the system under investigation. The number of basis functions calculated through the previous formula gives the number of equivalent grid points [12]. The field coefficients in [8] are offset by half a cell to build a scheme in space that is similar to the conventional Yee-FDTD scheme.

It should be noted that like in FDTD, the update equations in MRTD are fully explicit. This means they are easy to implement but, on the other hand, the time step must again be chosen below a stability limit, as will be discussed in detail in this chapter. In particular, discretisation in time, based on pulse functions, results in a leapfrog arrangement analogous to the one in FDTD, with E and H fields that are offset by half a time step. Once the field offset is chosen, the MRTD update scheme is obtained by applying the method of moments.

8.2.3   Method of Moments

As one of the finite-difference-based time-domain techniques, MRTD can be derived from the MoM when specific expansion and testing functions are applied [13]. The

MoM was proposed in the past in order to solve equations that can be expressed in the following compact form

$$L\left(\vec{E}, \vec{H}\right) - f\left(\vec{J}, \rho\right) = 0 \tag{8.9}$$

A full explanation of the method can be found in [14]. It consists in representing first the unknown function as a sum of unknown coefficients multiplied by known basis functions. The second step is error testing by means of a series of chosen test functions. With the number of test functions equal to the number of unknowns, a system of linear equations whose solutions allow us to determine the wanted coefficients is generated.

A brief explanation of the procedure for a 1D problem will be considered

$$f(x) = \frac{\partial b(x)}{\partial x} \tag{8.10}$$

where $f(x)$ and $b(x)$ represent the unknown and known functions, respectively.

First, the function $f(x)$ is expanded as follows:

$$f(x, t) = \sum_{n=0}^{N} a_n c_n(x) \tag{8.11}$$

where a_n are unknown expansion coefficients and $c_n(x)$ are the basis functions, which are known. Next, a set of testing functions $\omega_n(x)$ is chosen and the following inner product is considered

$$\langle \omega_n, f \rangle = \langle \omega_n, b' \rangle = \int \omega_n(x) f(x) \, dx \tag{8.12}$$

A system of linear equations can be generated as

$$\begin{bmatrix} a_1 \langle \omega_1, c_1 \rangle & a_2 \langle \omega_1, c_2 \rangle & \cdots & a_N \langle \omega_1, c_N \rangle \\ a_1 \langle \omega_2, c_1 \rangle & a_2 \langle \omega_2, c_2 \rangle & \cdots & a_N \langle \omega_2, c_N \rangle \\ \vdots & \vdots & \ddots & \vdots \\ a_1 \langle \omega_N, c_1 \rangle & a_2 \langle \omega_N, c_2 \rangle & \cdots & a_N \langle \omega_N, c_N \rangle \end{bmatrix} = \begin{bmatrix} \langle \omega_1, b' \rangle \\ \langle \omega_2, b' \rangle \\ \vdots \\ \langle \omega_N, b' \rangle \end{bmatrix} \tag{8.13}$$

and solved to calculate the unknown coefficients, a_n.

When the following property for the testing functions holds

$$\langle \omega_n, c_n \rangle = \delta_{m,n} \tag{8.14}$$

the matrix becomes diagonal, the scheme is explicit and each coefficient can be calculated as

$$a_n = \langle \omega_n, b' \rangle \tag{8.15}$$

In particular, when the testing functions are the same as the basis functions and Equation (8.14) is true, (the set of basis functions is orthonormal), the method is named Galerkin's procedure.

8.3 MRTD Update Scheme

8.3.1 Approximation in Time: Testing with Pulse Functions

The MRTD update in time is defined by the choice of scaling/testing functions that give the localisation of the expansion coefficients in time. In order to build an explicit scheme, the expansion in time is performed through pulse functions whose derivatives yield Dirac delta functions located at the edges of the pulses

$$\frac{\partial h_{n+1/2}(t)}{\partial t} = \delta(t - n\Delta t) - \delta(t - (n+1)\Delta t) \tag{8.16}$$

Representation of this series of derivatives as Delta functions is given in Figure 8.1.

If these derivatives at time steps $n + 1/2$ represent B (and H) coefficients, from Maxwell's equations we can represent D (and E) at time steps, n, with the pulse functions at the bottom of Figure 8.1. The expansion of the E field in time is

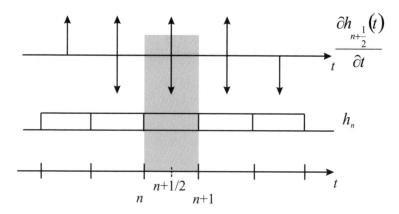

Figure 8.1 Representation of the derivatives of the basis functions in time when h_n are pulsed functions.

expressed by

$$E(\vec{r}, t) = \sum_{n=0}^{N} h_n(t)_n E(\vec{r}) \tag{8.17}$$

where $_n E(\vec{r})$ is the wavelet/scaling discretisation in space at time step n. For the B field the expression is similarly written as

$$B(\vec{r}, t) = \sum_{n=0}^{N} h_{n+1/2}(t)_{n+1/2} B(\vec{r}) \tag{8.18}$$

From the inner products with the pulses, h_n, the following are obtained

$$\langle h_n(t), E(\vec{r}, t) \rangle = \Delta t_n E(\vec{r}) \tag{8.19}$$

$$\left\langle h_n(t), \frac{\partial B(\vec{r}, t)}{\partial t} \right\rangle = {}_{n+1/2} B(\vec{r}) - {}_{n-1/2} B(\vec{r}) \tag{8.20}$$

By applying these inner product operations to both sides of Maxwell's equations, the update relations of the fields can be derived. For instance, for the component B_x, the following is written

$$\left\langle h_n, \frac{\partial B_x}{\partial t} \right\rangle = \left\langle h_n, \frac{\partial E_y}{\partial z} - \frac{\partial E_z}{\partial y} \right\rangle \tag{8.21}$$

$${}_{n+1/2} B_x(\vec{r}) - {}_{n-1/2} B_x(\vec{r}) = \Delta t \left[\frac{\partial_n E_y(\vec{r})}{\partial z} - \frac{\partial_n E_z(\vec{r})}{\partial y} \right] \tag{8.22}$$

Then, solving for the new term, $_{n+\frac{1}{2}} B_x(\vec{r})$, the following is found:

$${}_{n+1/2} B_x(\vec{r}) = {}_{n-1/2} B_x(\vec{r}) + \Delta t \left[\frac{\partial_n E_y(\vec{r})}{\partial z} - \frac{\partial_n E_z(\vec{r})}{\partial y} \right] \tag{8.23}$$

which represents the update equation for the component B_x and calculates the new field at time $n + 1/2$ from the value at previous time step $n - 1/2$.

8.3.2 Approximation in Space: Testing with Wavelet/Scaling Functions

The basis functions in 3D space are expressed as $\Gamma(x)\Gamma(y)\Gamma(z)$, where Γ represents either φ_l scaling functions, or $\psi_{l,p}^r$ wavelet functions, with $l = i, j, k$ as the directional index in the three space directions. In order to perform an approximation of the derivatives in the space domain, the test functions are applied together with the time differentiated functions in Equation (8.23).

It is convenient to represent the discretisation of the field in vectorial notation [15] that is a generalised form for any wavelet basis: all the wavelet/scaling coefficients are written as

$$
_{n+1/2}B_{x,i,j,k} =
\begin{bmatrix}
{}_{n+1/2}B_{i,j,k}^{x,\varphi\varphi\varphi} \\
{}_{n+1/2}B_{i,j,k,0,0}^{x,\psi\varphi\varphi} \\
\vdots \\
{}_{n+1/2}B_{i,j,k,r_{\max.z},2^{r_{\max.z}}-1}^{x,\varphi\varphi\psi} \\
{}_{n+1/2}B_{i,j,k,0,0,0,0}^{x,\psi\psi\varphi} \\
\vdots \\
{}_{n+1/2}B_{i,j,k,r_{\max.y},2^{r_{\max.y}}-1}^{x,\varphi\psi\psi} \\
{}_{n+1/2}B_{i,j,k,0,0,0,0,0,0}^{x,\psi\psi\psi} \\
\vdots \\
{}_{n+1/2}B_{i,j,k,r_{\max.x},2^{r_{\max.x}}-1,r_{\max.y},2^{r_{\max.y}}-1,r_{\max.z},2^{r_{\max.z}}-1}^{x,\psi\psi\psi}
\end{bmatrix}
\tag{8.24}
$$

Another vector, Γ, is defined as follows

$$
\Gamma_{i,j,k} =
\begin{bmatrix}
\varphi_i(x)\,\varphi_j(y)\,\varphi_k(z) \\
\psi_{i,0}^0(x)\,\varphi_j(y)\,\varphi_k(z) \\
\vdots \\
\varphi_i(x)\,\varphi_j(y)\,\psi_{k,2^{r_{\max.z}}-1}^{r_{\max.z}}(z) \\
\psi_{i,0}^0(x)\,\psi_{j,0}^0(y)\,\varphi_k(z) \\
\vdots \\
\varphi_i(x)\,\psi_{j,2^{r_{\max.y}}-1}^{r_{\max.y}}(y)\,\psi_{k,2^{r_{\max.z}}-1}^{r_{\max.z}}(z) \\
\psi_{i,0}^0(x)\,\psi_{j,0}^0(y)\,\psi_{k,0}^0(z) \\
\vdots \\
\psi_{i,2^{r_{\max.x}}-1}^{r_{\max.x}}(x)\,\psi_{j,2^{r_{\max.y}}-1}^{r_{\max.y}}(y)\,\psi_{k,2^{r_{\max.z}}-1}^{r_{\max.z}}(z)
\end{bmatrix}
\tag{8.25}
$$

thus

$$
_{n+1/2}B_x(\vec{r}) = \sum_{i,j,k} \Gamma_{i,j,k}^T {}_{n+1/2}B_{x,i,j,k}
\tag{8.26}
$$

The update equation for the B component can be derived by taking an inner product of Equation (8.23) with each wavelet/scaling coefficient. The update equation becomes

$$
_{n+1/2}B_{x,i,j,k} = {}_{n-1/2}B_{x,i,j,k} + \frac{\Delta t}{\Delta x \Delta y \Delta z}
$$

$$
\left[\sum_m U^{B_x}_{E_y,mn} E_{y,i,j,k+m} + \sum_m U^{B_x}_{E_z,mn} E_{y,i,j+m,k} \right] \tag{8.27}
$$

where U represents the matrix of the inner products between the E basis functions and the B basis functions. It has to be noted that E and B components are offset only along the direction of differentiation of the E field. The index, m, refers to the position in space amongst all the neighbouring cells whose total number depends on the choice of the basis functions.

The size of U matrices is equal to $2^{3+r_{max,x}+r_{max,y}+r_{max,z}} \times 2^{3+r_{max,x}+r_{max,y}+r_{max,z}}$ and it can be calculated before the actual start of the simulation. Its generic form is expressed by the following

$$
U^{F_1}_{F_2,m} = \begin{bmatrix} \left\langle {}_{F_1}\Gamma_1, \frac{\partial {}_{F_2}\Gamma_1|_m}{\partial n} \right\rangle & \left\langle {}_{F_1}\Gamma_1, \frac{\partial {}_{F_2}\Gamma_2|_m}{\partial n} \right\rangle & \cdots & \left\langle {}_{F_1}\Gamma_1, \frac{\partial {}_{F_2}\Gamma_L|_m}{\partial n} \right\rangle \\ \left\langle {}_{F_1}\Gamma_2, \frac{\partial {}_{F_2}\Gamma_2|_m}{\partial n} \right\rangle & \left\langle {}_{F_1}\Gamma_2, \frac{\partial {}_{F_2}\Gamma_2|_m}{\partial n} \right\rangle & & \\ \vdots & & \ddots & \\ \left\langle {}_{F_1}\Gamma_L, \frac{\partial {}_{F_2}\Gamma_L|_m}{\partial n} \right\rangle & & & \left\langle {}_{F_1}\Gamma_L, \frac{\partial {}_{F_2}\Gamma_L|_m}{\partial n} \right\rangle \end{bmatrix} \tag{8.28}
$$

where F_1 is the field that has to be updated and F_2 is the field on which this update depends. m represents the offset in the direction of differentiation, $\partial/\partial n$ is the derivative in space with $n = y, z$, if Equation (8.27) is considered as example. In this case the (2,2) entry in Equation (8.28) is made explicate as follows when the field offsets in x-, y- and z- directions are called s_x, s_y and s_z, respectively

$$
U^{B_{x1}}_{E_y,m,2,2} = \left\langle \psi^0_{i,0}(x)\varphi_{j+s_y}(y)\varphi_{k+s_z}(z), \frac{\partial}{\partial z}\psi^0_{i,0}(x)\varphi_{j+s_y}(y)\varphi_k(z) \right\rangle =
$$

$$
= \iiint \psi^0_{i,0}(x)\varphi_{j+s_y}(y)\varphi_{k+s_z}(z) \frac{\partial}{\partial z}\left(\psi^0_{i,0}(x)\varphi_{j+s_y}(y)\varphi_k(z)\right)\partial x \partial y \partial z \tag{8.29}
$$

Separating the integral by direction

$$
\iiint \psi^0_{i,0}(x)\varphi_{j+s_y}(y)\varphi_{k+s_z}(z) \frac{\partial}{\partial z}\left(\psi^0_{i,0}(x)\varphi_{j+s_y}(y)\varphi_k(z)\right)\partial x \partial y \partial z =
$$

$$
= \int \psi^0_{i,0}(x)\psi^0_{i,0}(x)\partial x \int \varphi_{j+s_y}(y)\varphi_{j+s_y}(y)\partial y \int \varphi_{k+s_z}(z)\frac{\partial \varphi_k(z)}{\partial z}\partial z \tag{8.30}
$$

It is known the orthogonality of the basis functions, therefore

$$\int \psi_{i,0}^{0}(x)\,\psi_{i,0}^{0}(x)\,\partial x \int \varphi_{j+s_y}(y)\,\varphi_{j+s_y}(y)\,\partial y \int \varphi_{k+s_z}(z)\,\frac{\partial \varphi_k(z)}{\partial z}\,\partial z$$

$$= \Delta x \cdot \Delta y \cdot \int \varphi_{k+s_z}(z)\,\frac{\partial \varphi_k(z)}{\partial z}\,\partial z \qquad (8.31)$$

From Equation (8.31), it can be seen that only one integral has to be evaluated. Its value depends on the choice of the basis functions so it can be solved either analytically or numerically. However, these values can be tabulated referring to each basis functions family and thus they don't have to be calculated for each simulation. Through this procedure all the update equations for the components B and D can be similarly derived.

8.3.3 Media Discretisation

After updating the B and D components, the actual H and E fields have to be calculated by means of the constitutive relationships and depend on the media characteristics. Generally material properties are a function of position inside the computational domain. In this chapter, isotropic materials, in some applications also dispersive and nonlinear, are considered. Assuming a medium with magnetic permeability $\mu = \mu_0$, the only constitutive relationship to be taken into account is the one to obtain the E field from an update of D and it can be written as

$$E(\vec{r}, t) = \frac{1}{\varepsilon(\vec{r}, t)} D(\vec{r}, t) \qquad (8.32)$$

where $\varepsilon(\vec{r}, t)$ represents the space- and time- dependent permittivity tensor of the medium. Equation (8.32) can be discretised by means of the scaling and pulse functions in space and in time respectively, following the Galerkin's method already described in the previous paragraphs. The permittivity tensor, ε, that links the flux density vector to the electric field is given by

$$\varepsilon(\vec{r}, t) = \begin{bmatrix} \varepsilon_x(\vec{r}, t) & 0 & 0 \\ 0 & \varepsilon_y(\vec{r}, t) & 0 \\ 0 & 0 & \varepsilon_z(\vec{r}, t) \end{bmatrix} \qquad (8.33)$$

The constitutive relation is written in the form of scalar equations as

$$D_x = \varepsilon_x(\vec{r}, t)\,E_x \qquad (8.34a)$$

$$D_y = \varepsilon_y(\vec{r}, t)\,E_y \qquad (8.34b)$$

$$D_z = \varepsilon_z(\vec{r}, t)\,E_z \qquad (8.34c)$$

Discretisation of these equations is considered for the case of expansion in scaling functions in space only, for simplicity. The field expansions are expressed by

$$F_x \left(\vec{r}, t \right) = \sum_{k,l,m,n=-\infty}^{+\infty} {}_k F^{\varphi_x}_{l+\frac{1}{2},m,n} h_k \left(t \right) \varphi_{l+\frac{1}{2}} \left(x \right) \varphi_m \left(y \right) \varphi_n \left(z \right) \tag{8.35a}$$

$$F_y \left(\vec{r}, t \right) = \sum_{k,l,m,n=-\infty}^{+\infty} {}_k F^{\varphi_y}_{l,m+\frac{1}{2},n} h_k \left(t \right) \varphi_l \left(x \right) \varphi_{m+\frac{1}{2}} \left(y \right) \varphi_n \left(z \right) \tag{8.35b}$$

$$F_z \left(\vec{r}, t \right) = \sum_{k,l,m,n=-\infty}^{+\infty} {}_k F^{\varphi_z}_{l,m,n+\frac{1}{2}} h_k \left(t \right) \varphi_l \left(x \right) \varphi_m \left(y \right) \varphi_{n+\frac{1}{2}} \left(z \right) \tag{8.35c}$$

where $F_r \left(\vec{r}, t \right) = \left[E_r \left(\vec{r}, t \right), D_r \left(\vec{r}, t \right) \right]$, with $r = x, y, z$. The coefficients ${}_k F^{\varphi_r}_{l,m,n}$ are the field expansion coefficients in terms of scaling functions. The indices l, m, n and k describe the localisation in space and time. In time, the function $h_k(t)$ is given by

$$h_k \left(t \right) = h \left(\frac{t}{\Delta t} - k \right) \tag{8.36}$$

with

$$h \left(t \right) = \begin{cases} 1 & \text{for} \quad |t| < 1/2 \\ 1/2 & \text{for} \quad |t| = 1/2 \\ 1 & \text{for} \quad |t| > 1/2 \end{cases} \tag{8.37}$$

In space, the scaling function is defined as

$$\varphi_m \left(x \right) = \varphi \left(\frac{x}{\Delta x} - m \right) \tag{8.38}$$

After substituting the field expansions into Equations (8.34a–c) and sampling them with pulse functions in time and scaling functions in space, it is assumed that

$$\varepsilon_r \left(\vec{r}, t \right) = \varepsilon_r \left(x \right) \varepsilon_r \left(y \right) \varepsilon_r \left(z \right) \varepsilon_r \left(t \right) \tag{8.39}$$

Sampling Equation (8.34a) with $\varphi_{l+1/2} \left(x \right)$, $\varphi_m \left(y \right)$, $\varphi_n \left(z \right)$ and $h_k \left(t \right)$ brings to

$${}_k D^{\varphi_x}_{l+1/2,m,n} = \sum_{k',l',m',n'=-\infty}^{+\infty} \varepsilon \left(x \right)^{\varphi_x}_{l+1/2,l'+1/2} \varepsilon \left(y \right)^{\varphi_x}_{m,m'} \varepsilon \left(z \right)^{\varphi_x}_{n,n'} \varepsilon \left(t \right)^x_{k,k'} E^{\varphi_x}_{l'+1/2,m',n'}$$

$$\tag{8.40}$$

where the coefficients $\varepsilon\,(r)_{m,m'}^{\varphi_x}$ and $\varepsilon\,(t)_{k,k'}^{x}$ represent the integrals

$$\varepsilon\,(r)_{m,m'}^{\varphi_x} = \frac{1}{\Delta r} \int\limits_{-\infty}^{+\infty} \varphi_m\,(r)\,\varepsilon_x\,(r)\,\varphi_{m'}\,(r)\,dr \tag{8.41}$$

$$\varepsilon\,(t)_{k,k'}^{x} = \frac{1}{\Delta t} \int\limits_{-\infty}^{+\infty} h_k\,(t)\,\varepsilon_x\,(t)\,h_{k'}\,(t)\,dt \tag{8.42}$$

8.3.4 Numerical Dispersion

In order to choose the space and time-step size properly, numerical dispersion and stability need to be considered for MRTD.

Due to numerical dispersion, there is a variation of the wavenumber, k, with the angular frequency, ω. Under the assumption of a linear, nondispersive, isotropic medium, by substitution of the solution for a monochromatic plane wave into Maxwell's equations, the following dispersion relationship is derived

$$k = \pm\frac{\omega}{c} \tag{8.43}$$

with

$$c = \frac{1}{\sqrt{\mu\varepsilon}} \tag{8.44}$$

In a 3D space, a wavevector, $\vec{k}$, is defined in a Cartesian system as

$$\vec{k} = k_x\hat{i} + k_y\hat{j} + k_z\hat{k} \tag{8.45}$$

where

$$k = \sqrt{k_x^2 + k_y^2 + k_z^2} \tag{8.46}$$

From these parameters, phase velocity v_{p}, and group velocity v_{g} are defined

$$v_{\mathrm{p}} = \pm\frac{\omega}{k} = \pm c \tag{8.47}$$

$$v_{\mathrm{g}} = \pm\frac{\partial\omega}{\partial k} = \pm c \tag{8.48}$$

From Equation (8.48), it can be seen that wavelength and frequency have a linear relationship, and that phase and group velocity are independent of frequency.

Referring to a numerical scheme, these relationships become more complex as an effect of the discretisation adopted in space and time. Time is considered as a sequence of discrete time steps and space is a multiple of cells that, although small, make the waves propagate in the directions defined by the grid rather than in any other direction. This makes the propagation velocity dependent on direction and frequency. Compared to the dispersion equation for FDTD, (that is equivalent to MRTD with Haar scaling functions only), it is demonstrated that increasing the resolution by one level effectively doubles the resolution of the scheme. Thus, the generic relationship can be obtained simply dividing the space step by $2^{r_{max}+1}$, as follows [12]

$$
\left[\frac{1}{c\Delta t} \sin\left(\frac{\omega \Delta t}{2} \right) \right]^2 = \left[\frac{2^{r_{max,x}+1}}{\Delta x} \sin\left(\frac{k_x \Delta x}{2^{r_{max,x}+2}} \right) \right]^2 + \left[\frac{2^{r_{max,y}+1}}{\Delta y} \sin\left(\frac{k_y \Delta y}{2^{r_{max,y}+2}} \right) \right]^2
$$

$$
+ \left[\frac{2^{r_{max,z}+1}}{\Delta z} \sin\left(\frac{k_z \Delta z}{2^{r_{max,z}+2}} \right) \right]^2
$$

(8.49)

The given dispersion analysis can be generalised for any wavelet basis and resolution of wavelet with the following

$$
\left[\frac{1}{c\Delta t} \sin\left(\frac{\omega \Delta t}{2} \right) \right]^2 = \left[\frac{2^{r_{max,x}+1}}{\Delta x} \left(\sum_{l=0}^{L_s-1} a\,(l) \sin\left(\frac{k_x \Delta x}{2^{r_{max,x}+2}} \right) \right) \right]^2
$$

$$
+ \left[\frac{2^{r_{max,y}+1}}{\Delta y} \left(\sum_{l=0}^{L_s-1} a\,(l) \sin\left(\frac{k_y \Delta y}{2^{r_{max,y}+2}} \right) \right) \right]^2
$$

(8.50)

$$
+ \left[\frac{2^{r_{max,z}+1}}{\Delta z} \left(\sum_{l=0}^{L_s-1} a\,(l) \sin\left(\frac{k_z \Delta z}{2^{r_{max,z}+2}} \right) \right) \right]^2
$$

8.3.5 Numerical Stability

For the case MRTD with expansion in scaling functions only (scaling-MRTD), the condition for numerical stability is [9]

$$
\Delta t \leq \frac{1}{\sum\limits_{i=0}^{L_s-1} |a\,(l)| \sqrt{\frac{1}{\mu\varepsilon} \left(\frac{1}{\Delta x^2} + \frac{1}{\Delta y^2} + \frac{1}{\Delta z^2} \right)}}
$$

(8.51)

Considering the 2D case in the xz-plane, the limit becomes

$$\Delta t \leq s \frac{\Delta}{c_0} \tag{8.52}$$

where

$$s = \frac{1}{\sqrt{2} \sum_{l=0}^{L_s - 1} |a(l)|} \tag{8.53}$$

where $\Delta = \Delta x = \Delta z$ in an uniform mesh, c_0 is the speed of the light and the Courant number, s, represents the stability factor in two dimensions and depends on the order of the adopted basis functions [16]. The result is that the time limit for the MRTD scheme is smaller than for FDTD with the same cell grid size. However, although the single time step is smaller, it should be noted that generally MRTD allows a much coarser grid than FDTD so that the computational efficiency improves overall.

It has been demonstrated in [12] that the generic formula, valid for any MRTD basis and any level of wavelet resolution, is expressed as

$$\Delta t \leq \frac{1}{\sum_{i=0}^{L_s-1} |a(l)| \sqrt{\frac{1}{\mu \varepsilon} \left(\frac{1}{\left(\frac{\Delta x}{2^{r_{\max,x}+1}}\right)^2} + \frac{1}{\left(\frac{\Delta y}{2^{r_{\max,y}+1}}\right)^2} + \frac{1}{\left(\frac{\Delta z}{2^{r_{\max,z}+1}}\right)^2} \right)}} \tag{8.54}$$

8.4 Scaling MRTD

8.4.1 Choice of Basis Functions: Cohen-Daubechies-Feauveau Family

The main purpose of the MRTD scheme is to minimise the number of unknowns, which is the number of grid points per wavelength, without deteriorating the accuracy of results. Practically, this means producing a better approximation of the field through accurate wavelet expansion, which minimises the numerical dispersion error of the algorithm for a given grid resolution. If $\psi_{l,m}$ is the wavelet at level l and shifted by $m/2^l$ units, while $\tilde{\psi}_{l,m}$ represents its dual, they both must hold the orthogonality relationship

$$\langle \psi_{l,m}, \tilde{\psi}_{l,m} \rangle = \delta \left(l - l' \right) \delta \left(m - m' \right) \tag{8.55}$$

When $\psi_{l,m} = \tilde{\psi}_{l,m}$, the wavelets are called orthonormal. The common requirements are for the wavelet basis to be regular (smooth) with regard to the degree of differentiability of a function, and with vanishing moments defined by

$$m_1(n) = \int x^n \psi(x) \, dx \qquad (8.56)$$

where n indicates the generic n^{th} moment [17].

In particular, it has been found that a good approximation demands $\psi_{l,m}$ to have as many vanishing moments as possible, and $\tilde{\psi}_{l,m}$ to be as smooth as possible. It is also required that the scaling/wavelet functions have minimal support in order to reduce the number of computations needed. If specifically the set of orthonormal wavelet families is considered, the two conditions of regularity and minimal support are found to be in conflict [17].

The oldest and simplest scaling/wavelet family is the Haar basis functions. When these basis functions are applied to an MRTD scheme, an algorithm very similar to the Yee FDTD scheme can be generated [18]. The main advantages in the use of Haar functions refer to their finite domain, so that they don't overlap from one cell to the next one, and the simplicity of performing derivatives and integral calculations due to their pulse nature. Besides that, their main disadvantage is the lack of smoothness that brings a higher numerical dispersion compared to other existing wavelet families. The MRTD method that Krumpholz and Katehi proposed in 1995 is based on Battle–Lemarie scaling/wavelet functions that are derived by B-spline functions [8]. They have been shown to have very good regularity properties, but suffer from having noncompact support. There is a theoretically infinite number of terms in the update equations and thus, a truncation of the sequence of coefficients in the summation calculations (usually 8–12 on each side) is needed with consequences in terms of arithmetic precision that could even destroy properties, such as zero moments and orthogonality [19]. The property of having an exact number of interpolating coefficients or compact support is achieved by Daubechies orthogonal wavelets [20,21], where the number of vanishing moments is maximised. Cohen–Daubechies–Feauveau (CDF) biorthogonal scaling functions have been chosen to develop the MRTD scheme here. In the literature, it was found that this family of functions satisfies the requirements of the MRTD scheme as it shows the maximum number of vanishing moments for a given support, good regularity and compact support [16]. Thanks to these properties, the sequence of MRTD coefficients in the update equations is rigorously finite, while a good level of regularity is still kept. The CDF notation (α, β) is adopted to indicate the lengths of the reconstruction and decomposition filters of the family. Through this work, φ for the field expansions of order (2,4) have been chosen from the CDF family. This order of functions relies on a total number of coefficients equal to five (compact support) and shows a good compromise between higher-order accuracy and the increased number of operations required.

In [16], an extensive study of the numerical dispersion characteristics of the CDF-MRTD scheme compared to other wavelet families is given. Dogaru and Carin showed that numerical dispersion depends on many factors, such as the spatial resolution, the Courant number, the number of level of wavelets used to expand the fields and the angle of electromagnetic propagation. However, generally, it is found that the MRTD scheme allows a grid resolution at least two times coarser than FDTD when the same level of accuracy is required. Another important aspect is that dispersion performances are heavily influenced by the adopted Courant number. The choice of a Courant number smaller than the required limit for stability means better results in terms of numerical dispersion. As a consequence of this, when the same Courant number is taken, the low-order CDF (2,4) family can achieve better accuracy than the case in which both the scaling and the first level of wavelet are included, which implies a stricter stability limit on the time-step size.

8.4.2   Derivation of Update Scheme

For the analysis of a 2D problem in the xz-plane, assuming the y-axis as the homogeneous direction and the x-axis as the propagation direction, the fundamental electromagnetic components of a TE mode are E_y, H_x, H_z, and from Maxwell's equations, the following 2D scalar equations can be derived:

$$\frac{\partial H_x}{\partial t} = \frac{1}{\mu_0} \frac{\partial E_y}{\partial z} \tag{8.57a}$$

$$\frac{\partial H_z}{\partial t} = -\frac{1}{\mu_0} \frac{\partial E_y}{\partial x} \tag{8.57b}$$

$$\frac{\partial E_y}{\partial t} = \frac{1}{\varepsilon_0 \varepsilon_r} \left(\frac{\partial H_x}{\partial z} - \frac{\partial H_z}{\partial x} \right) \tag{8.57c}$$

where μ_0 is the permeability of the free space and ε_r is the relative dielectric constant of the medium.

With respect to the unit cell shown in Figure 8.2, the electromagnetic fields are expanded as a combination of scaling functions in space and Haar functions in time using

$$H_x(x, z, t) = \sum_{n,i,j=-\infty}^{+\infty} {}_n H_{i+1/2,j}^{x,\varphi} \varphi_{i+1/2}(x)\, \varphi_j(z) h_n(t) \tag{8.58a}$$

$$H_z(x, z, t) = \sum_{n,i,j=-\infty}^{+\infty} {}_n H_{i,j+1/2}^{z,\varphi} \varphi_i(x)\, \varphi_{j+1/2}(z) h_n(t) \tag{8.58b}$$

$$E_y(x, z, t) = \sum_{n,i,j=-\infty}^{+\infty} {}_{n+1/2} E_{i+1/2,j+1/2}^{y,\varphi} \varphi_{i+1/2}(x)\, \varphi_{j+1/2}(z) h_{n+1/2}(t) \tag{8.58c}$$

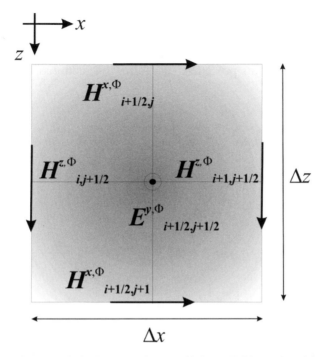

Figure 8.2 Electric and magnetic expansion coefficients field as placed inside the MRTD unit cell in the case of only scaling functions (S-MRTD). (Reproduced with permission from Letizia, R. and Obayya, S.S.A. (2009) Efficient second harmonic generation through selective photonic crystal-microcavity coupling. *IEEE J. Lightwave Technol.*, **27** (21), 4763–4772. © 2009 IEEE.)

where n, i, j, are the discrete indexes in time and in space, respectively, φ is the scaling function chosen from the CDF(2,4) family, h is the Haar function and represents the sampling in time, $_{n+1/2}E_{i+1/2,j+1/2}^{y,\varphi}, _{n}H_{i+1/2,j}^{x,\varphi}, _{n}H_{i,j+1/2}^{z,\varphi}$ are the expansion coefficients. The discretisation in space follows a scheme that is very similar to Yee's scheme. As shown in Figure 8.2, the components placed in a cell, on which the update iteration takes place, are expansion coefficients. The actual field at the time step t_0 in a point (x_0, z_0) has to be calculated from

$$E_y(x_0, z_0, t_0) = \sum_{i',j'=-\infty}^{+\infty} {}_{n'}E_{i',j'}^{y,\phi}\varphi_{i'}(x_0)\,\varphi_{j'}(y_0) \qquad (8.59)$$

where i', j' and n' are the indexes in space and time, respectively. Due to the finite support of CDF scaling functions, only a few terms of the previous summation have to be considered.

Substituting the field expansions in the form of Equations (8.58a–c) into the scalar Maxwell's Equations (8.57a–c) and testing them with the dual of the biorthogonal scaling function, $\tilde{\varphi}_m$ (with $m = i, j$) in space and pulse functions in time, and following the method-of-moments procedure, the following field-update equations can be obtained

$$_{n+1}H^{x,\varphi}_{i+1/2,j} = {}_n H^{x,\varphi}_{i+1/2,j} - \frac{\Delta t}{\mu_0 \Delta z}\left(\sum_{l=-L_s}^{L_s-1} a\,(l)_{n+1/2}E^{y,\varphi}_{i+1/2,j-l-1/2} \right) \qquad (8.60a)$$

$$_{n+1}H^{z,\varphi}_{i,j+1/2} = {}_n H^{z,\varphi}_{i,j+1/2} + \frac{\Delta t}{\mu_0 \Delta x}\left(\sum_{l=-L_s}^{L_s-1} a\,(l)_{n+1/2}E^{y,\varphi}_{i-l-1/2,j+1/2} \right) \qquad (8.60b)$$

$$_{n+1/2}E^{y,\varphi}_{i+1/2,j+1/2} = {}_{n-1/2}E^{y,\varphi}_{i+1/2,j+1/2} + \frac{\Delta t}{\varepsilon_0 \varepsilon_{i+1/2,j+1/2}}$$

$$\times \left(\sum_{l=-L_s}^{L_s-1} a\,(l)\left(-\frac{1}{\Delta z}{}_n H^{x,\varphi}_{i+1/2,j+1/2} + \frac{1}{\Delta x}{}_n H^{z,\varphi}_{i-l-1,j+1/2} \right) \right) \qquad (8.60c)$$

where L_s, called the *stencil size*, represents the effective support of the basis function, which determines the number of expansion coefficients considered in the summation and it is equal to 5 for CDF (2,4), the connection coefficients $a(l)$, [22], are obtained numerically by Equation (8.61) and their values for the assumed scaling functions are given in Table 8.1 [16].

The CDF functions have the advantage of making use of a compact support (finite number of nonzero coefficients in the MRTD scheme) so $a(l)$ is exactly equal to zero for $l > L_s - 1$ and $l < L_s$. For $l < 0$, the $a(l)$ values are known from the symmetry

Table 8.1 Connection coefficients and courant number at the stability limit in two dimensions. (Reproduced with permission from Letizia, R. and Obayya, S.S.A. (2008) Efficient multiresolution time-domain analysis of arbitrarily shaped photonic devices. *IET Optoelectron.*, **2** (6), 241–253. © 2008 IET.)

l	CDF(2,2)	CDF(2,4)
1	1.229 166 7	1.291 813 4
2	−0.093 750 0	−0.137 134 8
3	0.028 761 7	0.028 761 7
4	0	−0.003 470 1
5	0	0.000 008 0
s	0.5300	0.4839

relation $a(-l) = -a(l-1)$

$$\int_{-\infty}^{+\infty} \varphi_i(x) \frac{\partial \varphi_{i'+\frac{1}{2}}(x)}{\partial x} dx = \sum_{l=-L_s}^{L_s-1} a(l)\delta_{i+l,i'} \tag{8.61}$$

By virtue of their interpolation property, as biorthogonal wavelets, the expansion coefficients for CDF families on which the updating takes place, can be taken as physical field values with negligible error [16]. For instance, the field E_y at an arbitrary point in space $(x_0 = i\,\Delta x, z_0 = j\,\Delta z)$ at time $t_0 = n\,\Delta t$ is given by

$$E_y(x_0, z_0, t_0) = \int\int E_y(x, z, t)\delta(x - x_0)\delta(z - z_0)\delta(t - t_0)dx\,dz\,dt = {}_n E_{i,j}^{y,\phi} \tag{8.62}$$

that allows the field coefficient to be taken as the field value.

This allows the building of a simple algorithm in which the computational overhead of the total field reconstruction is reduced.

Figure 8.3 shows a sketch of the update process in a 1D space: the component ${}_n E^{\varphi}$ at position i is calculated from a number of components ${}_{n-\frac{1}{2}} H^{\varphi}$ in the range $\left[i - L_s + 1/2; i + L_s - 1/2\right]$ that is determined by the stencil size L_s and are weighted by the connection coefficients $a(l)$.

In order to ensure the numerical stability of the MRTD scheme, the time step Δt has to be smaller than a certain limit as

$$\Delta t \leq s\frac{\Delta}{c_0}, \tag{8.63}$$

$$s = \frac{1}{\sqrt{2}\sum_{l=0}^{L_s-1} |a(l)|} \tag{8.64}$$

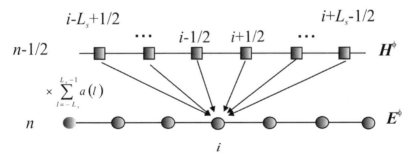

Figure 8.3 Scheme of the update process in a 1D space for S-MRTD with compact basis functions determined by the stencil size L_s.

where $\Delta = \Delta x = \Delta z$ in an uniform mesh, c_0 is the speed of the light and the Courant number, s, represents the stability factor in two dimensions and is equal to 0.4839 for S-MRTD with CDF (2,4) (Table 8.1). Numerical experiments have revealed that although this value is enough to guarantee the stability of the algorithm, it cannot guarantee a good accuracy of the results with coarser meshes. Smaller values of s, typically five times less the stability limit, can significantly improve the accuracy of the method, even for coarser spatial discretisation. In the spirit of the numerical comparison performed in [18], it has been found that the value $s = 0.1$ is the most suitable to ensure numerical stability and also a high level of accuracy.

8.4.3   UPML in S-MRTD

One of the most important challenges in the field of computational modelling has been the introduction of absorbing boundary conditions (ABCs) at the grid boundary, allowing the simulation of an infinite region. Amongst all the possible techniques, here the 2D-S-MRTD scheme is presented in conjunction with the UPML absorbing boundary condition which efficiently terminates the boundaries of the space domain in an absorbing artificial material medium. Starting from the traditional scalar equations of UPML reported in the literature for FDTD [5,22], and already introduced in Chapter 7 for the termination of the FVTD scheme, they are now implemented in the framework of the MRTD technique. Therefore, UPML derivatives in space are discretised by testing them with the scaling functions through Galerkin's procedure, while a second-order central difference scheme is adopted for the derivatives in time. This procedure leads to a two-step update scheme for each field component. Considering the TE propagation and the geometry of the grid (Figure 8.2), the following discretised equations are derived

$$_nB_{i+1/2,j}^{x,\varphi} = {}_{n-1}B_{i+1/2,j}^{x,\varphi} - \Delta t \sum_{l=-L_s}^{L_s-1} a(l) \left(\frac{{}^{n-1/2}E_{i+1/2,j+l+1/2}^{y,\varphi}}{\Delta z} \right) \tag{8.65a}$$

$$_nH_{i+1/2,j}^{x,\varphi} = \left(\frac{2\varepsilon_0 - \sigma_z \Delta t}{2\varepsilon_0 + \sigma_z \Delta t} \right) {}_{n-1}H_{i+1/2,j}^{x,\varphi} + \left(\frac{2\varepsilon_0 + \sigma_x \Delta t}{2\varepsilon_0 + \sigma_z \Delta t} \right) \frac{1}{\mu_0} {}_nB_{i+1/2,j}^{x,\varphi}$$
$$- \left(\frac{2\varepsilon_0 - \sigma_x \Delta t}{2\varepsilon_0 + \sigma_z \Delta t} \right) \frac{1}{\mu_0} {}_{n-1}B_{i+1/2,j}^{x,\varphi} \tag{8.65b}$$

$$_nB_{i,j+1/2}^{z,\varphi} = \left(\frac{2\varepsilon_0 - \sigma_x \Delta t}{2\varepsilon_0 + \sigma_x \Delta t} \right) {}_{n-1}B_{i,j+1/2}^{z,\varphi} + \left(\frac{2\varepsilon_0 \Delta t}{2\varepsilon_0 + \sigma_x \Delta t} \right) \sum_{l=-L_s}^{L_s-1}$$

$$\times a\,(l) \left(\frac{\frac{1}{n-\frac{1}{2}} E^{y,\varphi}_{i+l+1/2,\,j+1/2}}{\Delta x} \right) \tag{8.66a}$$

$$_{n}H^{z,\varphi}_{i,\,j+1/2} = {}_{n-1}H^{z,\varphi}_{i,\,j+1/2} + \left(\frac{2\varepsilon_0 + \sigma_z \Delta t}{2\varepsilon_0} \right) \frac{1}{\mu_0}{}_{n}B^{z,\varphi}_{i,\,j+1/2} - \left(\frac{2\varepsilon_0 - \sigma_z \Delta t}{2\varepsilon_0} \right)$$

$$\times \frac{1}{\mu_0}{}_{n-1}B^{z,\varphi}_{i,\,j+1/2} \tag{8.66b}$$

$$_{n+1/2}D^{y,\varphi}_{i+1/2,\,j+1/2} = \left(\frac{2\varepsilon_0 - \sigma_z \Delta t}{2\varepsilon_0 + \sigma_z \Delta t} \right) {}_{n}D^{y,\varphi}_{i+1/2,\,j+1/2} + \left(\frac{2\varepsilon_0 \Delta t}{2\varepsilon_0 + \sigma_z \Delta t} \right)$$

$$\sum_{l=-L_s}^{L_s-1} a\,(l) \left(-\frac{{}_{n}H^{x,\varphi}_{i+1/2,\,j-l-1}}{\Delta z} + \frac{{}_{n}H^{z,\varphi}_{i-l-1,\,j+1/2}}{\Delta x} \right) \tag{8.67a}$$

$$_{n+\frac{1}{2}}E^{y,\varphi}_{i+\frac{1}{2},\,j+\frac{1}{2}} = \left(\frac{2\varepsilon_0 - \sigma_x \Delta t}{2\varepsilon_0 + \sigma_x \Delta t} \right) {}_{n}E^{y,\varphi}_{i+1/2,\,j+1/2} + \left(\frac{2\varepsilon_0}{\varepsilon_0 \varepsilon_{i+1/2,\,j+1/2}\,(2\varepsilon_0 + \sigma_x \Delta t)} \right)$$

$$\left({}_{n+1/2}D^{y,\varphi}_{i+1/2,\,j+1/2} - {}_{n-1/2}D^{y,\varphi}_{i+1/2,\,j+1/2} \right) \tag{8.67b}$$

where ε is the permittivity of the medium, σ_x and σ_z are the electric conductivity of the UPML layers, whose profile takes the following form

$$\sigma_i\,(i) = \frac{\sigma_{\text{max}} i^m}{d^m} \tag{8.68}$$

where $i = x, z$, d is the depth of the UPML and m stands for the order of the polynomial variation. The choice of σ_{max} that minimises the reflection from boundaries is [15]

$$\sigma_{\text{max}} \approx \frac{(m+1)}{150\pi \Delta \sqrt{\varepsilon_r}} \tag{8.69}$$

where Δ is the uniform spatial discretisation adopted.

References

[1] Locatelli, A., Pigozzo, F.M., Modotto, D. *et al.* (2002) Bidirectional beam propagation method for multilayered dieletrics with quadratic nonlinearity. *IEEE J. Sel. Topics Quantum Electron.*, **8**, 440–447.

[2] D'Orazio, A., De Sario, M., Petruzzelli, V. and Prudenzano, F. (2003) All-optical amplification via the interaction among guided and leaky propagation modes in lithum niobate waveguides exploiting the cascaded second order nonlinearity. *Opt. Qunatum Electron.*, **35**, 47–68.

[3] Martorell, J., Vilaseca, R. and Corbalan, R. (1997) Pseudo-metal reflection at the interface between a linear and a nonlinear material. *Optics Commun.*, **144**, 65–69.

[4] D'Orazio, A. (1998) Method of lines for the analysis of nonlinear phenomenon due to cascaded second order nonlinearity. *IEEE Photon. Technol. Lett.*, **10**, 1584–1586.

[5] Taflove, A. (1995) *Computational Electrodynamics: The Finite-Difference Time-Domain Method*, Artech House, Norwood, MA.

[6] Joseph, R.M., Hagness, S.C. and Taflove, A. (1991) Direct time integration of Maxwell's equations in linear dispersive media with absorption for scattering and propagation of femtosecond electromagnetic pulses. *Optics Lett.*, **16**, 1412–1414.

[7] Körner, T.O. and Fichtner, W. (1997) Auxiliary differential equation: efficient implementation in the finite-difference time-domain method. *Optics Letters*, **22**, 1586–1589.

[8] Krumpholz, M. and Katehi, L.P.B. (1996) MRTD: New time domain schemes based on multiresolution analysis. *IEEE Trans. Microwave Theory Tech.*, **44**, 555–571.

[9] Bushyager, N. and Tentzeris, E.M. (2006) *MRTD (Multi Resolution Time Domain) Method in Electromagnetics*, Morgan & Claypool Publishers.

[10] Petropoulos, P.G. (1994) Stability and phase error analysis of FD-TD in dispersive dieletrics. *IEEE Trans. Antennas Propag.*, **42**, 62–69.

[11] Mallat, S.G. (1989) A theory for multiresolution signal decomposition: The wavelet representation. *IEEE Trans. Pattern Anal. Mach. Intell.*, **11**, 674–693.

[12] Sarris, C. and Katehi, L.P.B. (2001) Fundamental gridding-related dispersion effects in multiresolution time-domain schemes. *IEEE Trans. Microwave Theory Tech.*, **49**, 2248–2257.

[13] Chen, Z.D. and Luo, S. (2006) Generalization of the finite-difference-based time-domain methods using the method of moments. *IEEE Trans. Antennas and Propag.*, **54**, 2515–2524.

[14] . Harrington, R. (1968) *Field Computations by Moment Method*, Macmillan, New York.

[15] Dogaru, T. and Carin, L. (2002) Application of Haar-wavelet-based multiresolution time-domain schemes to electromagnetic scattering problems. *IEEE Trans. Antennas Prop.*, **50**, 774–784.

[16] Dogaru, T. and Carin, L. (2001) Multiresolution time-domain using CDF biorthogonal wavelets. *IEEE Trans. Microwave Theory Tech.*, **49**, 685–691.

[17] Daubechies, I. (1992) *Ten Lectures on Wavelets*, SIAM, Philadelphia, PA.

[18] Carat, G., Gillard, R., Citerne, J. and Wiart, J. (2000) An efficient analysis of planar microwave circuits using a DWT-based Haar MRTD scheme. *IEEE Trans. Microwave Theory Tech.*, **48**, 2261–2270.

[19] Gotze, J., Odegard, J.E., Rieder, P. and Burrus, C.S. (1996) Approximate Moments and Regularity of Efficiently Implemented Orthogonal Wavelet Transform. *Proc. IEEE Int. Circuits Syst. Symp. Dig.*, pp. 405–408.

[20] Cheong, Y.W., Lee, Y.M., Ra, K.H. *et al.* (1999) Wavelet-Galerkin scheme of time-dependent inhomogeneous electromagnetic problems. *IEEE Microwave Guided Wave Lett.*, **9**, 297–299.

[21] Fujii, M. and Hoefer, W.J.R. (2000) Dispersion of time domain wavelet Galerkin method based on Daubechies compactly supported scaling functions with three and four vanishing moments. *IEEE Microwave Guided Wave Lett.*, **10**, 125–127.

[22] Gedney, S.D. (1996) An anisotropic perfectly matched layer absorbing media for the truncation of FDTD lattice. *IEEE Trans. Antennas Propagat.*, **44**, 1630–1639.

9

MRTD Analysis of PhC Devices

9.1 Introduction

Validation of the MRTD scheme developed and presented in the previous chapter is given here for the analysis of linear photonic devices. First of all, the rigorous implementation of the uniaxial perfectly matched layer (UPML) scheme at the lattice boundaries has been considered for validation in order to ensure efficient truncation of the computational domain with negligible artificial reflections. To this aim, the reflection coefficient from the UPML boundary is studied in a planar waveguide first, and then in a PhC-based waveguide. The obtained results are then compared to those performed by the same boundary scheme when implemented into the conventional FDTD method. Once the absorbing boundary condition has been validated and the required parameters have been set, the numerical accuracy and efficiency of the MRTD scheme are thoroughly investigated for Bragg resonators (1D-PhC) and PhC-optical filters for linear applications. Specifically, a comparison between MRTD performance and FDTD results is given for a Bragg resonator with different grid resolutions. This particular case proves the computational efficiency of MRTD over FDTD, which allows significant reduction in the required CPU running time when the same level of accuracy is required.

9.2 UPML-MRTD: Test and Code Validation

9.2.1 UPML in Planar Waveguide

As a first example, the simple slab waveguide shown in Figure 9.1 [1], has been considered, where the width is $w = 0.450$ μm and the core and the cladding have refractive indices of $n_{core} = 3.6$ and $n_{cl} = 3.42$, respectively. The propagation of the fundamental TE_y mode in the x-direction is performed with the aim of demonstrating the robustness of the UPML boundary condition incorporated into S-MRTD. The

Computational Photonics   Salah Obayya
© 2011 John Wiley & Sons, Ltd

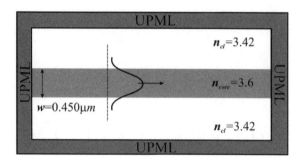

Figure 9.1 Schematic diagram of the planar waveguide simulated where $n_{cl} = 3.42$, $n_{core} = 3.6$ and $w = 0.45$ μm, [1].

source field is expressed as

$$E_{y,\text{source}}(z, t) = E_{\text{TE}_0}(z) \sin(2\pi f_0 t) e^{-\frac{(t-t_0)^2}{T_0^2}} \tag{9.1}$$

where E_{TE_0} is the profile of the fundamental TE_0 mode, $T_0 = 15$ fs is the bandwidth of the Gaussian profile in time, $t_0 = 60$ fs represents its delay and $\lambda_0 = c_0/f_0 = 0.86$ μm is the central wavelength.

Figure 9.2 shows the time evolution of the pulse energy, normalised to the input pulse, when the waveguide is assumed to be surrounded by a 20-cell UPML. The results obtained from varying the parameter m in the many simulations performed in this work show that, in most of the cases, for a 20-cell UPML medium, the best value for m is 2.5. The results are obtained by adopting uniform mesh with cell size $\Delta x = \Delta z = \Delta = 30$ nm and time-step size $\Delta t = 0.01$ fs. As may be noticed from Figure 9.2 and from Figure 9.3, which shows the electric-field patterns at different time intervals, once the pulse has completely been injected, the field propagates with constant power along the waveguide until it reaches the UPML edge in the longitudinal direction (x). At this point, it rapidly decreases to a negligible value, showing that the UPML boundary condition performs stably and rigorously.

Then, the reflection coefficient due to a 15- and 20-cell UPML boundary truncating the output of the waveguide has been calculated and compared to the one obtained by running simulations for UPML in a conventional FDTD scheme with the same mesh Δ [2]. The time variation of the electric field has been stored at the reference point (Figure 9.1). By means of an FFT of both the incident and reflected fields at this point, their spectral distribution ratio has been calculated to obtain the reflection coefficient variation with the wavelength. As shown in Figure 9.4, over the range of wavelengths around the input λ_0, the reflection detected from S-MRTD simulations is significantly lower than the one obtained with FDTD in both 15- and 20-cell UPMLs. In particular, it should be noted that the developed UPML scheme in the S-MRTD

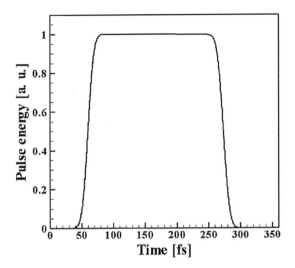

Figure 9.2 Time evolution of the pulse energy using UPML-MRTD scheme. (Reproduced with permission from Letizia, R. and Obayya, S.S.A. (2008) Efficient multiresolution time-domain analysis of arbitrarily shaped photonic devices. *IET Optoelectron.*, **2** (6), 241–253. © 2008 IET.)

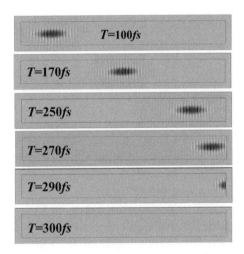

Figure 9.3 Time evolution of the electric-field pattern along the waveguide when the UPML boundary condition is used to truncate the computational window. (Reproduced with permission from Letizia, R. and Obayya, S.S.A. (2008) Efficient multiresolution time-domain analysis of arbitrarily shaped photonic devices. *IET Optoelectron.*, **2** (6), 241–253. © 2008 IET.)

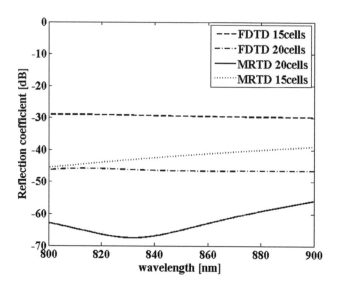

Figure 9.4   Variation of the UPML reflection coefficient with the wavelength using both FDTD and MRTD methods for 15- and 20-cell UPMLs. (Reproduced with permission from Letizia, R. and Obayya, S.S.A. (2008) Efficient multiresolution time-domain analysis of arbitrarily shaped photonic devices. *IET Optoelectron.*, **2** (6), 241–253. © 2008 IET.)

algorithm can bring reflections from computational domain boundaries down to less than -60 dB.

9.2.2   UPML in PhC Waveguide

In order to test the UPML scheme in a more complicated structure than the planar waveguide seen in the previous chapter, a PhC waveguide is considered next. The structure, whose schematic is shown in Figure 9.5, is a pillar-type PhC waveguide in which pillars made of GaAs ($n = 3.4$) are surrounded by air. The arrangement has periodicity $a = 0.58$ μm and filling factor $r/a = 0.18$, where r is the radius of the pillars. The TE Gaussian pulse given in Equation (9.1) with delay $t_0 = 30$ fs, bandwidth $T_0 = 90$ fs, central wavelength of $\lambda_0 = c_0/f_0 = 1.5$ μm and where E_{TE_0} represents the profile of the fundamental TE_0 mode in the waveguide that has been injected. The wavelength range of the source signal belongs to the photonic bandgap of the periodic structure, thus it can propagate along the line defect.

As shown in Figure 9.5, a detector is used to record the incident electric field and the reflected field from the UPML boundary. The FFT of the recorded temporal data and the ratio between reflected and incident field give the value of the reflection coefficient from the boundary. In the case under investigation here, reflections coming back from the UPML boundary at the right end of the waveguide are calculated for

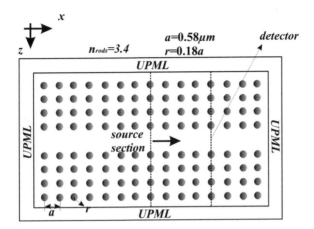

Figure 9.5 Schematic diagram of a pillar-type PhC-waveguide excited with a Gaussian pulse at 1.5 μm in order to test the efficiency of UPML-MRTD scheme. (Reproduced with permission from Letizia, R. and Obayya, S.S.A. (2008) Efficient multiresolution time-domain analysis of arbitrarily shaped photonic devices. *IET Optoelectron.*, **2** (6), 241–253. © 2008 IET.)

different cases. Figure 9.6 shows the evolution of the reflection coefficient obtained by varying the depth of the UPML boundaries. From this figure, it can be seen that for a UPML depth of 1.74 μm, the reflection coefficient is less than −40 dB at the central wavelength λ_0.

In Figure 9.7, the performance of the UPML is analysed for different values of σ_{max} adopted in the polynomial scaling. For $\sigma_{max} \Delta = 0.0175/\pi$, reflection coefficient is as low as −40 dB.

9.3 MRTD versus FDTD for the Analysis of Linear Photonic Devices

9.3.1 Bragg Resonators: Comparison with Analytical Approach

After studying the stability of the proposed method, the Bragg resonator (BR) problem is considered next. The Bragg grating reflector is an example of a 1D-PhC that serves as a reflector for frequencies within a photonic bandgap. In particular, it finds application in devices such as the distributed Bragg reflector (DBR) laser and the distributed feedback (DFB) laser.

Considering a medium of refractive index, n, containing a region of length, L, in which the perturbed refractive index varies periodically with spatial period, Λ, and indices in the range $n \pm n_1$ for $n_1/n \ll 1$. This region comprises a distributed Bragg phase grating that couples oppositely travelling waves with propagation constants,

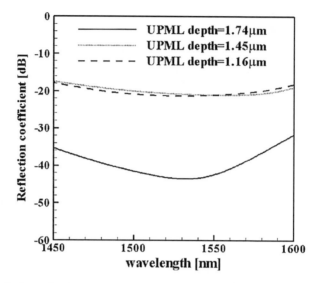

Figure 9.6 Variation of reflection coefficient with the depth of the UPML boundaries. (Reproduced with permission from Letizia, R. and Obayya, S.S.A. (2008) Efficient multiresolution time-domain analysis of arbitrarily shaped photonic devices. *IET Optoelectron.*, **2** (6), 241–253. © 2008 IET.)

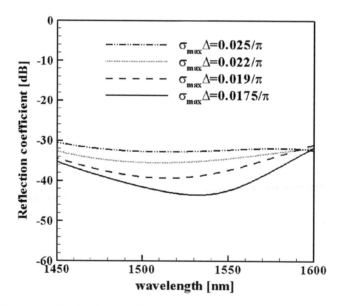

Figure 9.7 Variation of reflection coefficient with the value of σ_{max} adopted. (Reproduced with permission from Letizia, R. and Obayya, S.S.A. (2008) Efficient multiresolution time-domain analysis of arbitrarily shaped photonic devices. *IET Optoelectron.*, **2** (6), 241–253. © 2008 IET.)

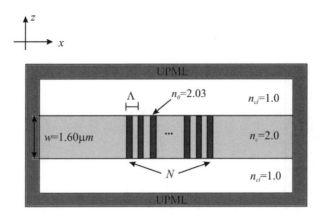

Figure 9.8   Schematic diagram of Bragg resonators (BR) with $\Lambda = 0.25$ μm and $N = 20$ [4].

$\pm\beta$, at angular frequency, ω. The Bragg propagation constant is defined by

$$2\beta_{\mathrm{B}} = K = \frac{2\pi}{\Lambda} \qquad (9.2)$$

where K represents the grating number. The Bragg condition is satisfied when $\beta = \beta_{\mathrm{B}}$. The Bragg wavelength is $\lambda_{\mathrm{B}} = 2\hat{n}\Lambda$ and the Bragg frequency is $\omega_{\mathrm{B}} = cK/2\hat{n}$ where $\hat{n}$ represents the modal index for the planar guide. At this frequency, the coupling between oppositely travelling waves is maximised, and the transmission is minimised, as the many small reflections from each phase perturbation add constructively in the backward direction. The coupling produces a transmission stopband, centred at ω_{B}, in which transmission is forbidden [3].

Figure 9.8 shows the schematic diagram of the BR considered, where the waveguide is designed with a core of refractive index, $n_c = 2$, and a width, $w = 1.60$ μm, surrounded by air ($n_{\mathrm{cl}} = 1.0$), [4]. The added Bragg resonators consist of $N = 20$ periods and for a central wavelength $\lambda_0 = 0.99$ μm, the periodicity Λ is equal to 0.25 μm and $n_0 = 2.03$ is the refractive index of the BR. This structure has been discretised with $\Delta = \Delta x = \Delta z = \Lambda/4$ and the adopted time step is $\Delta t = 0.02$ fs. A Gaussian pulse modulated by the fundamental TE_0 mode profile, Equation (9.1), is injected as excitation with parameters $T_0 = 5$ fs and $t_0 = 15$ fs.

The reflection coefficient spectra from the BR has been calculated and compared to the results obtained by conventional FDTD for the analysis of the BR, as shown in Figure 9.9. It may be observed from Figure 9.9 that the results from S-MRTD are in good agreement with their FDTD counterparts [4], even though a resolution of only four cells per period Λ has been used, whereas FDTD has required a resolution of 10 cells per period in order to achieve the same level of accuracy. A comparison of the

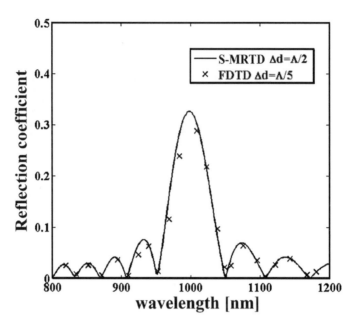

Figure 9.9 Variation of the TE reflection coefficient with the wavelength using both MRTD method and FDTD with grid resolution $\Delta d = \Delta x = \Delta z$ of $\Lambda/2$ and $\Lambda/5$ respectively. (Reproduced with permission from Letizia, R. and Obayya, S.S.A. (2008) Efficient multiresolution time-domain analysis of arbitrarily shaped photonic devices. *IET Optoelectron.*, **2** (6), 241–253. © 2008 IET.)

CPU running time shows that for the case of MRTD, there is a saving equal to 1/2 of that required by FDTD for the same level of accuracy.

Figure 9.10(a) shows the electric field pattern at the interface between the waveguide and the BR when a sinusoidal continuous wave has been injected and has reached a steady state. The wavelength of the sinusoidal wave is fixed at $\lambda = 0.99$ μm, which corresponds to the maximum reflection coefficient for the BR, λ_B. The source is injected at a length of 19 μm and the BR interface is placed at 32 μm. Figure 9.10(b) shows the field pattern as obtained in the case of a continuous wave input at a wavelength of $\lambda = 1.18$ μm, at which the reflection coefficient has a value of 10^{-3}. The amplitude of the electric field in this case has been normalised to the maximum amplitude recorded in the case of maximum reflection. It can be seen that the standing-wave pattern is still present, but with peaks that are much lower than in the previous case. The accuracy of the proposed approach has increased over the conventional FDTD as a result of the high-order spatial finite difference employed. Multiresolution analysis, combined with the second-order leapfrog finite-differences in time, based on Haar scaling functions, allowes for a much coarser spatial discretisation. In order to quantify the more linear dispersion characteristic of the S-MRTD scheme, the variation of the maximum reflection coefficient and its wavelength λ_B with respect

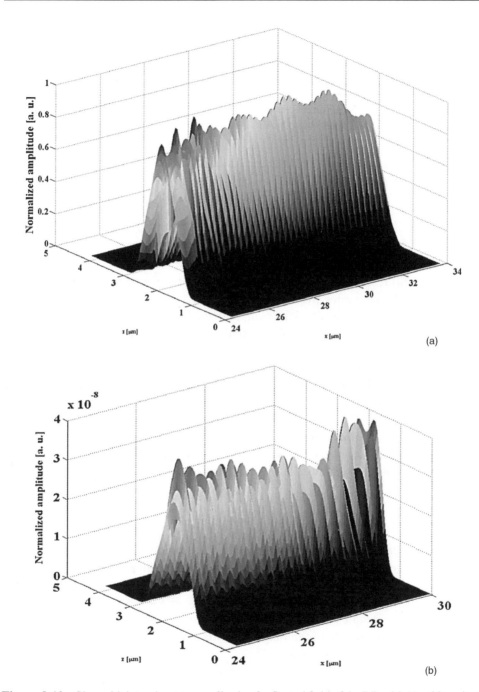

Figure 9.10 Sinusoidal steady-state amplitude of reflected field of the BR with $N = 20$ excited with a continuous wave: of wavelength $\lambda = 0.99$ μm, (a); and excited with a continuous wave of wavelength $\lambda = 1.18$ μm, (b). (Reproduced with permission from Letizia, R. and Obayya, S.S.A. (2008) Efficient multiresolution time-domain analysis of arbitrarily shaped photonic devices. *IET Optoelectronics*, **2** (6), 241–253. © 2008 IET.)

Table 9.1 Relative Error (%) for MRTD and FDTD compared to analytical solution at different spatial discretisations. (Reproduced with permission from Letizia, R. and Obayya, S.S.A. (2008) Efficient multiresolution time-domain analysis of arbitrarily shaped photonic devices. *IET Optoelectron.*, **2** (6), 241–253. © 2008 IET.)

$\Lambda/\Delta d$	Maximum Reflection Coefficient Error		Wavelength Peak Error	
	MRTD	FDTD	MRTD	FDTD
10	1.5	3	0.3	1.1
8	3.8	4.5	0.8	1.4
6	3	7.8	1.6	1.9
4	12	13.7	0.9	3.4

to different grid resolution values has been investigated for both the MRTD and the FDTD. The cell size varies from a resolution of 10 cells per period to one of four cells per period. The results obtained from both the S-MRTD and the FDTD schemes are reported in Table 9.1, where the analytical values of the maximum reflection coefficient, calculated at $\lambda = 0.99$ μm, is taken as a reference value to express the relative error. As expected, for both methods, when adopting a bigger stencil in space, the error increases. However, the S-MRTD code still shows greater accuracy than the FDTD in each case. Moreover, a slight shift of the peak reflection coefficient from the BR wavelength of 0.99 μm has been observed and reported in Table 9.1 for each cell size value. Again, the proposed method shows better accuracy compared to FDTD. In particular, this difference becomes more evident and significant as the grid resolution decreases towards coarser values.

9.3.2 PhC-Based Optical Filter

Here, the design of a PhC-based optical filter is considered. The structure, whose schematic diagram is given in Figure 9.11, consists of a single microcavity coupled to a straight waveguide based on a PBG structure [5–15]. The periodic square array consists of dielectric rods with refractive index, $n_{\text{rods}} = 3.4$ in air, a lattice constant,t $a = 0.58$ μm, and rod radius, $r_{\text{rods}} = 0.18a$. The crystal shows the photonic bandgap only for TE modes that extends in the range of $0.302 < \omega a/2\pi c < 0.443$ where $\omega a/2\pi c$ represents the normalised frequency [15].

A source field of expression similar to Equation (9.1) is injected into the waveguide with parameters $T_0 = 15$ fs, $t_0 = 50$ fs and central wavelength $\lambda_0 = c_0/f_0 = 1.45$ μm. The geometry of the structure is discretised with a mesh of cell size $\Delta x = \Delta z = \Delta = 29$ nm that leads to a simulation time interval $\Delta t \approx 0.01$ fs. Placing a detector at the output port, the time variation of the electric field has been recorded and by means of

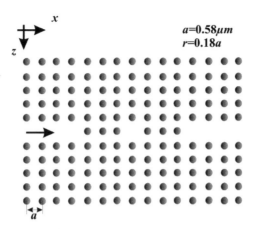

Figure 9.11 Schematic diagram of the PhC microcavity coupled to a waveguide. (Reproduced with permission from Letizia, R. and Obayya, S.S.A. (2008) Efficient multiresolution time-domain analysis of arbitrarily shaped photonic devices. *IET Optoelectron.*, **2** (6), 241–253. © 2008 IET.)

FFT, its spectral distribution has been calculated and normalised to the input. Figure 9.12 shows the resulting transmission coefficient obtained in three different cases: the microcavity is formed in a lattice of [1 × 1], [2 × 2] or [3 × 3] rods with a single-rod-missing defect in the middle of the waveguide. It can be noted from Figure 9.12 that from all the frequencies that propagate inside the waveguide, only the one in which the microcavity shows resonance (at a normalised frequency of 0.39) can pass through to be observed at the output port. The others are forbidden to propagate and reflect back to the input port. Thus, the structure realises an optical filter tuned at the resonance frequency of the PhC-microcavity. By increasing the number of rods that assist the cavity, it can be seen that the transmission becomes sharper and more selective as the quality factor increases as well.

Next, in order to investigate the tuning capability of the optical filter at a particular frequency, the effect of varying the single-defect radius (r_d) that forms the microcavity shown in Figure 9.13, while the number of rods surrounding it is kept to three, will be analysed. Figure 9.14 shows the variation of the resonance frequency and quality factor, Q, calculated by the formula in Equation (6.3), of the cavity with r_d.

$$Q = 2\pi \frac{|E_t|^2}{|E_t|^2 - |E_{t+T}|^2} \tag{9.3}$$

It can be seen that by increasing the size of the defect rod, a shift towards lower frequencies is possible, whereas the quality factor rises to higher values. In particular, it the radius of the defect rod under the actual filling-factor value of the PhC was

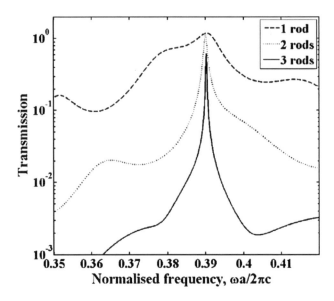

Figure 9.12 Variation of the transmission-field characteristic with the normalised frequency when the cavity is separated from the waveguide by three rods (solid line), two rods (dotted line) and one rod (dashed line). (Reproduced with permission from Letizia, R. and Obayya, S.S.A. (2008) Efficient multiresolution time-domain analysis of arbitrarily shaped photonic devices. *IET Optoelectron.*, **2** (6), 241–253. © 2008 IET.)

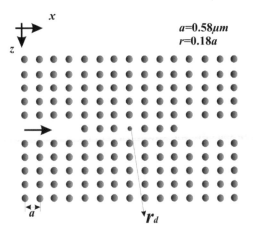

Figure 9.13 Schematic diagram of the PhC microcavity coupled to the waveguide when the single defect consists of a rod of smaller radius r_d. (Reproduced with permission from Letizia, R. and Obayya, S.S.A. (2008) Efficient multiresolution time-domain analysis of arbitrarily shaped photonic devices. *IET Optoelectron.*, **2** (6), 241–253. © 2008 IET.)

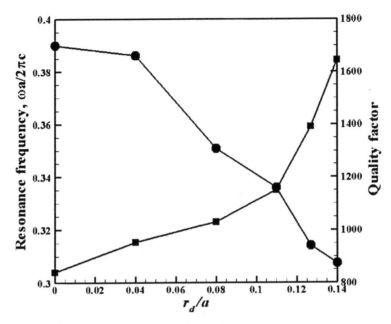

Figure 9.14 Variation of the resonance frequency and quality factor Q with r_d/a, where r_d is the defect radius and a is the lattice period. (Reproduced with permission from Letizia, R. and Obayya, S.S.A. (2008) Efficient multiresolution time-domain analysis of arbitrarily shaped photonic devices. *IET Optoelectron.*, **2** (6), 241–253. © 2008 IET.)

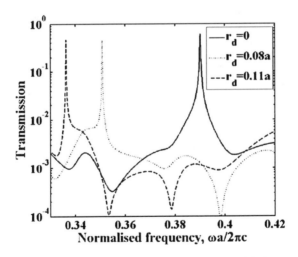

Figure 9.15 Variation of the transmission-field characteristic with the normalised frequency for different values of r_d : $r_d = 0$ (solid line), $r_d = 0.08a$ (dotted line), $r_d = 0.11a$ (dashed line). (Reproduced with permission from Letizia, R. and Obayya, S.S.A. (2008) Efficient multiresolution time-domain analysis of arbitrarily shaped photonic devices. *IET Optoelectron.*, **2** (6), 241–253. © 2008 IET.)

varied in order to ensure the resonance mode had the same mode profile inside the cavity. In Figure 9.15 the shift of the transmission spectra peak with the value of r_d, increasing this value from zero (rod-missing) to $0.08a$ and then $0.11a$, is reported. As it can be seen, the transmission resonance has shifted to a normalised frequency of 0.3508 and 0.336, respectively.

References

[1] Obayya, S.S.A. (2004) Novel finite element analysis of optical waveguide discontinuity problems. *IEEE J. Lightwave Technol.*, **22**, 1420–1425.

[2] Taflove, A. (1995) *Computational Electrodynamics: The Finite-Difference Time-Domain Method*, Artech House, Norwood, MA.

[3] Tamir, T. (1990) *Guided-wave Optoelectronics*, Springler-Verlgo, Berlin.

[4] Pinto, D., Obayya, S.S.A., Rahman, B.M.A. and Grattan, K.T.V. (2006) FDTD analysis of nonlinear Bragg grating based optical devices. *Opt. Quantum Electron.*, **38**, 1217–1235.

[5] Joannopoulos, J.D., Johnson, S.G., Winn, J.N. and Maede, R.D. (2008) *Photonic Crystals, Molding the Flow of Light*, 2nd edn, Princeton University Press.

[6] Borel, P.I., Frandsen, L.H., Harpøth, A. *et al.* (2004) Bandwidth engineering of photonic crystal waveguide bends. *Electron. Lett.*, **40**, 1263–1264.

[7] Costa, R., Melloni, A. and Martinelli, M. (2003) Bandpass resonant filters in photonic-crystal waveguides. *IEEE Photon. Technol. Lett.*, **15**, 401–403.

[8] Pistono, E., Ferrari, P., Duvillaret, L. *et al.* (2003) Tunable bandpass microwave filters based on defect commandable photonic bandgap waveguides. *Electron. Lett.*, **39**, 1131–1133.

[9] Villeneuve, P.R., Fan, S. and Joannopoulos, J.D. (1996) Microcavities in photonic crystals: Mode symmetry, tunability, and coupling efficiency. *Phys. Rev. B, Condens. Matter*, **54**, 7837–7842.

[10] Chong, H.M.H. and De La Rue, R.M. (2004) Tuning of photonic crystal waveguide microcavity by thermooptic effect. *IEEE Photon. Technol. Lett.*, **16**, 1528–1530.

[11] Mekis, A., Meier, M., DodaBalapur, A. *et al.* (1999) Lasing mechanism in two dimensional photonic crystal lasers. *Appl. Phys., A Mater. Sci. Process.*, **69**, 111–114.

[12] Painter, O., Lee, R.K., Scherer, A. *et al.* (1999) Two-dimensional photonic band-gap defect mode laser. *Science*, **284**, 1819–1821.

[13] Yablonovitch, E. (1994) Photonic crystals. *J. Modern Opt.*, **41**, 173–194.

[14] Chong, H.M.H. and De La Rue, R.M. (2004) Tuning of photonic crystal waveguide microcavity by thermooptic effect. *IEEE Photon. Technol. Lett.*, **16**, 1528–1530.

[15] Mekis, A., Chen, J.C., Kurland, I., *et al.* (1996) High transmission through sharp bends in photonic crystal waveguides. *Phys. Rev. Lett.*, **77**, 3787–3790.

10

MRTD Analysis of SHG PhC Devices

10.1 Introduction

In recent years, there has been a great deal of attention dedicated to nonlinear optical conversion in photonic-crystal-based devices as they have been found to be very promising for the future of modern telecommunications. Nonlinear processes such as second harmonic generation (SHG), can be exploited, in addition to the possibilities offered by PhC technology to open a new way of generating short and coherent wavelengths, not directly covered by laser sources, for innovative ultra-fast all-optical circuits. FDTD modelling of nonlinear frequency-conversion processes requires that the spatial interval in the propagation direction be a small fraction of the shortest generated wavelength to limit the phase velocity error typical of FDTD schemes. This constraint often results into an excessive cost in memory capacity and computational time. Moreover, in SHG problems, the energy coupling between the propagating EM field (fundamental wave) and the generated field (second harmonic wave) strongly depends on the phase shift between the two interacting fields. If this phase difference is not accurately estimated, the calculation of the amount of coupled energy and other relevant parameters will also be inaccurate. Therefore, in the field of nonlinear photonics, it is crucial to rely on higher-order approximation schemes such as MRTD that are capable of better accuracy than standard FDTD, while the computational burden is also minimised.

Keeping in mind these considerations, this chapter starts with the validation of the proposed nonlinear MRTD code in the cases of SHG in PhC waveguides and compound PhC structures for selective frequency doubling. Next, a new and more efficient design for selective SHG in nonlinear 2D photonic crystal-based devices is proposed. It will be shown that fine selection of the SH wave is obtained by exploiting the filtering properties of waveguide-microcavity coupling solutions.

Computational Photonics   Salah Obayya
© 2011 John Wiley & Sons, Ltd

10.2   Second Harmonic Generation in Optics

10.2.1   Introduction

A wide range of designs for all-optical signal processing using the PBG property of PhCs can be found in the literature, [1–5]. It is known that in a PhC, the electromagnetic-field distribution can be manipulated in order to create local field enhancement in one dielectric or in another. In nonlinear optics, this localisation of the fields can be exploited to enhance nonlinear effects that strongly depend on the strength of the local field. Near the photonic bandgap, low-frequency modes concentrate the energy in the high-refractive-index regions, whereas the high-frequency modes concentrate their energy in the low-refractive-index regions [1]. Therefore, if a periodic pattern is excited with a strong fundamental light source with wavelength close to the low-frequency photonic bandgap edge, the field will concentrate in the high-index medium, which can show a large value of nonlinear susceptibility. This strong field localisation can significantly assist nonlinear interactions of the fundamental field with the photonic crystal. Recently published work shows that the use of PhC-based devices for nonlinear processes, such as the generation of a second harmonic, is one of the most promising applications of this new design technology [6–11, 12].

10.2.2   Nonlinear Polarisation Vector

When a dielectric material is excited by an electric field, a small (compared to atomic dimensions) displacement of positive and negative charges is caused. Every molecule of the material is characterised by an induced dipole moment and the material is 'polarised'. This induced array of dipoles radiates its own electric field that is added to the incident field. If the frequency of the external field is very far from the resonance frequencies of the material, the polarisation, defined as the sum of all the dipole moments, depends linearly upon the external field. The polarisation vector, P, is expressed as function of the external electric field, E, by means of the susceptibility, χ, as follows

$$P_i(\omega) = \varepsilon_0 \cdot \sum_j \chi_{ij}(\omega) \cdot E_j(\omega), \quad i, j = (x, y, z) \tag{10.1}$$

in which the three components in space have been separated and the frequency dependence is expressed. From Equation (10.1) it can be seen that if the tensor, χ, is not diagonal, generally P is not parallel to the electric field imposed. If the electric field becomes strong enough, the relationship (10.1) doesn't hold anymore and it needs to be generalised into the following

$$P(\omega) = \varepsilon_0 \cdot \chi^{(1)}(\omega) \cdot E(\omega) + P^{\mathrm{NL}}(\omega) \tag{10.2}$$

in which a nonlinear polarization term, $P^{NL}(\omega)$, has been included in the linear relation. The relationship between the vectors P and E defines the system and describes all the properties of the material to which it refers. Generally, in nonlinear optics, this optical response is described by expressing the polarization, P, as a power series of the field strength, E, as

$$P = \varepsilon_0 \chi^{(1)} E + \varepsilon_0 \chi^{(2)} E^2 + \varepsilon_0 \chi^{(3)} E^3 + \cdots \qquad (10.3)$$

where ε_0 is the dielectric constant in free space and $\chi^{(n)}$ where $n = 1, 2, 3 \ldots$, represents the susceptibility of order n.

Nonlinear susceptibilities are also strictly related to the structural symmetry of the material. As a consequence, all materials having a centre of inversion symmetry show all elements of all even-order susceptibility tensors identically equal to zero. Thus, it is not possible to realise even-order nonlinear processes in these types of materials.

For simplicity, the propagation of a sinusoidal wave in an isotropic and nonlinear medium, for which Equation (10.3) holds, is considered. The electric field is expressed as

$$E = E_0 \cos(\omega t) = \frac{E_0}{2} \left(e^{j\omega t} + e^{-j\omega t} \right) \qquad (10.4)$$

If terms depending on susceptibilities of order greater than three are neglected, the induced polarisation can be written as

$$P(t) = \varepsilon_0 \left(\chi^{(1)} \frac{E_0}{2} \left(e^{j\omega t} + e^{-j\omega t} \right) + \chi^{(2)} \frac{E_0^2}{4} \left(e^{2j\omega t} + e^{-2j\omega t} + 2 \right) \right.$$
$$\left. + \chi^{(3)} \frac{E_0^3}{8} \left(e^{3j\omega t} + e^{-3j\omega t} + 3e^{2j\omega t} e^{-j\omega t} + 3e^{-2j\omega t} e^{j\omega t} \right) \right) \qquad (10.5)$$

thus

$$P(t) = \varepsilon_0 \left[\frac{1}{2} \chi^{(2)} E_0^2 + \left(\chi^{(1)} + \frac{3}{4} \chi^{(3)} E_0^2 \right) E_0 \cos(\omega t) \right.$$
$$\left. + \frac{1}{2} \chi^{(2)} E_0^2 \cos(2\omega t) + \frac{1}{4} \chi^{(3)} E_0^3 \cos(3\omega t) \right] \qquad (10.6)$$

Equation (10.6) shows that the polarisation, P, parallel to the electric field, E, as the medium is assumed to be isotropic, consists of four terms with angular frequencies of 0, ω, 2ω and 3ω, respectively.

Therefore, it is possible to represent the terms of which the polarisation vector consists in the frequency domain as follows

$$P(0) = \varepsilon_0 \frac{1}{2} \chi^{(2)} E_0^2 \tag{10.7a}$$

$$P(\omega) = \varepsilon_0 \left(\chi^{(1)} + \frac{3}{4} \chi^{(3)} E_0^2 \right) E_0 \cos{(\omega t)} \tag{10.7b}$$

$$P(2\omega) = \frac{1}{2} \varepsilon_0 \chi^{(2)} E_0^2 \cos{(2\omega t)} \tag{10.7c}$$

$$P(3\omega) = \frac{1}{4} \chi^{(3)} E_0^3 \cos{(3\omega t)} \tag{10.7d}$$

where $P(0)$ refers to the term with zero frequency that does not lead to the generation of electromagnetic radiation, but to a process of *optical rectification* (also called the inverse Pockels effect), in which a static electric field is created within the nonlinear crystal. The terms $P(2\omega)$ and $P(3\omega)$ instead represent the sources of the second and third harmonics, respectively. Finally, the term $P(\omega)$ consists of both the linear response at the frequency of the input signal, represented by $\varepsilon_0 \chi^{(1)} E_0 \cos{(\omega t)}$, and a term that depends on the third-order susceptibility, $\chi^{(3)}$. The latter term can be used to vary the refractive index of the medium, making possible self-focusing of the beam.

The system just assumed is very simplified compared to reality. To generalise the model, we need to consider the vector nature of the fields and that the dielectric susceptibilities are tensors in anisotropic materials. In the generic case of anisotropic and nonlinear medium, the relationship between P and E becomes

$$P_i = \varepsilon_0 \left(\sum_j \chi_{ij} E_j + \sum_{j,k} \chi_{ijk}^{(2)} E_j E_k + \sum_{j,k,l} \chi_{ijkl}^{(3)} E_j E_k E_l + \cdots \right), \ i, j, k, l = x, y, z \tag{10.8}$$

where P_i represents the *i*-component of polarisation vector, P, and E_j the *j*- component of electric-field vector, E. In anisotropic materials, the nonlinearity, $\chi^{(n)}$ of order n is a tensor with rank $n + 1$ and consists of 3^{n+1} elements.

The polarisation vector plays a crucial role in the description of nonlinear processes. By varying with time, in fact the polarisation can act as source for new components of the electromagnetic field. If, for instance, the wave equation is considered for the case of nonlinear optical medium, we have

$$\nabla^2 \vec{E}(t) - \frac{n^2}{c_0^2} \cdot \frac{\partial^2 \vec{E}(t)}{\partial t^2} = \frac{4\pi}{c_0^2} \cdot \frac{\partial^2 \vec{P}^{NL}}{\partial t^2} \tag{10.9}$$

where n is the linear refractive index and c_0 represents the speed of light in free space. This expression can be interpreted as an inhomogeneous wave equation in which the term $\vec{P}^{NL}$, which is related to the nonlinear response of the medium, controls the electromagnetic propagation. In particular, it says that when the term $\partial^2 \vec{P}^{NL}/\partial t^2$ is nonzero, the charges are accelerated and, according to Larmor's theorem, they generate electromagnetic radiation.

Typically, every nonlinear light–matter interaction can be expressed in terms of the nonlinear contribution to the polarisation vector, as described by Equation (10.2).

10.2.3 Physics of Second Harmonic Generation

The demonstration of optical second harmonic generation by the irradiation of a quartz crystal with a ruby laser in 1961 [13] marked the beginning of a new field of nonlinear optics. Nonlinear polarisation allows power to be exchanged between waves at different frequencies. Thus, SHG can be seen as a particular type of frequency mixing. A single pump wave, the fundamental wave at frequency ω, is incident on a nonlinear medium that exhibits second-order nonlinearity and generates a wave at double the frequency, the second harmonic 2ω. Second harmonic generation is described schematically in Figure 10.1.

Frequency conversion in optics can even be understood as the modulation of the refractive index by an electric field at a given optical frequency, or frequencies, through a second-order nonlinearity. The modulated index then produces sidebands at the optical frequency, yielding harmonics, and sum and difference frequencies [14].

In Figure 10.2 the difference between linear and nonlinear response is shown at a steady state: when the response is nonlinear, the polarisation keeps the same period

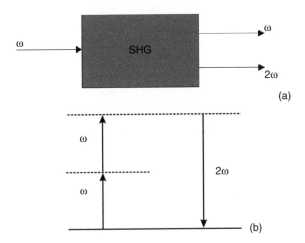

Figure 10.1 Schematic description of second harmonic generation process: (a) geometry of SHG; (b) energy-level diagram describing SHG.

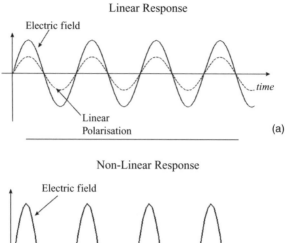

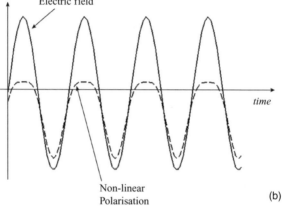

Figure 10.2 (a) Sinusoidal incident electric field and consequent polarisation in case of linear crystal and (b) of a crystal lacking of inversion symmetry with second order nonlinearity.

as the electric field, but the sinusoidal variation is lost. It is possible to break down the polarisation vector into its Fourier components, as shown in Figure 10.3.

If the material is lacking a symmetry inversion property, it has nonzero second-order nonlinearity and thus it is possible to obtain a nonzero second-harmonic component with generation of a new electromagnetic field at frequency 2ω. The amplitude of this field is always orders of magnitude less than the incident or the transmitted fields. However, there are several methods that can be used to enhance the generation of a second harmonic. All these methods refer to the concept of the *phase matching* condition.

It can be demonstrated that conversion efficiency is proportional to a particular term according to the following [14]

$$\frac{P^\omega}{P^{2\omega}} \propto \frac{\sin^2(\Delta k L/2)}{(\Delta k L/2)^2}, \quad \Delta k = k^{2\omega} - 2k^\omega \qquad (10.10)$$

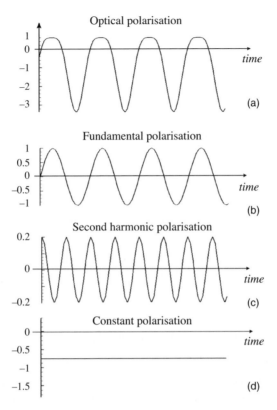

Figure 10.3 Analysis of the nonlinear polarisation wave: (a) total nonlinear polarisation; (b) first harmonic of the P expressed through Fourier series; (c) second harmonic at frequency 2ω; (d) constant component at frequency zero.

where k^ω is the wave vector of the incident field, $k^{2\omega}$ is the wave vector at the second harmonic, L is the interaction length of the dielectric material considered and P^ω and $P^{2\omega}$ represent the power of the fundamental and second harmonic waves, respectively. It is clear from Equation (10.10) that the optimal work condition to maximise the conversion efficiency is $\Delta k = 0$, and thus $k^{2\omega} = 2k^\omega$. If the latter condition does not hold, the phase mismatch between fundamental and second harmonic makes the two waves travel with different phase velocities. As a consequence of this, in some sections of the structure the fundamental wave is summed in phase with the generated second harmonic, whereas in others the overlap between the two waves is destructive. This oscillating response does not allow a consistent interaction between the fundamental and the generated waves. Therefore, the conversion efficiency is maximised when the two waves can travel at the same phase velocity, even if their frequency is different. Perfect phase matching for SHG with collinear beams leads to the following:

$$n(\omega) = n(2\omega) \tag{10.11}$$

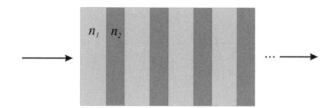

Figure 10.4 Schematic diagram of a 1D periodic layered structure (1D-PBG). Propagation is along the horizontal direction.

This condition can't be naturally achieved due to the chromatic dispersion effects that materials show: the refractive index is a monolithically growing function with frequency. With regard to this problem, through the last decades, research efforts have made it possible to define several techniques that allow phase matching to be satisfied. They will be presented in the following paragraphs.

10.2.4 Phase Matching Through Effective Index for Periodic Structures

A way to modify the topological dispersion of the medium to compensate chromatic dispersion is presented here. The basic idea is to introduce a periodic modulation of the linear refractive index of the medium to induce a phase-matching condition for SHG typically by means of a uniform Bragg grating. Generally, photonic bandgap structures offer this opportunity.

An enhancement of the SHG efficiency in periodic structures such as in Figure 10.4 was proposed for the first time in the early 1970s. The first generator of a second harmonic based on a PBG structure was proposed in 1997 by Scalora *et al.* [15], who showed that conversion efficiency was greatly enhanced, compared to that obtained in a bulk medium of the same length, by aligning the fundamental field with the maximum wavelength at the edge of the first-order bandgap and the second harmonic with the maximum wavelength at the edge of the second-order bandgap. These numerical results were experimentally demonstrated by Dumeige *et al.* who used a 1D PhC device constituting a stack of N alternated layers of AlGaAs and AlO_x [16]. Generally, a structure consisting of N levels which alternates two materials with high index contrast is considered. The levels are transparent both at the fundamental and the second-harmonic frequency and only one level has significant second-order nonlinearity. The incident field is perpendicular to the stack which is isotropic in the plane of the structure. Under this condition, the calculation of the transmission coefficient, T, and reflection coefficient, R, of the fundamental and SH waves is a scalar problem that was solved analytically by Sprung *et al.* [17]. T depends on N, on the transmission coefficient of the unit cell, T_u, and on the Bloch phase, β, which are properties of the infinite structure of the unit cell. The relationship is the following

$$T^{-1} = 1 + \left[\frac{\sin(N\beta)}{\sin(\beta)} \right]^2 (T_u^{-1} - 1) \tag{10.12}$$

The complex effective index method [18–20], based on the effective dispersion relationship, helps in understanding the properties of a finite multilayer stack and allows the calculation of the effective refractive index of the structure. To derive an explicit dispersion relationship for a structure of finite length, using the transfer matrix method, a generic complex transmission coefficient is defined [21]

$$t \equiv x + jy \equiv \sqrt{T} e^{j\varphi_t} \tag{10.13}$$

where $\sqrt{T}$ is the transmission amplitude, $\varphi_t = \tan^{-1}(y/x) \pm m\pi$ is the total phase accumulated as light propagates through the material and m represents an integer. Considering an analogy with the case of a homogeneous medium, the total phase associated with the transmitted field is expressed as

$$\varphi_t = k(\omega) D = (\omega/c) n_{\text{eff}}(\omega) D, \tag{10.14}$$

with $k(\omega)$ representing the effective wave vector and n_{eff} the effective refractive index associated to the multilayer stack of length D.

As a result of a transmission spectrum with the presence of gaps, the effective index is expected to be complex. The imaginary part should be large inside the gap to allow scattering losses. Thus, the following expression for $\sqrt{T}$ is assumed

$$\sqrt{T} = |t| = e^{-\gamma D} \tag{10.15}$$

where $\gamma = (\omega/c) n_i$ and n_i is the imaginary component of the field. The complex transmission matrix is written as $t = e^{\ln \sqrt{T}} e^{i\varphi_t} = e^{i\varphi} = x + jy$. Therefore the following expression is derived

$$i\varphi = i\varphi_t + \ln \sqrt{T} = i \left(\frac{\omega}{c} \hat{n}_{\text{eff}} D \right) \tag{10.16}$$

where still $\varphi_t = \tan^{-1}(y/x) \pm m\pi$. Equation (10.16) becomes

$$\hat{n}_{\text{eff}} = (c/\omega D) \left[\varphi_t - (i/2) \ln(x^2 + y^2) \right] \tag{10.17}$$

From Equation (10.17), it can be understood that at resonance, where $T = x^2 + y^2 = 1$, the imaginary part of the index is identically zero. It is also possible to define the effective index as the ratio between the speed of light in vacuum and the effective phase velocity of the wave in the medium as follows

$$\hat{k}(\omega) = \frac{\omega}{c}\hat{n}_{\text{eff}}(\omega) \tag{10.18}$$

Equation (10.18) expresses the effective dispersion relationship for the structure and it generally holds without referring to a specific periodicity [21].

For periodic structures, phase-matching conditions are automatically satisfied if the fundamental and SH fields are tuned at the resonance peaks on the right side of the respective transmission spectrum [15]. The effective index can be expressed for a periodic structure consisting of N periods as

$$\hat{n}_{\text{eff}} = \frac{c}{\omega N a}\left\{\tan^{-1}\left[z\tan(N\beta)\cot(\beta)\right] + \text{int}\left[\frac{N\beta}{\pi} + \frac{1}{2}\right]\pi\right\} \tag{10.19}$$

where β is the phase constant of an infinite structure with the same unit cell as the finite structure under examination. Equation (10.19) also tells something more about the localisation of the resonances at which phase-matched mixing of three waves is possible. A structure consisting of 20 periods of layers $\lambda/4 - \lambda/2$ is considered as example. This design allows the fields at the band edges to be tuned in order to achieve the PM condition and, at the same time, it results in high mode densities for all the fields. Transmittivity is a periodic function with Bloch phase, β, with period equal to π/N and with N resonances belonging to a pass band in the range $\beta \in [0, \pi]$, for the first pass band and $\beta \in [\pi, 2\pi]$ for the second pass band, and so on. It is interesting to note that at the resonances of $\beta(\omega) = \pi l/N$, T is in fact independent of T_u and thus it does not depend on any of the geometric and optic local properties. In the case of the SHG process, if the fundamental field is allocated to the first peak on the right of the first-order bandgap in the transmission characteristic, the phase constant at λ_f is

$$\beta(\omega) = \pi\frac{(N-1)}{N} \tag{10.20}$$

Substituting Equation (10.20) into (10.19), the following is obtained

$$\text{int}\left[\frac{N\beta}{\pi} + \frac{1}{2}\right] = \pi(N-1) \tag{10.21}$$

which represents the phase of the fundamental field.

The phase-matching condition for SHG leads to

$$2k^{\text{eff}}(\omega) = k^{\text{eff}}(2\omega) \tag{10.22}$$

where k^{eff} represents the effective wave vector. Equation (10.22) can be written for the phase constants as

$$2\beta\,(\omega) = \beta\,(2\omega) \tag{10.23}$$

that, with regards to the second harmonic field phase, leads to

$$\pi\,(N-1) = \frac{1}{2}\tan^{-1}\left[z\tan\,(N\beta)\cot\,(\beta)\right] + \operatorname{int}\left[\frac{N\beta}{\pi} + \frac{1}{2}\right]\frac{\pi}{2} \tag{10.24}$$

From Equation (10.24), the following is derived

$$\tan^{-1}\left[z\tan\,(N\beta)\cot\,(\beta)\right] = \pi\,(2N-2-M) \tag{10.25}$$

where M is the integer part on the RHS in Equation (10.24). The tangent function is defined in the range $[-\pi/2; \pi/2]$, thus Equation (10.25) is satisfied under the condition $M - 2N + 2 = 0$.

Therefore, the phase constant is given by

$$\beta\,(2\omega) = \frac{\pi\,(2N-2)}{N} \tag{10.26}$$

Relation (10.26) describes the phase-matched condition for the second harmonic field (SHF): it should be tuned at resonance $(2N-2)$, that is the second resonance far from the second band edge.

The reasons for the enhancement in SHG efficiency are found first of all in the increased modal density near to the band edges; furthermore there is a strong overlap of modes and the small group velocities allow for longer interactions [22].

In general, the approach presented above can be helpful when phase matching is required in a wave-mixing nonlinear process.

In Figure 10.5, the numerical and experimental results of SHG in a 1D PBG structure obtained in 2001 by Centini *et al.* are illustrated as an example [23].

It was shown that enhancement of the second-order interaction is possible by exploiting the high density of modes, the strong field localisation and the increased coherence length near the photonic band edge. Pioneering work in this sense was first done by Trull *et al.* in 1995 [24] and Martorell and Corbalan in 1997 [25], who studied SHG in defective PhCs, both theoretically and experimentally. However, in this initial study, SH was obtained by the nonlinear interfaces between layers that would not allow conversion efficiency high enough to bring the design into practical applications. Later, in 2001, Shi *et al.* proposed a defective PhC with nonlinear layers for which numerical simulations registered an increment of SHG efficiency up to five orders of magnitude compared to the past [26]. In this kind of defective PhC, the

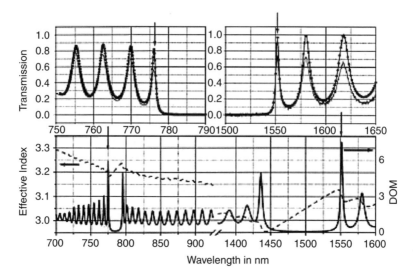

Figure 10.5 Calculated (circles) and measured (solid lines) transmission spectra around fundamental and SH wavelength (indicated by vertical narrows), (top view). Density of modes (DOM) (solid line) and effective refractive index (dashed line), (bottom view) [23].

phase condition for SHG is different from the one in a bulk nonlinear crystal or in a periodic dielectric system. From the transmission spectra it is evident that there is a resonance peak localised at the incident frequency (fundamental) belonging to the PhC bandgap. This can be explained by the existence of some localised states inside the defect that correspond to specific frequencies allowed inside the forbidden gap; because of the finite periods of the PhC, these states couple with the surrounding free space and decay as resonance states.

The results of SHG efficiency from this structure are compared to the ones obtained by a periodic structure with waves tuned at the band edges. As it is localised at one of the defect states inside the bandgap, the fundamental field intensity was found to be increased as happens at the edges of the bandgap inside a periodic structure. However, the phase condition in these two cases is very different. In the case of a periodic structure with waves tuned to the band edges, the phase-matching condition is crucial and must be satisfied for an efficient conversion process. In Shi's design this condition is not essential to achieve efficient SHG. A detailed analysis of a 2D PhC structure with a defect for SHG was proposed by Shi in [12] with the fundamental wave propagating parallel to the plane of the structure. In this case, the 2D PhC structure consists of 17 periods with $a = 0.5$ μm and a fundamental wave, $\lambda_f = 1.362$ μm, that corresponds to a defect mode inside the PhC. As a consequence, the fundamental wave is well confined inside the defect (PhC microcavity). Furthermore, it can be noted that the coherence length of the structure is longer than the total dimensions of the PhC in the propagation direction, thus no phase-matching condition is necessary to achieve efficient SHG.

Up to now, research work on SHG in PhC-based structures has shown that conversion efficiency in these structures is still quite low if it is compared to results achievable in bulk or waveguides that require dimensions on the scale of millimetres or centimetres. However, the possibilities offered by PhC technology for the development of SHG devices, making nonlinear effects available for integration onto microchips, have strongly driven the research efforts in this direction. Different designs for SHG in PhC waveguides and microcavities are currently under intense investigation for the enhancement of generation efficiency [27–32].

10.3   Extended S-MRTD for SHG Analysis

10.3.1   TE/TE Coupling

The nonlinear process coupling the TE fundamental wave/ TE second harmonic in the S-MRTD scheme is considered first. Assuming the 2D problem in the xz-plane, the y-axis as the homogeneous direction and the x-axis as the direction of propagation, the fundamental electromagnetic components of the TE mode are E_y, H_x and H_z. From the discretised Maxwell's equations, we find the following set of equations

$$
_n H^{x,\varphi}_{i+1/2,j} = {}_{n-1} H^{x,\varphi}_{i+1/2,j} - \frac{\Delta t}{\mu_0 \Delta z} \left(\sum_{l=-L_s}^{L_s-1} a\,(l)\, {}_{n-1/2} E^{y,\varphi}_{i+1/2,j-l-1/2} \right)
$$

$$\text{(10.27a)}$$

$$
_n H^{z,\varphi}_{i,j+1/2} = {}_{n-1} H^{z,\varphi}_{i,j+1/2} + \frac{\Delta t}{\mu_0 \Delta x} \left(\sum_{l=-L_s}^{L_s-1} a\,(l)\, {}_{n-1/2} E^{y,\varphi}_{i-l-1/2,j+1/2} \right)
$$

$$\text{(10.27b)}$$

$$
_{n+1/2} D^{y,\varphi}_{i+1/2,j+1/2} = {}_{n-1/2} D^{y,\varphi}_{i+1/2,j+1/2}
$$
$$
+ \Delta t \left(\sum_{l=-L_s}^{L_s-1} a\,(l) \left(-\frac{1}{\Delta z}\, {}_n H^{x,\varphi}_{i+1/2,j-l-1} + \frac{1}{\Delta x}\, {}_n H^{z,\varphi}_{i-l-1,j+1/2} \right) \right)
$$

$$\text{(10.27c)}$$

Equation (10.27c) allows the calculation of $D^{y,\varphi}$ at time step $(n + 1/2)$ from the values of $H^{x,\varphi}$ and $H^{z,\varphi}$ at time step (n). The constitutive relationship then, accounting for second-order nonlinearity, is

$$
D_y = \varepsilon_0 \varepsilon_\infty E_y + P_{\mathrm{NL}y} \tag{10.28}
$$

Assuming a $\chi^{(2)} = \chi_{22}^{(2)}$ nonlinear medium and instantaneous second-order non-linearity, the only y-component of the nonlinear polarisation, P_{NLy}, is nonzero and given by

$$P_{NLy} = \varepsilon_0 \chi_{22}^{(2)} E_y^2 \tag{10.29}$$

The nonlinear effect can then be included inside the permittivity of the dielectric medium as follows. Using Equations (10.28) and (10.29), the constitutive relationship can be rewritten as

$$D_y = \varepsilon_0 \varepsilon_r E_y \tag{10.30}$$

where

$$\varepsilon_r = \left[\varepsilon_\infty + \chi_{22}^{(2)} E_y \right] \tag{10.31}$$

Substituting Equation (10.31) in (10.30), the following relationship for the E-field is obtained

$$E_y = \frac{D_y}{\varepsilon_0 \left[\varepsilon_\infty + \chi_{22}^{(2)} E_y \right]} \tag{10.32}$$

Adopting Equation (10.32), the following updating equation is derived and inserted into the S-MRTD algorithm in order to calculate the E-field component

$$_{n+1/2}E_{i+1/2,j+1/2}^{y,\varphi} = \frac{_{n+1/2}D_{i+1/2,j+1/2}^{y,\varphi}}{\varepsilon_0 \left[\varepsilon_\infty + \chi_{i+1/2,j+1/2}^{(2)} \cdot {}_{n-1/2}E_{i+1/2,j+1/2}^{y,\varphi} \right]} \tag{10.33}$$

where

$$\chi_{i+1/2,j+1/2}^{(2)} = \chi_{22}^{(2)} \Big|_{i+1/2,j+1/2} \tag{10.34}$$

and it represents the nonlinear susceptibility value at the grid point of coordinates $(i + 1/2, j + 1/2)$.

It should be noted that in Equation (10.33), the latest value of $E^{y,\varphi}$ is obtained by iteration using the new value of $D^{y,\varphi}$ and the old value of $E^{y,\varphi}$ itself.

10.3.2 TE/TM Coupling ($\chi^{(2)}$ Tensor)

A TE polarised input wave is considered with E-field components in the xz-plane and the fundamental H component in the longitudinal y-direction. SHG from TM-polarised harmonic waves can be achieved with epitaxial growth of the $Al_{30\%}Ga_{70\%}As$ crystal on [001]-orientated GaAs substrate by aligning the x-axes along the principal axes of the crystal [110]. The $\chi^{(2)}$ tensor has the following components

$$\chi^{(2)} = \begin{bmatrix} 0 & 0 & 0 & 0 & 0 & 0 \\ 0 & 0 & 0 & \chi_{24}^{(2)} & 0 & 0 \\ 0 & \chi_{32}^{(2)} & 0 & 0 & 0 & 0 \end{bmatrix} \tag{10.35}$$

where it is assumed that $\chi_{32}^{(2)} = \chi_{24}^{(2)} = 1$ (arbitrary units). Due to the form of the nonlinear susceptibility tensor, the only nonzero component of the nonlinear polarisation vector is P_{NLz} and it is given by

$$P_{NLz} = \varepsilon_0 \chi^{(2)} E_y^2 \tag{10.36}$$

and realises a TE/TM coupling of the fundamental/SH waves, respectively.

The discretised equation for the update of the nonlinear polarisation states

$$_{n+1}P_{i,j}^{NLz} = \varepsilon_0 \chi_{i,j}^{(2)} \left(_{n+1}E_{i,j}^{y,\varphi} \right)^2 \tag{10.37}$$

From Equation (10.37) the following equations for the D-field components can be written

$$_{n+1/2}D_{i+1/2,j+1/2}^{x,\varphi} = \varepsilon_0 \varepsilon_{\infty i, j n+1/2} E_{i+1/2,j+1/2}^{x,\varphi} \tag{10.38a}$$

$$_{n+1/2}D_{i+1/2,j+1/2}^{z,\varphi} = \varepsilon_0 \varepsilon_{\infty i, j n+1/2} E_{i+1/2,j+1/2}^{z,\varphi} + \varepsilon_0 \chi_{i+1/2,j+1/2}^{(2)} \left(_{n+1/2}E_{i+1/2,j+1/2}^{y,\varphi} \right)^2 \tag{10.38b}$$

Using Equations (10.38), the update equations for the E- field components for the generated TM polarisation can be derived as

$$_{n+1/2}E_{i+1/2,j+1/2}^{x,\varphi} = \frac{1}{\varepsilon_0 \varepsilon_{\infty,i+1/2,j+1/2}} \left(_{n+1/2}D_{i+1/2,j+1/2}^{x,\varphi} \right) \tag{10.39a}$$

$$_{n+1/2}E_{i+1/2,j+1/2}^{z,\varphi} = \frac{1}{\varepsilon_0 \varepsilon_{\infty,i+1/2,j+1/2}} \left(_{n+1/2}D_{i+1/2,j+1/2}^{z,\varphi} \right)$$
$$- \varepsilon_0 \chi_{i+1/2,j+1/2}^{(2)} \left(_{n+1/2}E_{i+1/2,j+1/2}^{y,\varphi} \right)^2 \tag{10.39b}$$

From this point, the updates of the components D_x D_z and H_y follows the S-MRTD equations for the case of TM polarised field propagation and are written as

$$_{n+1}D_{i+1/2,j}^{x,\varphi} = {}_nD_{i+1/2,j}^{x,\varphi} + \frac{\Delta t}{\Delta z}\left(\sum_{l=-L_s}^{L_s-1} a\,(l)\,_{n+1/2}H_{i+1/2,j-l-1/2}^{y,\varphi}\right) \qquad (10.40a)$$

$$_{n+1}D_{i+1/2,j}^{z,\varphi} = {}_nD_{i+1/2,j}^{z,\varphi} - \frac{\Delta t}{\Delta x}\left(\sum_{l=-L_s}^{L_s-1} a\,(l)\,_{n+1/2}H_{i-l-1/2,j+1/2}^{y,\varphi}\right) \qquad (10.40b)$$

$$_{n+1/2}H_{i+1/2,j+1/2}^{y,\varphi} = {}_{n-1/2}H_{i+1/2,j+1/2}^{y,\varphi}$$
$$- \frac{\Delta t}{\mu_0}\left(\sum_{l=-L_s}^{L_s-1} a\,(l)\left(-\frac{1}{\Delta z}\,_nE_{i+1/2,j-l}^{x,\varphi} + \frac{1}{\Delta x}\,_nE_{i-l,j+1/2}^{z,\varphi}\right)\right)$$
$$(10.40c)$$

while the set of Equations (10.27), combined with Equation (10.41) below, solves the fields for TE polarisation

$$_{n+1/2}E_{i+1/2,j+1/2}^{y,\varphi} = \frac{_{n+1/2}D_{i+1/2,j+1/2}^{y,\varphi}}{\varepsilon_0\varepsilon_\infty} \qquad (10.41)$$

The solution of the equations is obtained through the following iterative process:

1. Calculation of the updated values of H_x and H_z with Equations (10.27a–b), using the E-field value at the previous time step.
2. Substitution of the updated magnetic-fields component into Equation (10.27.c) to calculate the new value of D_y.
3. Next, calculation of the E_y component using Equation (10.41).

Once the fields for TE polarisation have been updated, the following steps are iterated to simulate the coupling with the TM-polarised harmonic:

1. Updating of nonlinear term of the polarisation, P_{NLz}, through Equation (10.37).
2. At the same time step, calculation of components D_x and D_z from Equations (10.38a–b) respectively.
3. Updating of electric-field components E_x and E_z with Equations (10.39a–b).
4. Finally, through Equations (10.40a–c), the set of new components for TM polarisation is calculated.

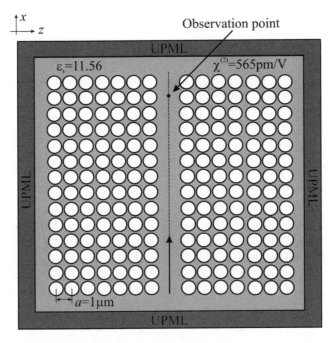

Figure 10.6 Schematic diagram of 2D straight nonlinear PhC waveguide. The lattice parameters are: $a = 1$ μm, $\varepsilon_s = 11.56$ and radius $= 0.475a$. (Reproduced with permission from Letizia, R. and Obayya, S.S.A. (2009) Efficient second harmonic generation through selective photonic crystal-microcavity coupling. *IEEE J. Lightwave Technol.*, **27** (21), 4763–4772. © 2009 IEEE.)

10.4 SHG in PhC-Waveguides

10.4.1 PBG Property: Calculating the PhC Dispersion Diagram

First of all, in order to test the validity of the presented MRTD approach, the nonlinear PhC waveguide analysed through condensed node spatial network method in [33] will be considered.

The structure, shown in Figure 10.6, is an air-hole-type photonic crystal waveguide in which circular holes, arranged in a square lattice, are drilled into a GaAs slab ($\varepsilon_s = 11.56$) which presents a nonlinear coefficient $\chi_{22}^{(2)} = 565$pm/V. The photonic crystal has periodic lattice of periodicity $a = 1$ μm and a filling factor $r/a = 0.475$ where r is the radius of the holes.

Before proceeding with the analysis of the PhC waveguide through the nonlinear MRTD scheme, the PBG of the periodic lattice has been calculated around the normalised frequency ($\omega a/2\pi c$) of 0.281 for TE polarisation modes. In order to calculate the dispersive diagram of the periodic structure when no defects are included, an FDTD algorithm combined with periodic boundary conditions (PBC) using a wrap-around approach has been used [34]. As the periodic arrangement is

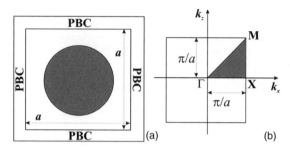

Figure 10.7 (a) Unit cell, and (b) irreducible Brilluoin zone for a rectangular lattice with period a.

rectangular, the total computational domain consists of the lattice unit cell, Figure 10.7(a), whose irreducible Brillouin zone is shown in Figure 10.7(b).

Due to the 2D translational symmetry of PhCs, the field quantities can be expressed from Bloch's theorem as

$$F(x + a, z) = e^{-jk_x a} F(x, z) \tag{10.42}$$
$$F(x, z + a) = e^{-jk_z a} F(x, z) \tag{10.43}$$

where F represents the field components for E and H. The PBCs are enforced by taking the values of the field at the lower/left edge, multiplying them by a complex factor, $k_{x,z} a$, that represents the required phase shift in the x and z-directions respectively, and assigning the result to the upper/right boundary.

Figure 10.8 shows the band diagram for TE polarisation obtained by simulating a unit cell with periodic boundary conditions applied. As may be observed from this figure, there is a photonic bandgap for the TE modes in the normalised frequency range from 0.2142 to 0.306. Starting from this periodic structure, a waveguide is obtained by removing a row of circular holes (single-missing-hole line defect). The computational domain of $[15a \times 15a]$ is discretised with an average $\Delta x, z = 40$ nm and a temporal step size of $\Delta t = 0.05$ fs has been adopted to ensure the code stability. To generate an SH inside this waveguide structure, a sinusoidal TE wave at the normalised frequency of 0.281 with amplitude 10^9 V/m modulated by a raised cosine in space, has been injected as a source. It should be noted that the fundamental frequency belongs to the photonic bandgap so that it is tightly confined inside the line defect and it can propagate along the PhC.

10.4.2 Propagation Properties and SHG Efficiency

Figure 10.9 shows the time evolution of the electric field at an observation point placed at the output end of the PhC waveguide. By using the FFT of these

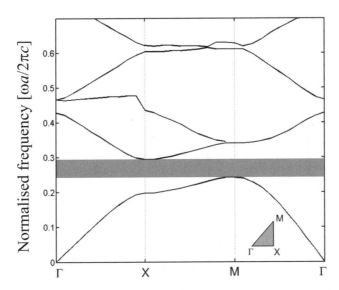

Figure 10.8 Transmission coefficient of the periodic lattice of Figure 10.16 where the PBG extends from 0.2142 to 0.306 in the normalised frequency domain. (Reproduced with permission from Letizia, R. and Obayya, S.S.A. (2009) Efficient second harmonic generation through selective photonic crystal-microcavity coupling. *IEEE J. Lightwave Technol.*, **27** (21), 4763–4772. © 2009 IEEE.)

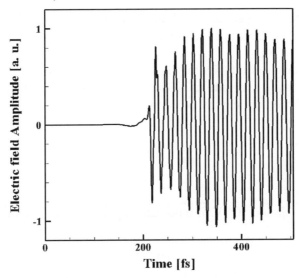

Figure 10.9 Time variation of the electric field recorded at the observation point inside the PhC waveguide. (Reproduced with permission from Letizia, R. and Obayya, S.S.A. (2009) Efficient second harmonic generation through selective photonic crystal-microcavity coupling. *IEEE J. Lightwave Technol.*, **27** (21), 4763–4772. © 2009 IEEE.)

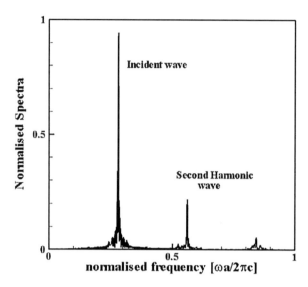

Figure 10.10 Normalised spectral distribution of both fundamental and SH waves inside the PhC waveguide. (Reproduced with permission from Letizia, R. and Obayya, S.S.A. (2009) Efficient second harmonic generation through selective photonic crystal-microcavity coupling. *IEEE J. Lightwave Technol.*, **27** (21), 4763–4772. © 2009 IEEE.)

time-domain data, the spectral distribution of the electric field is calculated and normalised to the fundamental spectrum peak input. The result of this procedure is shown in Figure 10.10.

As maybe noted from this figure, the second harmonic of the input frequency is generated inside the waveguide as result of the nonlinear interaction between the electric field and the nonlinearity of the medium. The SHG efficiency is calculated using

$$\text{Eff} = \frac{|E_{\text{sh}}|^2_{\text{max}}}{|E_{\text{f}}|^2_{\text{max}}} \qquad (10.44)$$

where $|E_{\text{sh}}|_{\text{max}}$ and $|E_{\text{f}}|_{\text{max}}$ represent the maximum value of electric field for the fundamental and SH waves, respectively. The calculated efficiency is about 19% and agrees well with the efficiency results shown in [33]. In this case, the second harmonic field (SHF) is mixed with the fundamental field (FF), which is also still propagating in the waveguide. Furthermore, the second harmonic wave is not well guided inside the structure as its frequency does not belong to the photonic bandgap of the waveguide. This is confirmed in Figure 10.11, where the electric field-profiles of both the fundamental and SH waves are shown along the transverse section of the PhC waveguide.

It can be noted from Figure 10.11 that the fundamental frequency mode is strongly guided inside the core (line defect), whereas the SH wave penetrates for some rows inside the periodic pattern at the edges as the photonic bandgap guiding principle does

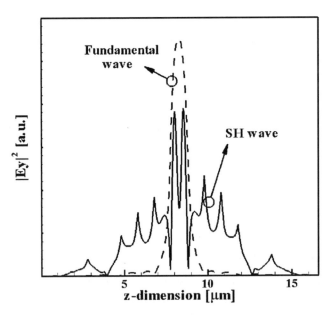

Figure 10.11 Electric-field profile of both the fundamental and the SH waves along the transversal section of the PhC waveguide. (Reproduced with permission from Letizia, R. and Obayya, S.S.A. (2009) Efficient second harmonic generation through selective photonic crystal-microcavity coupling. *IEEE J. Lightwave Technol.*, **27** (21), 4763–4772. © 2009 IEEE.)

not hold at its frequency. This results in a weak guidance of the SH wave and in the spread of some of the total SH energy through the PhC cladding.

10.5   Selective SHG in Compound PhC-Based Structures

10.5.1   Introduction

Due to the relatively small size of PhC devices, it has been clearly demonstrated that achieving high conversion efficiency in PhC devices is relatively difficult [6–11, 12]. In particular, the generation of harmonics in a PhC waveguide can encounter some problems due to the absence of a photonic bandgap at high frequencies. This will result in very weak guiding of the generated wave. As the photonic bandgap guiding process is compromised at these frequencies, instead of being confined, it tends to spread outside the core into the periodic arrangement.

   However, as an attempt to enhance the conversion efficiency and selectivity of the SH wave, Satoh *et al.* [10] have proposed, for the first time, a compound nonlinear PhC structure that can convert an input frequency into its second harmonic by using the nonlinear property of the medium. Two different nonlinear PhC waveguides, built in two different periodic lattices, are coupled together through a tapered section to achieve selection of only the second harmonic frequency at the output of the second waveguide. The compound arrangement first of all provides the PhC waveguide with the ability to guide the second harmonic and secondly, the stopband characteristics forbid the fundamental wave to propagate. Thus, selection of the only the second harmonic at the output of the generation structure is realised.

Simulation results from the 2D frequency converter proposed in [10] will be given as further validation of the numerical code.

However, this approach suffers from a low value of frequency conversion efficiency due to the radiation losses encountered in the tapered waveguide region, and that significantly reduces the power of the second harmonic at the output of the PhC waveguide. Moreover, the fundamental frequency is not completely suppressed at the output waveguide.

10.5.2 Selective SHG: Coupling PhC Waveguides Through a Tapered Section

The structure for the 2D frequency converter proposed in [10] by Satoh's group is considered to further investigate the validity of the developed code. The structure, in Figure 10.12, consists of two different pillar-type PhC waveguides placed together to allow only the SH wave as output. The parameters are the same as the structure in Figure 10.6 for the input region, used now to assist the harmonic generation, while the lattice period is changed to half in the output waveguide ($a_2 = a/2$). This, due to the PBG property, will prevent the fundamental wave propagating in the output

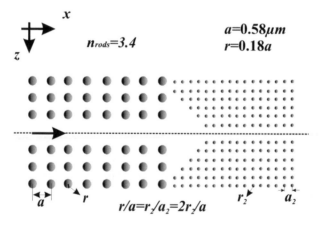

Figure 10.12 Schematic diagram of a frequency converter consisting of two different PhC waveguides coupled through a tapered section. The output waveguide takes place in a lattice having $a_2 = a/2$. The input waveguide is excited with a Gaussian pulse at $\omega a/2\pi c = 0.381$. (Reproduced with permission from Letizia, R. and Obayya, S.S.A. (2008) Efficient multiresolution time-domain analysis of arbitrarily shaped photonic devices. *IET Optoelectron.*, **2** (6), 241–253. © 2008 IET.)

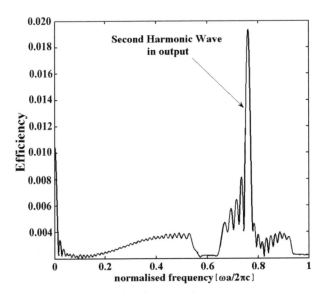

Figure 10.13 Spectral distribution of the electric field recorded at the detector point in output at $x = 22$ μm from the input section. Efficiency of selective SHG is calculated at about 2%. (Reproduced with permission from Letizia, R. and Obayya, S.S.A. (2008) Efficient multiresolution time-domain analysis of arbitrarily shaped photonic devices. *IET Optoelectron.*, **2** (6), 241–253. © 2008 IET.)

waveguide. In order to assist the coupling between the two PhC waveguides, a tapered section is arranged at the interface.

The structure is excited with a Gaussian pulse TE polarised of $t_0 = 30$ fs and $T_0 = 90$ fs at a normalised frequency of 0.381 and an amplitude of 10^9 V/m. A detector is placed inside the output waveguide at $x = 22$ μm from the input source. In Figure 10.13, the FFT of the recorded time data in this point is shown. It is apparent that only the generated second harmonic wave can be found in the output as the PBG characteristic of the second PhC waveguide acts like a mirror for the fundamental frequency. This can be clearly noted in Figures 10.14 and 10.15, where the evolution of both the fundamental and second harmonic waves is shown along the longitudinal direction of the frequency converter. The SHG efficiency calculated is about 2%, again, this value agrees well with the results reported in [10].

10.6   New Design for Selective SHG: PhC Microcavity Coupling

10.6.1   Introduction

The coupling of two PhC waveguides through a tapered section enables the collection of the SH from the generation waveguide and address it into the output waveguide

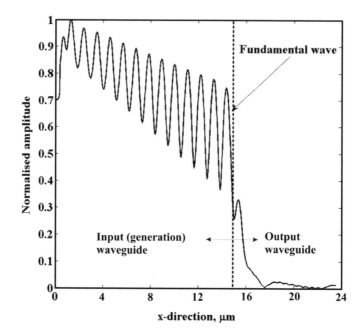

Figure 10.14 Fundamental electric-field variation along the longitudinal direction of the two waveguides that make up the frequency converter. The fundamental frequency is prevented from propagating inside the output photonic crystal waveguide. (Reproduced with permission from Letizia, R. and Obayya, S.S.A. (2008) Efficient multiresolution time-domain analysis of arbitrarily shaped photonic devices. *IET Optoelectron.*, **2** (6), 241–253. © 2008 IET.)

while the fundamental wave is reflected back. However, due to the radiation losses encountered in the taper, the whole SHG efficiency deteriorates [10]. Moreover, the fundamental wave can still partially propagate in the output waveguide section leading to a low-quality selection of the SH wave. Alternatively, a new and more efficient design for an SH converter, whose schematic is shown in Figure 10.20, is suggested here. The basic idea is that in addition to the original waveguide where SH conversion takes place, there is another side-coupled PhC waveguide, designed to allow only the propagation of the SH wave.

10.6.2 Stopband Properties

As seen in the previous chapters, the photonic bandgap property is directly related to the lattice constant of the periodic structure. Therefore, by scaling down the lattice constant from 1 to 0.5 µm, the PBG of the new periodic structure is expected to be in the frequency band of the SH wave. By following this concept, the side-coupled PhC waveguide is designed in a lattice of periodicity $a_2 = 0.5$ µm. By carrying out a simple FDTD analysis to find out the dispersion diagram for the new unit cell with

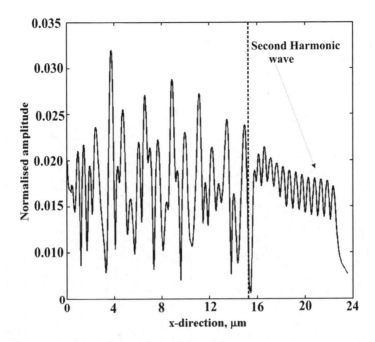

Figure 10.15 Second harmonic electric-field variation along the longitudinal direction of the two waveguides that make up the frequency converter. The output photonic crystal waveguide assists the propagation of the SH wave. (Reproduced with permission from Letizia, R. and Obayya, S.S.A. (2008) Efficient multiresolution time-domain analysis of arbitrarily shaped photonic devices. *IET Optoelectron.*, **2** (6), 241–253. © 2008 IET.)

period a_2, it can be easily verified that a PBG lies in the range $0.4 < \omega a/2\pi c < 0.55$. Therefore, the generated SH wave belongs to the PBG of the second periodic lattice and hence it will have a better confinement inside its line defect. As shown in Figure 10.16, this second PhC waveguide is placed in parallel to the first PhC waveguide to form a compound structure that generates SH in two stages as follows:

First stage – the SH wave is generated in the first PhC waveguide as a result of nonlinear interaction of second order.

Second stage – the SH wave is filtered and dropped from the first waveguide to the second waveguide in order to achieve as high selectivity and efficiency as possible.

The dimensions of the new structure are $[11a \times 11a]$ for the first region plus $[12a_2 \times 22a_2]$ for the second region, where $a_2 = a/2$, with $a = 1$ μm. Spatial average discretisation is chosen to be $\Delta x, z = 50$ nm and time interval is fixed at $\Delta t = 0.06$ fs. Initially, the two PhC waveguides coupled in this compound structure are simulated

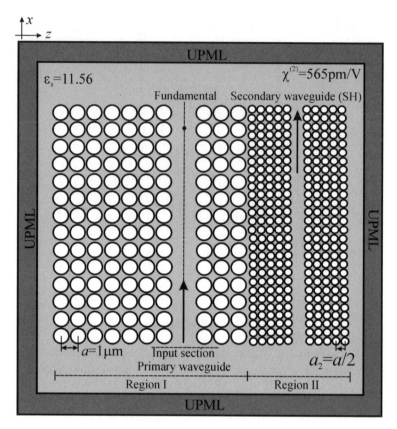

Figure 10.16 Schematic diagram of a frequency converter consisting of two different and coupled PhC waveguides. The secondary waveguide is in a lattice of period $a_2 = a/2$. The primary waveguide is excited with a Gaussian pulse at $\omega a/2\pi c = 0.2257$. (Reproduced with permission from Letizia, R. and Obayya, S.S.A. (2009) Efficient second harmonic generation through selective photonic crystal-microcavity coupling. *IEEE J. Lightwave Technol.*, **27** (21), 4763–4772. © 2009 IEEE.)

to understand how the fields are coupled into the secondary waveguide. A Gaussian pulse, with width in time $T_0 = 600$ fs and delay $t_0 = 1500$ fs at a centred normalised frequency of 0.2257 with amplitude 10^9 V/m, and modulated in space by a raised cosine, is injected as source. The time variation of the electromagnetic field has been recorded at a detector point placed at the output end of the secondary waveguide. Figure 10.17 shows the spectral distribution of both the fundamental and SH waves at the output, normalised to the spectrum of the fundamental wave launched into the primary waveguide. It is apparent that at the output waveguide the coupled SH signal is very weak and it is still mixed with a significant portion of the fundamental wave. Therefore, a specific design engineered to enhance the coupling of the SH wave

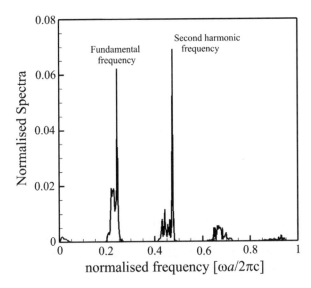

Figure 10.17 Transmission of the electric field at the output end of the secondary waveguide in the case with no cavity. (Reproduced with permission from Letizia, R. and Obayya, S.S.A. (2009) Efficient second harmonic generation through selective photonic crystal-microcavity coupling. *IEEE J. Lightwave Technol.*, **27** (21), 4763–4772. © 2009 IEEE.)

into the secondary waveguide and avoid interference from the fundamental frequency propagating in the primary waveguide, is needed.

10.6.3   Coupling Waveguides and Cavities

In order to design PhC based photonic integrated circuits, it is essential to understand and control the coupling of light between a PhC cavity and a PhC waveguide [35, 36] or between two PhC waveguides [37]. When a PhC resonant cavity is attached at a side oif a PhC waveguide, like a tuned stub is used in microwave engineering, the system behaves like wavelength filter. Following this concept, a pair of waveguides can be coupled to a cavity to build a very sharp filter through resonant tunnelling.

The filtering characteristics of all these systems are determined by the geometry of the cavity that is attached to the waveguide and correspond to a specific frequency response of the resonator itself. In order to drop a wavelength from a waveguide, it is necessary to couple a single-mode cavity that is tuned at that specific wavelength and that, through resonance, is able to drop the signal from the waveguide and couple it back to the same waveguide or send it to a second waveguide. In Figure 10.18, the process of resonance tunnelling is displayed: the signal which is dropped to waveguide 2 corresponds to the resonant single-mode of the cavity [38].

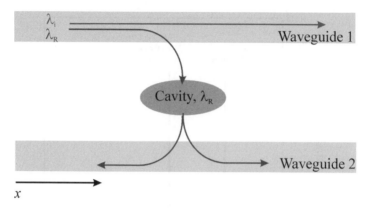

Figure 10.18 Sketch of selective coupling of two waveguides through a single-mode resonant cavity tuned at λ_R. The mode that propagates inside waveguide 1 at wavelength λ_R is dropped by the resonator and sent into waveguide 2. All the other wavelengths (λ_i) that are different from the resonance mode carry on propagating along waveguide 1.

Furthermore, Figure 10.19 shows two generic cases of coupling waveguides to resonators: the case of *side coupling* in Figure 10.19(a) with the cavity placed at the side of the infinite waveguide, and then the case of two half waveguides that are coupled together via *resonant tunnelling* through the centre cavity, Figure 10.19(b).

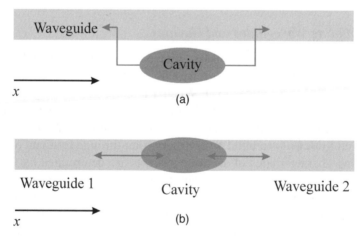

Figure 10.19 Sketch of two generic cases of coupling between waveguides and resonators: (a) side coupling, and (b) resonant coupling in which two waveguides are coupled by a high Q cavity.

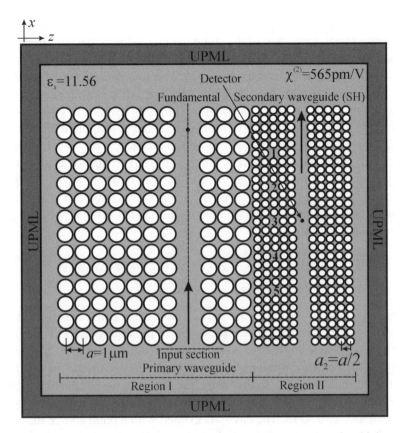

Figure 10.20 Schematic diagram of the frequency converter proposed, which consists of two different PhC waveguides and coupled through a row of microcavities. (Reproduced with permission from Letizia, R. and Obayya, S.S.A. (2009) Efficient second harmonic generation through selective photonic crystal-microcavity coupling. *IEEE J. Lightwave Technol.*, **27** (21), 4763–4772. © 2009 IEEE.)

10.6.4 Enhancement of Efficiency and Selectivity Through Selective PhC Microcavities

To enhance the performance of the suggested compound structure, a new arrangement is proposed and shown in Figure 10.20. In order to transfer the SH into the output PhC waveguide and avoid the losses due to its weak guidance inside the primary waveguide, a row of PhC microcavities is arranged as a coupling region between the two waveguides. These microcavities are realised by single-hole point defects inside the square lattice, finely tuned to resonate at the SH frequency. The aim of adding microcavities is to build an optical filter able to maximise the SH transfer to the secondary waveguide. Coupled resonator optical waveguides (CROWs) can be found in the literature in the design of narrow-band resonance filters or add/drop structures

[36–46]. The concept of coupling the waveguide to a tuned microcavity is applied here for the first time, to the best of the author's knowledge, to design a high-efficiency SHG-based frequency converter. The basic concept behind this kind of structure is that by placing a point defect, the microcavity, near to a PhC waveguide, a PhC-based resonance filter can be realised. The main advantage of this arrangement is that the dropping frequency of the filter can be tuned simply by engineering the single-defect microcavity. In the proposed converter the same concept is applied and the dropping frequency of the filter in this example is the SH frequency generated in the primary waveguide. The PhC microcavity resonating at the SH frequency is able to trap this frequency from the primary waveguide and drop it into the secondary waveguide where it can propagate by means of the PhC's engineered stopband characteristics. In particular, by chaining a row of side resonators instead of only one, the filter efficiency can be greatly improved.

10.6.5 PhC-Microcavity Design

A preliminary study of the cavity properties has been carried out to investigate the resonance frequency and the mode profile inside the cavity itself. The resonator is obtained by removing a single-hole point defect in a lattice of $[7a_2 \times 7a_2]$. By varying the point-defect geometric properties, the resonance frequency can be controlled and tuned to the desired value [2]. In this work, the lattice period of the PhC surrounding the single-hole defect, a_2, is varied in order to design a microcavity exactly tuned to the SH frequency of 0.4515. The cavity is excited with a point source that evolves in time as a Gaussian pulse of amplitude 1 V/m, $t_0 = 50.6$ fs and $T_0 = 20$ fs. By taking the FFT of the recorded time variation of the electric field at the centre of the cavity, the spectral distribution of the resonant mode has been calculated. Figure 10.21 shows the variation of the normalised resonance frequency with a_2/a, and it can be noted from this figure that when $a_2 = 0.5a$, the cavity resonates exactly at the SH frequency (normalised frequency 0.4515). For $a_2 = 0.5a$, the spectral energy distribution of the resonant mode is shown in Figure 10.22 and its contour profile in the plane of the structure is reported in Figure 10.23. The quality factor, Q, of the PhC microcavity is calculated using [47]

$$Q = 2\pi \frac{|E_t|^2}{|E_t|^2 - |E_{t+T}|^2} \tag{10.45}$$

where E_t is the electric field taken at a particular time step t, T is the time cycle of the resonant mode and E_{t+T} is the recorded electric field after one cycle. In this microcavity the time cycle is $T = 7$ fs and the quality factor has been calculated to be $Q = 222$.

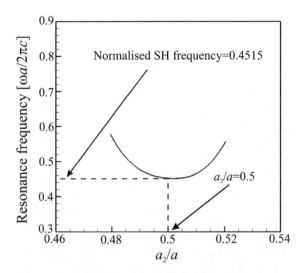

Figure 10.21 Variation of the resonant frequency with a_2/a where a_2 is the PhC microcavity period and $a = 1$ μm. (Reproduced with permission from Letizia, R. and Obayya, S.S.A. (2009) Efficient second harmonic generation through selective photonic crystal-microcavity coupling. *IEEE J. Lightwave Technol.*, **27** (21), 4763–4772. © 2009 IEEE.)

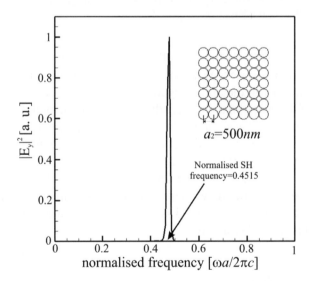

Figure 10.22 Spectral energy density of the electric field inside the microcavity. The cavity shows only a resonant mode at the SH normalised frequency of 0.4515. (Reproduced with permission from Letizia, R. and Obayya, S.S.A. (2009) Efficient second harmonic generation through selective photonic crystal-microcavity coupling. *IEEE J. Lightwave Technol.*, **27** (21), 4763–4772. © 2009 IEEE.)

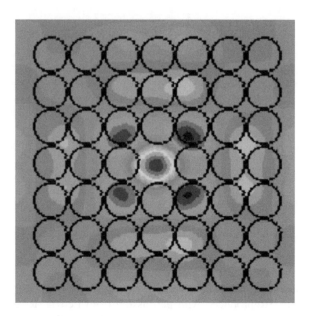

Figure 10.23 Contour profile of the resonant mode on the plane of the cavity. (Reproduced with permission from Letizia, R. and Obayya, S.S.A. (2009) Efficient second harmonic generation through selective photonic crystal-microcavity coupling. *IEEE J. Lightwave Technol.*, **27** (21), 4763–4772. © 2009 IEEE.)

In order to maximise the efficiency, the cavities are carefully placed along the propagation direction. It is expected that the location of the microcavities plays a crucial role in determining the efficiency at the output of the secondary waveguide as a consequence of the phase matching condition between the pump and the generated wave. Along the primary waveguide in fact, the velocity phase difference between the two waves will change, affecting the efficiency of nonlinear frequency conversion. Figure 10.24 shows the variation of the amplitude of the SHF and FF along the propagation direction in the primary waveguide when the total length is 42 μm. From this figure, it is clear that the SH wave reaches its maximum at distance $x = 6.3$ μm from the source section. Thus, cavity 3 is placed at this value along the longitudinal direction of the compound structure.

10.6.6 Simulation Results

Next, the effect of inserting one cavity, two cavities and a row of five cavities on the overall device performance will be investigated. A detector point is placed in the middle of the microcavity and from the FFT of the recorded time-domain data, the spectral distribution, normalised with respect to the injected mode energy, is calculated.

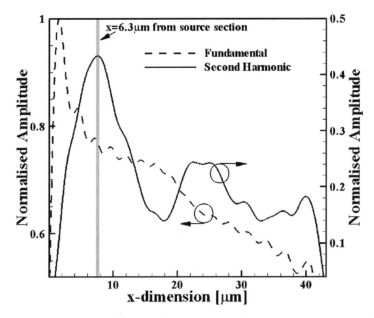

Figure 10.24 Variation of fundamental (dashed line) and second harmonic field amplitude (solid line) along the propagation direction for a PhC waveguide of length 42 μm. The SH wave reaches its maximum at distance $x = 6.3$ μm from the source. Cavity 3 in the compound structure is carefully placed at this distance along the propagation direction x. (Reproduced with permission from Letizia, R. and Obayya, S.S.A. (2009) Efficient second harmonic generation through selective photonic crystal-microcavity coupling. *IEEE J. Lightwave Technol.*, **27** (21), 4763–4772. © 2009 IEEE.)

Figure 10.25 shows the results when only one microcavity is placed, cavity 1. It can be noted that, compared to the previous case, without a cavity, the conversion efficiency has lowered to about 1.5% while, in terms of selectivity, no significant change is recorded. However, the situation changes when the order of filtering is increased by adding a second microcavity. Thus, both the fundamental and the SH normalised spectra are computed for the case of two resonators, cavity 1 and cavity 2, with a separation distance of three periods and the results are shown in Figure 10.26.

Comparing Figures 10.25 and 10.26, it is clearly demonstrated that inserting the second microcavity has resulted in an improvement in the selectivity of the filter and also an enhancement of the SH efficiency which has increased to about 2.8%.

Finally, the case of five microcavities is considered. Compared to the previous results, Figure 10.27 shows the variations of the fundamental and SH normalised spectra as observed at the centre of cavity 3. It can be noted that only 0.5% of the fundamental field can be detected in the resonator and the efficiency has significantly increased to about 8.5%. Therefore, placing a row of microcavities to side-couple the PhC waveguide can dramatically improve the selection of the SH frequency. This

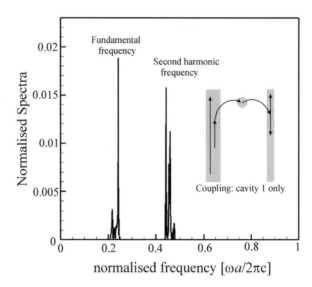

Figure 10.25 Normalised spectra of the fundamental and SH waves at the output in the case of one cavity (cavity 1). (Reproduced with permission from Letizia, R. and Obayya, S.S.A. (2009) Efficient second harmonic generation through selective photonic crystal-microcavity coupling. *IEEE J. Lightwave Technol.*, **27** (21), 4763–4772. © 2009 IEEE.)

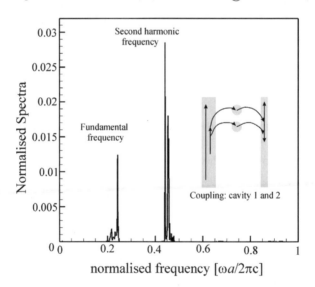

Figure 10.26 Normalised spectra of the fundamental and SH waves at the output in the case of two cavities (cavities 1 and 2). (Reproduced with permission from Letizia, R. and Obayya, S.S.A. (2009) Efficient second harmonic generation through selective photonic crystal-microcavity coupling. *IEEE J. Lightwave Technol.*, **27** (21), 4763–4772. © 2009 IEEE.)

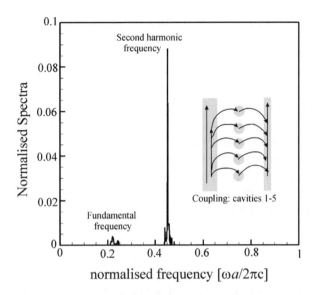

Figure 10.27 Normalised spectra of the fundamental and SH waves at the output in the case of a row of five cavities (from cavity 1 to cavity 5). (Reproduced with permission from Letizia, R. and Obayya, S.S.A. (2009) Efficient second harmonic generation through selective photonic crystal-microcavity coupling. *IEEE J. Lightwave Technol.*, **27** (21), 4763–4772. © 2009 IEEE.)

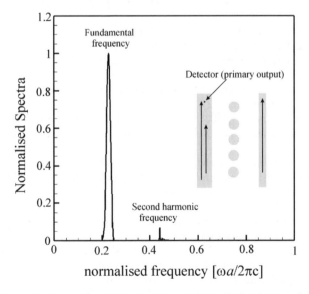

Figure 10.28 Normalised spectra of the fundamental and SH waves at the output of the primary waveguide in the case of a row of five cavities (from cavity 1 to cavity 5).

value of 8.5% SHG efficiency in photonic crystal waveguides is about three times higher than the best value reported in the literature [10].

Figure 10.28 shows that when the observation point is taken at the output of the primary waveguide, the pump wave propagates independently and that there is still a portion of the SH wave, about 5%, that will need a longer simulation time to be coupled. The separation between microcavities is a critical parameter in determining the overall selectivity and efficiency of SH at the output. If the separation between microcavities increases, the quality factor of individual cavities should increase; however, due to the relatively weak interaction between these microcavities, the overall selectivity would deteriorate. Based upon different numerical experiments that have been carried out through this work, it has been found that a separation of three periods gives the best filtering performance.

References

[1] Joannopoulos, J.D., Johnson, S.G., Winn, J.N. and Maede, R.D. (2008) *Photonic Crystals, Molding the Flow of Light*, 2nd edn, Princeton University Press.
[2] Villeneuve, P.R., Fan, S. and Joannopoulos, J.D. (1996) Microcavities in photonic crystals: Mode symmetry, tunability, and coupling efficiency. *Phys. Rev. B, Condens. Matter*, **54**, 7837–7842.
[3] Baylindir, M., Temelkuran, B. and Ozbay, E. (2000) Propagation of photons by hopping: A waveguiding mechanism through localized coupled cavities in three-dimensional photonic crystals. *Phys. Rev. B*, **61**, 11855–11858.
[4] Loncar, M., Nedeljkovic, D., Doll, T. *et al.* (2000) Waveguiding in planar photonic crystals. *Appl. Phys. Lett.*, **77**, 1937–1940.
[5] Jin, C.J., Cheng, B.Y., Man, B.Y. *et al.* (1999) Band gap and wave guiding effect in a quasiperiodic photonic crystal. *Appl. Phys. Lett.*, **75**, 1848–1850.
[6] Ranieri, F., Dumeige, Y., Levenson, A. and Letartre, X. (2002) Nonlinear decoupled FDTD code: phasematching in 2D defective photonic crystal. *Electron. Lett.*, **38**, 1704–1706.
[7] van Driel, H.M. (2003) "Nonlinear Optics in Photonic Crystals", Transparent Optical Networks. Proceedings of the 5th International Conference, vol. **1**, pp. 56–59.
[8] Cowan, A.R. and Young, J.F. (2001) Second Harmonic Generation in Planar Photonic Crystal Waveguides. Proceedings of the Quantum Electronics and Laser Science Conference, (QELS), pp. 44–45.
[9] Inoue, K., Sasada, M., Kawai, N. and Kawamata, J. (2000) Phase-matched second-harmonic generation in a two-dimensional photonic crystal. Proceedings of the Quantum Electronics and Laser Science Conference, (QELS), Technical Digest 7-12, p. 3.
[10] Satoh, H., Yoshida, N., Kitayama, S. and Konaka, S. (2006) Analysis of 2-D frequency converter utilizing compound nonlinear photonic-crystal structure by condensed node spatial network method. *IEEE Trans. Microwave Theory Tech.*, **54**, 210–215.
[11] Scalora, M., Bloemer, M.J., Manka, A.S. *et al.* (1997) Pulsed second-harmonic generation in nonlinear, one-dimensional, periodic structures. *Physical Rev. A*, **56**, 3166–3174.
[12] Shi, B., Jiang, Z.M., Zhou, X.F. and Wang, X. (2002) A two-dimensional nonlinear photonic crystal for strong second harmonic generation. *J. Appl. Phys.*, 91, 6769.
[13] Franken, P.A., Hill, A.E., Peters, C.W. and Weinreich, G. (1961) Generation of optical harmonics. *Phys. Rev. Lett.*, **7**, 118.

[14] Southerland, R.L. (2003) *Handbook of Nonlinear Optics*, Dekker.

[15] Scalora, M., Bloemer, M.J., Manka, A.S. *et al.* (1997) Pulsed second-harmonic generation in nonlinear, one-dimensional periodic structure. *Phys. Rev. A*, **56**, 3166–3174.

[16] Dumeige, Y., Monnier, P., Sagnes, I., *et al.* (2002) Second Order Nonlinear AlOx/AlGaAs 1D Photonic Crystal. ICTON '02: 4th International Conference on Transparent Optical Networks and European Symposium on Photonic Crystals, vol. **2**, pp. 168–171.

[17] Sprung, D.W.L., Wu, H. and Martorell, J. (1993) Scattering by a finite periodic potential. *Am. J. Phys.*, **61**, 1118–1124.

[18] Bloembergen, N. and Sievers, A.J. (1970) Nonlinear optical properties of periodic laminar structures. *Appl. Phys. Lett.*, **17**, 483.

[19] Jariv, A. and Yen, P. (1977) Electromagnetic propagation in periodic stratified media. II. Birefringence, phase matching, and X-ray lasers. *J. Opt. Soc. Am.*, **67**, 438.

[20] Dowling, J.P. and Bowden, C. (1994) Anomalous index of refraction in photonic bandgap materials. *J. Mod. Opt.*, **41**, 345.

[21] Centini, M., Sibilia, C., Scalora, M. *et al.* (1999) Dispersive properties of finite, one-dimensional photonic band gap structures: Applications to nonlinear quadratic interactions. *Physical Rev. E*, **60**, 4891–4898.

[22] Dumeige, Y., Sagnes, I., Monnier, P. *et al.* (2002) $\chi(2)$ semiconductor photonic crystals. *J. Opt. Soc. Am. B*, **19**, 2094–2101.

[23] Sibilia, C., Centini, M., D'Aguanno, G. and Scalora, M. (2001) Enhancement of second-harmonic generation in a one-dimensional semiconductor photonic band gap. *Appl. Phys. Lett.*, **78**, 3021–3023.

[24] Trull, J., Vilaseca, R., Martorell, J. and Corbalan, R. (1995) Second harmonic generation in local modes of a truncated periodic structure. *Opt. Lett.*, **20**, 1746–1748.

[25] Martorell, J., Vilaseca, R. and Corbalan, R. (1997) Second harmonic generation in a photonic crystal. *Appl. Phys. Lett.*, **70**, 702–704.

[26] Shi, B., Jiang, Z.M. and Wang, X. (2001) Detective photonic crystals with greatly enhanced second-harmonic generation. *Opt. Lett.*, **26**, 1194–1197.

[27] Antonucci, D., De Ceglia, D., D'Orazio, A. *et al.* (2007) Enhancement of the SHG efficiency in a doubly resonant 2D-photonic crystal microcavity. *Opt. Quant. Electron.*, **39**, 353–360.

[28] Luo, X. and Ishihara, T. (2004) Engineered second harmonic generation in photonic crystal-slabs consisting of centrosymmetric materials. *Adv. Funct. Mater.*, **14**, 905–912.

[29] Maymo, M. and Martorell, J. (2006) Visible second-harmonic light generated from a self-organized centrosymmetric lattice of nanospheres. *Opt. Express*, **14**, 2864–2872.

[30] Liscidini, M. and Andreani, L.C. (2006) Second-Harmonic generation in doubly resonant microcavities with periodic dielectric mirrors. *Phys. Rev. E*, **73**, 1–11.

[31] Tomita, I., Asobe, M., Suzuki, H. *et al.* (2006) Broadband quasi-phase-matched second-harmonic generation in a nonlinear photonic crystal. *J. Appl. Phys.*, **100**, 023120.

[32] Malvezzi, A.M., Vecchi, G., Patrini, M. *et al.* (2003) Resonant second-harmonic generation in a GaAs photonic crystal waveguide. *Phys. Rev. B*, **68**, 161306.

[33] Satoh, H., Yoshida, N. and Miyanaga, Y. (2002) Analysis of fundamental property of 2D photonic crystal optical waveguide with various medium conditions by condensed node spatial network. *Electron.Commun.Jpn.*, **85** (Part 2), 10–20.

[34] Chan, C.T., Yu, Q.L. and Ho, K.M. (1995) Order-N spectral method for electromagnetic waves. *Phys. Rev. B*, **51**, 16635–16642.

[35] Noda, S., Chutinan, A. and Imada, M. (2000) Trapping and emission of photons by a single defect in a photonic bandgap structure. *Nature*, **407**, 608–610.

[36] Seassal, C., Desieres, Y., Letartre, X., *et al.* (2002) Optical coupling between a two-dimensional photonic crystal-based microcavity and single-line defect waveguide on InP membranesk. *IEEE J.Quantum Electron.*, **38**, 811–815.

[37] Xu, Z., Wang, J., He, Q. *et al.* (2005) Optical filter based on contra-directional waveguide coupling in a 2D photonic crystal with square lattice of dielectric rods. *Opt. Express*, **13**, 5608–5613.
[38] Soltani, M., Haque, A., Momeni, B. and Adibi, A. (2003) Designing Complex Optical Filters Using Photonic Crystal Microcavities. Proceedings of the SPIE, Photonic Crystals Materials and Devices, vol. **5000**, 257–265.
[39] Fan, S., Villeneuve, P.R., Joannopoulos, J.D. and Haus, H.A. (1998) Channel drop tunnelling through localized states. *Phys. Rev. Lett.*, **80**, 960–963.
[40] Xu, Y., Lee, R.K. and Yariv, A. (2000) Propagation and second harmonic generation of electromagnetic waves in a coupled-resonator optical waveguide. *J. Opt. Soc. Am. B*, **17**, 387–400.
[41] Jin, C.J., Johnson, N.P., Chong, H.M.H. *et al.* (2005) Transmission of photonic crystal-resonator waveguide (PhCCRW) structure enhanced via mode matching. *Opt. Exp.*, **13**, 2295–2302.
[42] Mookherjea, S. and Yariv, A. (2002) Coupled resonator optical waveguide. *IEEE J. Sel.ect. Top. Quantum Electron.*, **8**, 448–456.
[43] Wang, Y., Wang, T. and Liu, J. (2006) Waveguide modes in coupled-resonator optical waveguides. *Phys. Lett. A*, **353**, 101–104.
[44] Imada, M., Noda, S., Chutinan, A. *et al.* (2002) Channel drop filter using a single defect in a 2-D photonic crystal slab waveguide. *IEEE J. Lightwave Technol.*, **20**, 845.
[45] Manolatou, C., Khan, M.J., Fan, S., *et al.* (1999) Coupling of modes analysis of resonant channel add-drop filters. *IEEE J. Quantum Electronics*, **35**, 1322–1332.
[46] Xu, Y., Li, Y., Lee, R.K. and Yariv, A. (2000) Scattering-theory analysis of waveguide-resonator coupling. *Phys. Rev. E*, **62**, 7389–7404.
[47] Rodriguez-Esquerre, V.F., Koshiba, M. and Figueroa, H. (2004) Finite-element time-domain analysis of 2-D photonic crystal resonant cavities. *IEEE Photon. Technol Lett.*, **16**, 816–818.

11

Dispersive Nonlinear MRTD for SHG Applications

11.1 Introduction

In this chapter, the MRTD approach based on the expansion in terms of only scaling functions (S-MRTD) is successfully extended for nonlinear SHG applications in dispersive media. Chromatic dispersion of materials makes the dielectric characteristics of the media vary with frequency. As a result, two or more waves at different wavelengths propagate into the medium with different phase velocities creating a phase mismatch that cannot be ignored in SHG processes; on the contrary it can significantly affect the entity of converted SH power. Therefore, in order to accurately describe SHG processes, the numerical scheme should include a model of the material dispersion of the medium and account for the deriving change in the refractive index.

In FDTD schemes, analysis of dispersive materials has been carried out through two main methods: the piecewise linear recursive convolution technique [1] and the auxiliary differential equation (ADE) scheme [2]. In this chapter, the ADE-MRTD scheme with Lorentz chromatic dispersion model to reproduce the characteristic linear susceptibility of the medium is proposed. First, the equations are derived in the case of TE/TE coupling between fundamental and second harmonic waves and then the case of TE/TM coupling, which also performs a polarisation conversion, is presented.

Validation of the developed ADE-MRTD scheme is given for the generation of the second harmonic in a planar asymmetric waveguide. A comparison between the ADE-FDTD and the proposed ADE-MRTD scheme shows that numerical dispersion of the latter scheme increases more slowly than in FDTD and, as a result, accuracy can be ensured, even though the mesh discretisation is coarse. Looking at the CPU running time required by the two different numerical codes, it is evident that the accuracy obtained by FDTD with a fine mesh can be equally well performed by using

Computational Photonics   Salah Obayya
© 2011 John Wiley & Sons, Ltd

fewer points in MRTD. This shows that the ADE-MRTD code is again more efficient overall than ADE-FDTD.

Next, a more challenging example is reported to show the potential and ability of nonlinear dispersive MRTD in a case of TE/TM coupling. Results show that the nonlinear ADE-MRTD scheme ensures the accurate analysis of SHG processes in photonic devices, while coarse meshes are allowed. Therefore, MRTD can be an efficient and promising alternative modelling technique to the more popular FDTD for the computation of nonlinear optical devices.

11.2   Dispersion Analysis

11.2.1   Introduction

In its original formulation, MRTD doesn't deal with the dispersive nature of materials. This can be negligible in some traditional electromagnetic problems such as microwave plane wave scattering at single frequencies from complicated structures. However, when the interaction of light with optical materials is studied, linear dispersion must be taken into account to ensure accurate results. This is especially true for TD simulations with pulsed excitations and a wideband frequency response, and for nonlinear problems involving two frequencies very distant one from each other, such as the SHG process. Thus, the variation of the material's permittivity and/or permeability with frequency occurring at low-intensity E and H fields must be included in the MRTD equations. With regard to this problem, the algorithm proposed from Chapter 8 to Chapter 10 has been extended for the modelling of Lorentz media by means of an ADE scheme, developed by Kashiwa and Fukai in 1990 [2]. The ADE technique consists in the solution of a differential equation deriving from the constitutive relationship and is second-order accurate in the space stencil Δx. The stability and phase-error properties of the ADE, when applied to conventional FDTD schemes, have been deeply investigated in the past few years [3]. It has been found that the overall scheme for the study of dispersive medium is generally more dispersive than the standard FDTD and their accuracy depends strongly on how well the chosen time step resolves the shortest timescale. Therefore, it is expected that for the analysis of dispersive media, the size of chosen time and space steps become even more crucial in determining the accuracy of the results.

11.2.2   Dispersive Materials: Lorentz Model

In general, the macroscopic propagation characteristics of a medium depend on the frequency, due to a phenomenon called chromatic dispersion that is experimentally measurable. The phase velocity of propagating waves varies with frequency as a result of the monotonically increasing variation of the refractive index.

The relationship linking the variation of optical properties in dielectric materials and semiconductors to the strength of the electric field is linear through first-order susceptibility, $\chi^{(1)}$, and is expressed as

$$\vec{P}(\omega) = \varepsilon_0 \chi^{(1)}(\omega) \vec{E}(\omega) \tag{11.1}$$

where P represents the polarisation vector at relatively low illumination intensity. With high-intensity light as a source, a nonlinear response of the material can occur and the chromatic dispersion will become the linear contribution at the total polarisation vector, P.

The Lorentz model [4] provides a definition for the linear susceptibility, $\chi^{(1)}$, so that the effect of P with frequency can be included in the determination of the E field by means of a constitutive relationship. Lorentz media define a complex function in the frequency domain that is characterised by one or more pairs of complex-conjugate poles. For the case of a single-pole-pair medium, the linear susceptibility becomes

$$\chi_p(\omega) = \frac{\Delta \varepsilon_p \omega_p^2}{\omega_p^2 + 2j\omega\delta_p - \omega^2} \tag{11.2}$$

where $\Delta \varepsilon_p = \varepsilon_{s,p} - \varepsilon_{\infty,p}$ represents the variation of the relative permittivity due to the Lorentz pole pair from the value at zero and infinite frequency, respectively, ω_p is the frequency of the pole pair and δ_p is the damping coefficient. By inverse Fourier transformation, the function $\chi(t)$ in time domain can be derived as

$$\chi_p(t) = \frac{\Delta \varepsilon_p \omega_p^2}{\sqrt{\omega_p^2 - \delta_p^2}} e^{-\delta_p t} \sin\left(\sqrt{\omega_p^2 - \delta_p^2}\, t\right) U(t) \tag{11.3}$$

From the generic expression of $\chi^{(1)}$ for a number, P, of pole pairs, the relative permittivity can be defined by the following

$$\varepsilon(\omega) = \varepsilon_\infty + \sum_{p=1}^{P} \frac{\Delta \varepsilon_p \omega_p^2}{\omega_p^2 + 2j\omega\delta_p - \omega^2} \tag{11.4}$$

11.2.3 Auxiliary Differential Equations

According to the constitutive relationship for the linear case, the following holds

$$D = \varepsilon_0 \varepsilon_\infty E + P_{\mathrm{L}} \tag{11.5}$$

where P_L is the linear polarisation term that is modelled through Lorentz scheme with a single pole pair as

$$P_L(\omega) = \varepsilon_0 \chi_L^{(1)}(\omega) E(\omega) \tag{11.6}$$

with

$$\chi_L^{(1)}(\omega) = \frac{(\varepsilon_s - \varepsilon_\infty)\omega_p^2}{\omega_p^2 + 2j\omega\delta_p - \omega^2} \tag{11.7}$$

where ε_s and ε_∞ refer to the relative permittivity at zero and infinite frequency, respectively.

Substituting the expression (11.7) into Equation (11.6), the ADE is finally obtained

$$\omega_p^2 P_L + 2\delta_p \frac{\partial P_L}{\partial t} + \frac{\partial^2 P_L}{\partial t^2} = \varepsilon_0 \Delta\varepsilon\omega_p^2 E, \quad \Delta\varepsilon = \varepsilon_s - \varepsilon_\infty \tag{11.8}$$

and it is discretised through central differences into

$$\omega_{p,i,j}^2 \, P_L|_{i,j}^n + 2\delta_{p.i.j} \frac{P_L|_{i,j}^{n+1} - P_L|_{i,j}^{n-1}}{2\Delta t} + \frac{P_L|_{i,j}^{n+1} - 2\,P_L|_{i,j}^n + P_L|_{i,j}^{n-1}}{\Delta t^2}$$
$$= \varepsilon_0 \Delta\varepsilon_{i,j}\omega_{p,i,j}^2 \, E|_{i,j}^n \tag{11.9}$$

Thus, the explicit relation for $P_L|_{i,j}^{n+1}$ is

$$P_L|_{i,j}^{n+1} = \alpha_{i,j} \, P_L|_{i,j}^n + \beta_{i,j} \, P_L|_{i,j}^{n-1} + \gamma_{i,j} \, E|_{i,j}^n \tag{11.10}$$

where

$$\alpha_{i,j} = \frac{2 - \Delta t^2 \omega_{p,i,j}^2}{1 + \delta_{p,i,j}\Delta t} \tag{11.11a}$$

$$\beta_{i,j} = \frac{\delta_{p,i,j}\Delta t - 1}{\delta_{p,i,j}\Delta t + 1} \tag{11.11b}$$

$$\gamma_{i,j} = \frac{\varepsilon_0 \Delta\varepsilon_{i,j}\omega_{p,i,j}^2 \Delta t^2}{1 + \delta_{p,i,j}\Delta t} \tag{11.11c}$$

11.3 SHG-MRTD Scheme for Dispersive Materials

11.3.1 TE/TE Coupling

The nonlinear process coupling the TE fundamental wave/TE second harmonic in S-MRTD scheme is considered first. Assuming the 2D problem in the xz-plane, the y-axis as homogeneous direction and x-axis as direction of propagation, the fundamental electromagnetic components of a TE mode are E_y, H_x and H_z. As seen in Chapter 8, Equations (8.70a–c) hold.

The constitutive relationship accounting for linear dispersion and second-order nonlinearity is modified as

$$D_y = \varepsilon_0 \varepsilon_\infty E_y + P_{Ly} + P_{NLy} \tag{11.12}$$

Using Equations (11.12) and (8.72), the constitutive relationship can be rewritten as

$$D_y = \varepsilon_0 \varepsilon_r E_y + P_{Ly} \tag{11.13}$$

where

$$\varepsilon_r = \left[\varepsilon_\infty + \chi_{22}^{(2)} E_y \right] \tag{11.14}$$

Substituting Equation (11.14) in (11.13), the following relationship for the E field is obtained

$$E_y = \frac{D_y - P_{Ly}}{\varepsilon_0 \left[\varepsilon_\infty + \chi_{22}^{(2)} E_y \right]} \tag{11.15}$$

Adopting Equation (11.15), the following updating equation is derived and inserted into the S-MRTD algorithm in order to calculate the E-field component

$$_{n+1/2}E_{i+1/2,j+1/2}^{y,\varphi} = \frac{_{n+1/2}D_{i+1/2,j+1/2}^{y,\varphi} - _{n+1/2}P_{i+1/2,j+1/2}^{Ly}}{\varepsilon_0 \left[\varepsilon_\infty + \chi_{i+1/2,j+1/2}^{(2)} \cdot _{n-1/2}E_{i+1/2,j+1/2}^{y,\varphi} \right]} \tag{11.16}$$

where

$$_{n+1}P_{i,j}^{Ly,\varphi} = \alpha_{i,jn} P_{i,j}^{Ly,\varphi} + \beta_{i,jn-1} P_{i,j}^{Ly,\varphi} + \gamma_{i,jn} E_{i,j}^{y,\varphi} \tag{11.17}$$

and

$$\chi_{i+1/2,j+1/2}^{(2)} = \chi_{22}^{(2)} \Big|_{i+1/2,j+1/2} \tag{11.18}$$

and it represents the nonlinear susceptibility value at the grid point of coordinates $(i + 1/2, j + 1/2)$.

11.3.2 TE/TM Coupling ($\chi^{(2)}$ Tensor)

A TE polarised input wave is considered with E-field components in the xz-plane and the fundamental H component in the longitudinal y-direction. An SHG with TM-polarised harmonic waves can be achieved with epitaxial growth of an $Al_{30\%}Ga_{70\%}As$ crystal on [001]-oriented GaAs substrate by aligning the x-axes along the principal axes of the crystal [110]. The $\chi^{(2)}$ tensor is shown in Equation (10.35) and the nonlinear polarisation vector is expressed by Equation (10.36).

From the discretised equation for the update of the nonlinear polarisation (10.37) and the constitutive relationship (11.12), the following equations for the D-field components can be written

$$_{n+1/2}D_{i+1/2,j+1/2}^{x,\varphi} = \varepsilon_0\varepsilon_{\infty i,j}\,_{n+1/2}E_{i+1/2,j+1/2}^{x,\varphi} + {}_{n+1/2}P_{i+1/2,j+1/2}^{Lx,\varphi} \qquad (11.19a)$$

$$_{n+1/2}D_{i+1/2,j+1/2}^{z,\varphi} = \varepsilon_0\varepsilon_{\infty i,j}\,_{n+1/2}E_{i+1/2,j+1/2}^{z,\varphi} + {}_{n+1/2}P_{i+1/2,j+1/2}^{Lz,\varphi}$$
$$+\varepsilon_0\chi_{i+1/2,j+1/2}^{(2)}\left(_{n+1/2}E_{i+1/2,j+1/2}^{y,\varphi}\right)^2 \qquad (11.19b)$$

where

$$_{n+1}P_{i,j}^{Lx,\varphi} = \alpha_{i,jn}P_{i,j}^{Lx,\varphi} + \beta_{i,jn-1}P_{i,j}^{Lx,\varphi} + \gamma_{i,jn}E_{i,j}^{x,\varphi} \qquad (11.20a)$$

$$_{n+1}P_{i,j}^{Lz,\varphi} = \alpha_{i,jn}P_{i,j}^{Lz,\varphi} + \beta_{i,jn-1}P_{i,j}^{Lz,\varphi} + \gamma_{i,jn}E_{i,j}^{z,\varphi} \qquad (11.20b)$$

Using Equation (11.20), the update equations for the E- field components for the generated TM polarisation can be derived as

$$_{n+1/2}E_{i+1/2,j+1/2}^{x,\varphi} = \frac{1}{\varepsilon_0\varepsilon_{\infty,i+1/2,j+1/2}}\left(_{n+1/2}D_{i+1/2,j+1/2}^{x,\varphi} - {}_{n+1/2}P_{i+1/2,j+1/2}^{Lx,\varphi}\right)$$
$$(11.21a)$$

$$_{n+1/2}E_{i+1/2,j+1/2}^{z,\varphi} = \frac{1}{\varepsilon_0\varepsilon_{\infty,i+1/2,j+1/2}}\left(_{n+1/2}D_{i+1/2,j+1/2}^{z,\varphi} - {}_{n+1/2}P_{i+1/2,j+1/2}^{Lz,\varphi}\right)$$
$$-\varepsilon_0\chi_{i+1/2,j+1/2}^{(2)}\left(_{n+1/2}E_{i+1/2,j+1/2}^{y,\varphi}\right)^2 \qquad (11.21b)$$

From this point, the update of the components D_x, D_z and H_y are expressed by the set of Equations (10.40.a–c) while the set of Equations (10.27a–c) solves the fields for

TE polarisation with fundamental electric component calculated by

$$_{n+1/2}E^{y,\varphi}_{i+1/2,j+1/2} = \frac{_{n+1/2}D^{y,\varphi}_{i+1/2,j+1/2} - _{n+1/2}P^{Ly}_{i+1/2,j+1/2}}{\varepsilon_0\varepsilon_\infty} \tag{11.22}$$

The solution of the equations is obtained through the following iterative process:

1. Calculation of the updated values of H_x and H_z with Equations (10.27a–b), using the E-field value at the previous time step.
2. Substitution of the updated magnetic-field components into Equation (10.27c) to calculate the new value of D_y.
3. At the same time step, P_{Ly} update through Equation (11.17).
4. Next, calculation of the E_y component with Equation (11.22).

Once the fields for TE polarisation have been updated, the following steps are iterated to simulate the coupling with the TM-polarised harmonic:

5. Updating of nonlinear polarisation term of polarisation, P_{NLz}, through Equation (10.37).
6. At the same time step, calculation of components D_x and D_z from Equations (11.19a–b) respectively.
7. Calculation of P_{Lx} and P_{Lz} through Equation (11.20a) and Equation (11.20b).
8. Updating of the electric-field components E_x and E_z with Equations (11.21a–b).
9. Finally, Equations (10.40a–c) are used to update the set of components for TM polarisation.

11.4 Simulation Results

11.4.1 SHG in a Planar Waveguide

In order to test the validity and efficiency of the nonlinear S-MRTD code proposed for the analysis of SHG, the asymmetric waveguide shown in Figure 11.1 is considered first. The device consists of a core that has width $L_{core} = 0.44\,\mu m$ and refractive index $n_{core} = 3.6$, between substrates with index $n = 3.1$ and with air as cladding ($n = 1$) [5].

The propagation of the fundamental TE_y mode in the x-direction is performed by launching a CW at a fundamental wavelength of $\lambda_f = 1.55\,\mu m$ modulated by the TE-mode profile in space. As a result of the nonlinearity of the dielectric material, $\chi^{(2)} = 200\,pm/V$, generation of the second harmonic is expected at wavelength $\lambda_{sh} = 0.775\,\mu m$.

By performing post-processing numerical filtering of the electric field recorded at the detector point, Figure 11.2 shows the variation of the electric field with time for

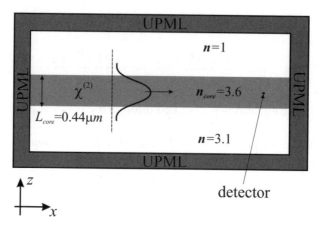

Figure 11.1 Schematic diagram of the planar waveguide simulated where $n_{core} = 3.6$, substrate has index $n = 3.1$, cladding is in air ($n = 1$) and $L_{core} = 0.44$ µm.

both the fundamental and the second harmonic waves, (continuous and bold lines, respectively). It can be noted from this figure that the numerical method is stable. From the FFT of these temporal data, the frequency response normalised to the fundamental spectrum peak is obtained and shown in Figure 11.3. The figure shows the second harmonic generated at wavelength 775 nm with efficiency $|E_{sh}|^2 \big/ |E_f|^2 \approx 30\%$,

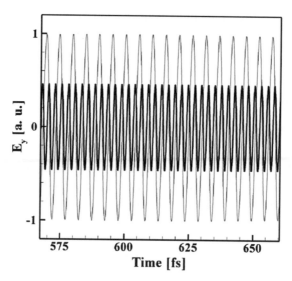

Figure 11.2 Time variation of fundamental (fine line) and second harmonic wave (bold line) recorded at the detector point of the waveguide. (Reproduced by permission from Obayya, S.S.A. (2010) Novel Auxiliary Differential Equation-Multiresolution Time Domain Scheme for Dispersive Nonlinear Photonic Devices. *J Quantum Electron.*, **46** (5), 837–845. © 2010 IEEE.)

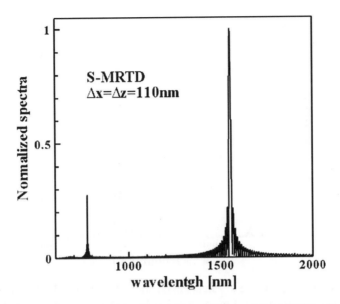

Figure 11.3 Normalised spectra of fundamental and SH wave for MRTD with discretisation $\Delta x = \Delta z = 110$ nm. (Reproduced by permission from Obayya, S.S.A. (2010) Novel Auxiliary Differential Equation-Multiresolution Time Domain Scheme for Dispersive Nonlinear Photonic Devices. *J Quantum Electron.*, **46** (5), 837–845. © 2010 IEEE.)

where E_{sh} and E_f represent the electric-field components of fundamental and SH waves, respectively.

Aiming to validate these results, the same structure has been simulated by using a nonlinear FDTD algorithm and a much finer mesh which applies to a cell size $\Delta x = \Delta z = 20$ nm. Frequency-domain variation of the electric field stored at the detector point is shown in Figure 11.4. As can be clearly seen from this figure, the second harmonic is generated from the input source with an efficiency that agrees well with the value obtained by means of the S-MRTD method combined with a coarser grid.

In order to investigate the improved computational efficiency of S-MRTD over conventional FDTD, an extensive study of the effect of varying the cell size has been carried out. Simulations have been performed varying the cell size in the range $[L_{core}/22; L_{core}/4]$ that in nm corresponds to [20 nm; 110 nm]. Figure 11.5 reports the value of SHG efficiency as calculated for each different mesh resolution with both FDTD and S-MRTD methods. The figure clearly shows a quite stable response of the MRTD code to the different grid sizes with very little effect on the efficiency of generation. On the other hand, the FDTD scheme suffers significantly from the choice of an inappropriate cell size leading to mistaken values of SHG efficiency. In particular, it has been seen that it is the amplitude of the generated SH found in the output that gradually decreases as the cell size is enlarged, whereas the fundamental wave keeps quite a similar value. This can be explained by recalling the requirement for a discretisation of $\lambda_{sh}/15$–20 in the FDTD scheme, where λ_{sh} represents the second

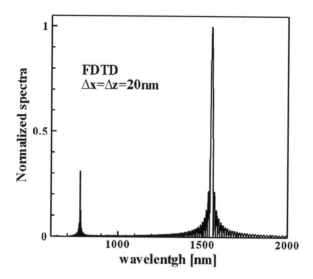

Figure 11.4 Normalised spectra of fundamental and SH wave for FDTD with discretisation $\Delta x = \Delta z = 20$ nm. (Reproduced by permission from Obayya, S.S.A. (2010) Novel Auxiliary Differential Equation-Multiresolution Time Domain Scheme for Dispersive Nonlinear Photonic Devices. *J Quantum Electron.*, **46** (5), 837–845. © 2010 IEEE.)

harmonic wavelength. When the cell size increases, in fact, high frequencies are the first to be affected by the FDTD numerical dispersion, thus a specific grid resolution can still be small enough to describe frequencies at the fundamental wavelength, but it is already over the limit for the second harmonic window.

To complete the study, CPU running time was recorded for each of the previous simulations and results are shown in Figure 11.6 to compare the computational costs

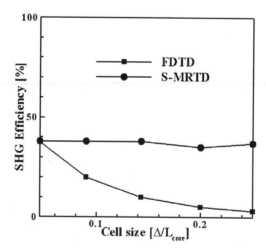

Figure 11.5 Comparison of SHG efficiency variation with unit cell size Δ for FDTD and S-MRTD scheme.

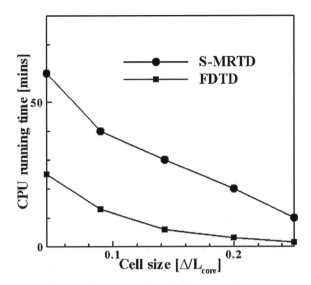

Figure 11.6 Comparison of CPU running time variation with unit cell size Δ for FDTD and S-MRTD scheme.

of S-MRTD and FDTD. As expected, the running time for the MRTD scheme is longer for each case, with a difference compared to FDTD that increased as the cell size becomes smaller. This fact is due to the limit on the choice of Δt imposed for stability reasons that is stricter in the MRTD scheme than in FDTD. However the situation changes if the results of Figures 11.5 and 11.6 are combined together. Accuracy comparable to FDTD with smaller cell sizes can be achieved by S-MRTD by using the largest one. This means that with the same level of accuracy, the CPU running time for S-MRTD becomes smaller than that required by conventional FDTD, proving the actual efficiency of the presented method, Table 11.1.

11.4.2   SHG in 1D Periodic Structure

Next, second harmonic generation is investigated through the ADE-S-MRTD scheme in the more challenging problem of a high-contrast grating added to the waveguide,

Table 11.1   Relative Error (%) for MRTD and FDTD vs CPU running time. (Reproduced by permission from Obayya, S.S.A. (2010) Novel Auxiliary Differential Equation-Multiresolution Time Domain Scheme for Dispersive Nonlinear Photonic Devices. *J Quantum Electron.*, **46** (5), 837–845. © 2010 IEEE.)

	Relative Error on SHG Efficiency	Δ/L_{core}	CPU Running Time (minutes)
FDTD	<3%	0.05	25
S-MRTD	<3%	0.25	12

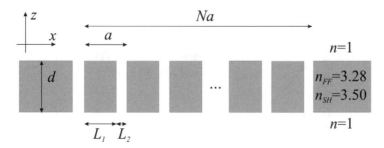

Figure 11.7 Schematic diagram of the 1D photonic bandgap structure with number of periods, N. (Reproduced by permission from Obayya, S.S.A. (2010) Novel Auxiliary Differential Equation-Multiresolution Time Domain Scheme for Dispersive Nonlinear Photonic Devices. *J Quantum Electron.*, **46** (5), 837–845. © 2010 IEEE.)

while dispersion analysis is added to allow a complete description of the process. This periodic structure was proposed in a modelling exercise in the COST-P11 project (http://w3.uniroma1.it/energetica/). This type of device consists practically in a 1D photonic bandgap structure and is used in the SHG process because of its ability to modify topologically the dispersion characteristic in such a way that the phase matching condition through the waveguide is achieved when the effective index of the device at the fundamental frequency is equal to the effective index at the second harmonic frequency [6, 7]. The result is a great enhancement of the SHG conversion efficiency. Furthermore, the intensity of the fields is increased by means of the resonances inside the 1D periodic structure [6].

The structure analysed to achieve SHG and TE/TM coupling , consists of a periodic grating in a planar waveguide, as shown in Figure 11.7. The dielectric structure made of $Al_{0.3}Ga_{0.7}As$ with refractive indices of $n_{ff} = 3.28$ at the fundamental wavelength $\lambda_f = 1.55$ μm and $n_{sh} = 3.50$ at the second harmonic wavelength $\lambda_{sh} = 0.775$ μm, is surrounded by air ($n = 1$).

The waveguide that couples the input/output of the device has width $d = 0.18$ μm and the grating consists of N periods, where each period $a = 0.18$ μm is formed by a dielectric part ($L_1 = 0.135$ μm) and air ($L_2 = 0.045$ μm). With this design the waveguide supports a single TE mode for the fundamental $\lambda_f = 1.55$ μm (only one electric component with out-of-plane direction, E_y), and a single TM mode for the second harmonic at $\lambda_{sh} = 0.775$ μm (only one magnetic component with out-of-plane direction, H_y). A normalised nonlinear coefficient is considered, $\chi^{(2)} = 1$, that allowsTE/TM coupling.

AlGaAs chromatic dispersion is accurately introduced by the Lorentz model for the linear susceptibility with parameters: $\varepsilon_\infty = (3.549)^2$, $\Delta\varepsilon = (\varepsilon_s - \varepsilon_\infty) = 4$, $\omega_p = 2\pi \cdot 108.51 \times 10^{12} \, \text{rad/s}$, and $\delta_p = 10^{-6}\omega_p$. The dispersion curve of the real part of the dielectric refractive index that is obtained with Lorentz model is shown in Figure 11.8.

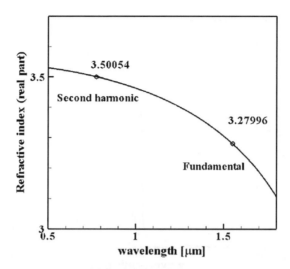

Figure 11.8 Lorentz model of dispersion curve for the real part of $Al_{0.3}Ga_{0.7}As$ refractive index. The curve reports $n_{sh} = 3.500\ 54$ at the second harmonic $\lambda_{sh} = 0.775$ μm and $n_{ff} = 3.279\ 96$ at fundamental $\lambda_f = 1.55$ μm. Both values approximate well the experimentally calculated indices, which are 3.50 and 3.28, respectively. (Reproduced by permission from Obayya, S.S.A. (2010) Novel Auxiliary Differential Equation-Multiresolution Time Domain Scheme for Dispersive Nonlinear Photonic Devices. *J Quantum Electron.*, **46** (5), 837–845. © 2010 IEEE.)

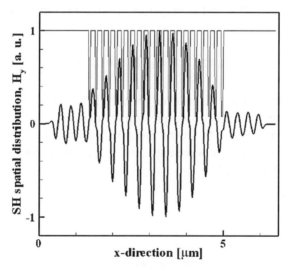

Figure 11.9 Spatial field distribution at the phase-matched SH wavelength for the case of $N = 20$ (thick line); the thin line represents the normalised profile of the refractive index along the *x*-direction. (Reproduced by permission from Obayya, S.S.A. (2010) Novel Auxiliary Differential Equation-Multiresolution Time Domain Scheme for Dispersive Nonlinear Photonic Devices. *J Quantum Electron.*, **46** (5), 837–845. © 2010 IEEE.)

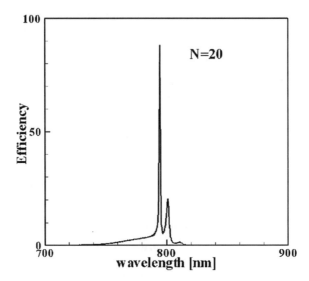

Figure 11.10   SHG conversion efficiency calculated for $N = 20$.

The structure is discretised with cell size $\Delta = \Delta x = \Delta z = d/8$ and the adopted time step is $\Delta t = 0.008$ fs. A Gaussian pulse modulated by the fundamental TE mode profile, $E_{TE}(z)$ in Equation (11.12), is injected as excitation with bandwidth in time $T_0 = 20$ fs, delay $t_0 = 80$ fs and $\lambda_f = c_0/f_0 = 1.55\ \mu$m is the central wavelength

$$E_{y,f\,source}(z,t) = E_{\text{TE}}(z)\sin(2\pi f_0 t)\,\mathrm{e}^{-\frac{(t-t_0)^2}{T_0^2}} \tag{11.23}$$

The spatial field distribution against the profile of the refractive index along the x-direction, at a phase-matched SH wavelength for the case of $N = 20$ is displayed in Figure 11.9. Next, Figure 11.10 shows the calculated SHG efficiency after $N = 20$ periods in the second harmonic wavelength range. Results are in agreement with what is reported in the literature [7].

References

[1] Kelley, D.F. and Luebbers, R.J. (1996) Piecewise linear recursive convolution for dispersive media using FDTD. *IEEE Trans. Antennas Propag.*, **44**, 792–797.
[2] Kashiwa, T. and Fukai, I. (1990) A treatment by FDTD method of dispersive characteristics associated with electronic polarization. *Microwave Opt. Technol. Lett.*, **3**, 1326–1328.
[3] Petropoulos, P.G. (1994) Stability and phase error analysis of FD-TD in dispersive dieletrics. *IEEE Trans. Antennas Propag.*, **42**, 62–69.
[4] Young, J.L. and Nelson, R.O. (2001) A summary and systematic analysis of FDTD algorithms for linearly dispersive media. *IEEE Antennas Propag.*, **43**, 61–77.

[5] Alsunaidi, M.A., Masoudi, H.M. and Arnold, J.M. (2000) A time-domain algorithm for the analysis of second-harmonic generation in nonlinear optical structures. *IEEE Phot. Technol. Lett.*, **12**, 395–397.

[6] Centini, M., Sibilia, C., Scalora, M., *et al.* (1999) Dispersive properties of finite, one-dimensional photonic band gap structures: Applications to nonlinear quadratic interactions. *Physical Rev. E*, **60**, 4891–4898.

[7] Maes, B., Bienstman, P., Baets, R. *et al.* (2008) Modeling comparison of second-harmonic generation in high-index-contrast devices. *Opt. Quant. Electron.*, **40**, 13–22.

Index